THE EARTH'S LAND SURFACE
LANDFORMS AND PROCESSES IN GEOMORPHOLOGY

Kenneth J. Gregory

Los Angeles | London | New Delhi
Singapore | Washington DC

SAGE Publications Ltd
1 Oliver's Yard
55 City Road
London EC1Y 1SP

SAGE Publications Inc.
2455 Teller Road
Thousand Oaks, California 91320

SAGE Publications India Pvt Ltd
B 1/I 1 Mohan Cooperative Industrial Area
Mathura Road, Post Bag 7
New Delhi 110 044

SAGE Publications Asia-Pacific Pte Ltd
33 Pekin Street #02-01
Far East Square
Singapore 048763

Library of Congress Control Number: 2010921012

British Library Cataloguing in Publication data

A catalogue record for this book is available from the British Library

ISBN 978-1-84860-619-7
ISBN 978-1-84860-620-3 (pbk)

Typeset by C&M Digitals (P) Ltd, Chennai, India
Printed in Great Britain by MPG Books Group, Bodmin, Cornwall
Printed on paper from sustainable resources

CONTENTS

PREFACE

In common with many others of my generation I first met geomorphology through Horrocks (1954) and then Wooldridge and Morgan (first published 1937, my copy 1954) as a school prize. Geomorphology then, in the 1950s, was so different – pre satellite, pre processes, pre techniques, pre computers, pre geomorphological journals. It is a paradox that now, when the emphasis in the environmental sciences has been towards a more holistic view, the trend in education has been to focus selectively upon particular areas. Thus the fissiparist reductionist approach contrasts starkly with the holistic imperative. Is this a time for reflection, in the light of the progress of the last 100 years, on how we make the journey, and whether we are going in the right direction? In this volume the emphasis upon landforms will be seen as retrograde by some – possibly as a return to rote learning or to "capes and bays". Whereas geomorphology started with a focus on landforms, it developed to the study of process and then may have become over-selective by concentrating on particular environments to the exclusion of others. But it is landforms that are the stuff of scenery and it is surely of interest to know how they have been studied and have produced reactions in landscape interpretations. Although detailed explanation of suggested origins cannot be given in the space available here, it can be explored using books, articles and the resources of the internet. Because the way in which we understand the land surface of the Earth depends upon those who have studied it, each chapter contains a short profile of at least one individual whose contributions have advanced our understanding of that land surface. Such profiles are included because to understand the epistemology of a discipline it is necessary to know something about the people who have influenced the epistemological development. Although not every one will agree with the choice of scientists, especially as they are all male (but see topic suggested at end of Chapter 1), they do provide a range of the types of influence – some within and some external to geomorphology, some from previous centuries and some comparatively modern. Readers can investigate for themselves the contributions of other influential individuals. In the past, larger books were required which were comprehensive and could be followed up by reference to articles in research journals. Now the internet resource suggests perhaps a shorter book is needed as a basis for rapid searching – many of the themes introduced in the following pages can initially be explored through the internet. As many graphic illustrations are now available on the internet figures have deliberately been kept as few as possible. Colour plates are located at the end of the book. Terms in the Glossary (p. 303) are shown in bold when first mentioned. Writing a book of this kind at the end of my career has been an attempt to reflect the enjoyment that I have derived from study of the land surface of the Earth. I hope that similar enjoyment may be experienced by some readers.

ACKNOWLEDGEMENTS

I much appreciate the advice received from several individuals who have commented on portions of the text, including Professor Vic Baker, Professor Tony Brown, Professor Anne Chin, Dr Peter Downs, Professor Andrew Goudie, Professor Will Graf, Professor Angela Gurnell, Professor Peter Haggett, Professor Vishwas Kale, Dr Colin Prowse, Professor Matti Seppala, Professor Mike Thomas and Professor Des Walling. I also appreciate Robert Rojek's suggestion that stimulated the idea for this book.

Every effort has been made to establish copyright, make acknowledgement and to obtain appropriate permissions for figures and plates but, if any have been missed, please let me know.

It is customary to acknowledge one's wife: Chris helped in so many ways, not only reading through the whole text and making helpful comments but she continues to provide unfailing support, which is greatly appreciated.

Ken Gregory

PART I
VISUALIZING THE
LAND SURFACE

1

RECOGNIZING THE LAND SURFACE

Terra firma, a Latin phrase for solid earth, is used to differentiate land from sea – thus connoting the land surface of the Earth. It is used specifically in the Amazon floral region where vegetation above flood level is called terra firma and that below, igapo or varzea. Terra firma in a search engine produces more than 11 million results, referring to tiles, travel firms, landscapes, rug company and many others but only one book (Orban, 2006). It is surprising that the land surface of the Earth has not received more explicit attention – perhaps because with hindsight it seems to have been hijacked by disciplines whose prime purpose lies elsewhere: a reason for what Tooth (2009) has described as invisible geomorphology where geomorphological awareness still needs to be raised by stressing the discipline's relevance and contribution to a host of environmental issues. For over a century, museums have been one way to inform about the surface of the Earth, but they emphasized geology, natural history and archaeology. In recent decades public awareness of the surface of the Earth has increased through press and media news reports and documentaries, cinema and the internet. It is ironic that the land surface of the Earth has not received more explicit attention but thought of geologically in terms of rock and tectonic control; biologically in terms of the vegetation types and ecosystems spread across the surface; economically and archaeologically in terms of the environments provided. This chapter introduces terra firma, what it is, how we think of it, and how we have divided it up.

1.1 WHAT IS THE LAND SURFACE?

Over the last five decades data collected by **remote sensing** from orbital satellites has revolutionized our valuation of the land surface of the blue planet, by

enabling definitive measures of its characteristics. Complementary recent progress
has been provided by computer analysis and **Geographical Information Systems
(GIS)**, by satellite navigation, and by advances in dating techniques. It is salutary
to recall that the shapes of the continents were not known until the eighteenth
and nineteenth centuries when national topographic surveys were undertaken.
Although we should now expect little variance in dimensions of the land
surface (see Table 1.1), it is surprising how much variation occurs in published
estimates – some because of conversions to the **International System of Units
(SI)**, or because they have not been verified or definitions have varied, but some
because changes of the land surface have occurred.

Table 1.1   DIMENSIONS FOR THE LAND SURFACE

Dimension	Value (based on several sources)	Notes
Spatial extent	148.847×10^6 km² representing 29.18% of the earth's surface.	Much more land area in the northern hemisphere than in the southern.
Relief	Highest point Mount Everest at 8848 m, lowest at –418m in the floor of the Dead Sea.	Hypsographic curve – see Figure 1.1.
Mean height of land above sea level	686 m.	Average depth of the oceans at 3795 m is much greater.
Ice extent	Area of 14,898,320 km² represents more than 10% of the land surface.	South polar region including Antarctica represents 84.6% and Greenland accounts for 11.6% of the ice surface area, together locking up more than 90% of the 33 million km³ of glacier ice in the world. Water equivalent of ice caps and glaciers is 26,000,000 km³ (Nace, 1969) equivalent to water needed to supply world rivers for nearly 900 years.
Freshwater, lakes	Lakes, described as wide places in rivers, together with inland seas have a surface area of 1,525,000 km², more than 1% of the Earth's surface. Lakes store 90,000 km³ of world's freshwater, 0.27% of world's water, about 80% in 40 large lakes.	Fresh water bodies in all continents, but majority occur in the northern hemisphere. World's 145 largest lakes estimated to contain over 95% of all lake freshwater. Lake Baikal, the world's largest, deepest and oldest lake, contains 27% of all lake freshwater.
Rivers	Contain 0.0064% of world's water. The total volume of water stored in rivers and streams is estimated at about 2,120 km³.	Because of their role in the hydrological cycle have a major influence on the land surface. An estimated 263 international river

Table 1.1   *(Continued)*

Dimension	Value (based on several sources)	Notes
		basins have drainage areas that cover about 45% (231 million km²) of the Earth's land surface (excluding polar regions). The world's 20 largest river basins have catchment areas ranging from 1 to 6 million km². The Amazon carries 15% of all the water returning to the world's oceans.
Coastline	356,000 km including 94 nations and other entities that are islands according to https://www.cia.gov/library/publications/the-world-factbook/fields/2060.html.	Allowing for the many inlets and estuaries total length may approximate 1 million km (Bird, 2000).
Mountains	Approximately 20% represented by mountains. Mountains, highlands and hills, collectively cover 36% of the land surface of the Earth (Fairbridge, 1968). See Figure 6.1.	No agreed definition of mountains but can be where relief greater than 1500 m, and according to Wikipedia >2500 m high or 1500–2499 m if their slope is >2°, or 1000–1499 m if their slope is >5°. On this basis 24% of earth's land surface could be regarded as mountainous with 64% of Asia mountainous.
Land cover	20% is dry land (desert, rock, ice and sand 42.0×10^6 km²), with 10% that doesn't have topsoil. Forests (48.5×10^6 km²) and woodland and scrubland (8.0×10^6 km²) account for 27%. 2% urban area. Irrigated 2,770,980 km² (CIA World Factbook, 2003, accessed 5 March 2009).	Forest is now c. 27% of land area. Since 1700, nearly 20% of the world's forests and woodlands have disappeared. Forest 39.52 $\times 10^6$ km² in 2005 (was 40.77×10^6 km² in 1990 and 39.89×10^6 km² in 2000) according to www.fao.org/forestry/site/32038/en (accessed 17 March 2009). In 2005, arable land accounted for 10.57%; permanent crops for 1.04%; and other 88.38% of land cover.
Human impact	Between one-third and one-half of the land surface has been transformed by human action. See Table 3.10.	13.31% of the land surface is arable with 4.71% covered by permanent crops whereas approximately 40% is used for crops or for pasture (13×10^6 km² cropland; 34×10^6 km² pastureland). 45,000 big dams now block the world's rivers, trapping 15% of all the water that used to flow from the land to the sea. Reservoirs now cover almost 1% of land surface.

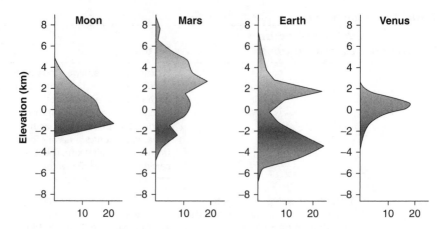

Figure 1.1   Comparative hypsometries (elevation is plotted against percentage of the Earth's surface area; based on http://comp.uark.edu/-sboss).

A way of demonstrating the overall distribution of the land surface is according to elevation above sea level. Such a hypsometric or hypsographic curve is a graph showing the proportion of the land mass occurring above a given level; it can be constructed for Earth as a whole or a part such as a drainage basin or glacier. Unlike the hypsometric curve for the moon or Venus that for the Earth (see Figure 1.1) is bimodal with one peak for the continents and one for the ocean basins, revealing more of the Earth's surface beneath the sea than above it.

Ice covers more than 10% of the land surface and frequent references to rates of melting of glaciers and ice sheets reflect the changes taking place so that regular auditing and monitoring of glacier and ice cap extent is necessary (see Table 1.2). Whereas ice-covered areas dominantly occur in high latitudes and high mountains, fresh water bodies are distributed over approximately 1% of the land surface in all continents. The largest closed basin lake is the Caspian Sea with a coastline approximately 7000 km long and a drainage basin covering 3.1 million km². Lake Baikal is the deepest single body of water with a volume of 22,000 km³ – nearly equivalent to the total volume in all five Great Lakes of North America. Some countries are particularly proud of their lakes. Finland, known as the Land of the Thousand Lakes, actually has more than 180,000. Manitoba claimed 100,000 lakes to beat Minnesota, which proclaimed 'Land of 10,000 Lakes' on its license plates. Lake Huron, the second largest lake in the world according to its surface area of 117,702 km², is one of the five Great Lakes of North America which are part of a system collectively containing some 22% of the world's freshwater – enough to cover the 48 contiguous states of the USA to a depth of 2.9 m, with a surface area nearly the same as that of the United Kingdom.

When seen from space, the surface of the blue planet, in addition to land, water and ice, shows the green of forest, and broad categories of land use (see Table 1.1). As much of the land surface has been substantially affected by human

Table 1.2   GLACIER INVENTORY: EXAMPLES OF PROGRESS

1   Plans for an inventory of global glacier features were included in the International
    Hydrological Decade (1965–1974), with the intention of repeating the survey every
    50 years to detect changes in glaciers. By 2008 about 37% of the estimated total
    glacier surface had been inventoried, available through the World Glacier
    Monitoring Service (WGMS) in Zürich and National Snow and Ice Data Center
    (NSIDC) in Boulder.
2   A simplified inventory method was developed in the early 1980s mainly based on
    satellite images, as outlined in Global Land Ice Measurements from Space (GLIMS)
    which covers 34% of the estimated glacier surface outside Greenland and Antarctica.
    See http://www.glims.org/ (accessed 17 March 2009).

action, dramatically increased since the nineteenth century, a new geological period, the Anthropocene, was suggested in 2000 by Paul Crutzen, a Nobel prize winner, because he regarded human impact on the Earth as so significant that it justified a new geological era to succeed the Pleistocene and Holocene of the Quaternary in the geological time scale (see Figure 5.1). A 2008 article (Zalasiewicz et al., 2008) was aptly entitled 'Are we now living in the Anthropocene?'.

The Earth's land surface is not static because change is, and always has been, a characteristic feature. Each year sees changes in detail – in ice extent for example. In addition to seasonal changes there are also changes from year to year, as drought years follow wetter years, and there are also evolutionary changes, for example as some coastlines progressively advance, complemented by others which are submerged. Some of the most dramatic changes arise from human impact, and most recently from global change on a scale which is substantial and potentially very serious. The level of the Caspian Sea fell over the last century so that, between 1927 and 1977, many economic activities, including oil exploration, oil field development, and pipeline construction, developed on the exposed lake floor. The lake level reached a record low of 29.0 m below mean sea level (Rodionov, 1990) in 1977, prompting an engineering solution response to bring water to the Sea from wetter parts of the Soviet Union (Glantz, 1995). After 1977 therefore the level of the Caspian began to rise rapidly (Rodionov, 1990) with consequential environmental problems including coastal inundation because of sea level rise; water pollution by raw sewage and oil production; fishing pressure and its impacts on fish populations (Glantz, 1995). The fall in level of the Aral Sea has also been dramatically affected by human activity: by 1989 the sea level had fallen by 14.3 m and the surface area had shrunk from 68,000 km² to 37,000 km². In the twenty-first century, as a result of conservation measures, the Sea level is rising again.

Interpreting the land surface requires a range of time scales analogous to the different resolutions of a microscope. A perceptive way of envisaging time scales (Schumm and Lichty, 1965) distinguished steady, graded and cyclic time (see Table 1.3). Any point on the land surface of the Earth can be thought of as a location in space and time so that to understand its characteristics we have to refer to changes occurring over time: some of the actual surface characteristics are the

Table 1.3   SOME CLASSIFICATIONS OF TIME

Authors	Subdivisions proposed		
Schumm and Lichty, 1965	*Steady state time*, typically of the order of a year or less when a true steady state situation may exist.	*Graded time*, may be hundreds of years during which a river develops a graded profile condition or a dynamic equilibrium exists involving progressive evolution of landform in relation to erosion and deposition.	*Cyclic or geological time*, encompasses millions of years as needed to complete an erosion cycle.
Udvardy, 1981	*Secular scale* spatial dimensions of about 100 km and time dimensions of about 100 years.	*Millennial scale* covering at least the last 10,000 years and spatial scales of up to 1000 km where climate and sea level change are the major factors operating.	*Evolutionary time or the phylogenetic scale* may be up to 500 million years and spatial extent may reach 40,000 km so that continental displacement may be important.
Driver and Chapman, 1996	*'Now time'*	*Generational time scale*, 10–100 years for sustainable development. · *Century time scale*, 100–1000 years.		*Late Quaternary time*, 1000–10,000 years.

result of relatively recent changes, some parts of the Earth retain characteristics developed over millions of years (see **Quaternary chronology**, Glossary p. 306). Where human impact is substantial then the surface characteristics may be just years or decades old. Time scales are inextricably combined with spatial scales so that one of the greatest challenges in studying the land surface of the Earth is to apply results gained from study at one time and spatial scale to one of the other time scales. Relating process measurements made over years or decades to development over centuries or millennia requires reconciling timeless and timebound scales.

1.2 ENVISAGING THE LAND SURFACE

How do we describe the land surface of the Earth? Language descriptions of characteristics of the surface are the obvious way, applying words to particular earth surface features. In English the nouns mountains, plains, valleys, plateaus have equivalents in other languages. Inevitably some languages include words for Earth surface characteristics according to the environment in their particular country, so that a language of place (Mead, 1953) reflects the fact that some vocabularies have words for particular features which do not have direct counterparts in other languages. In Russian there are words for types of valley, like *balki*, which cannot easily be translated into English because no comparable features occur as extensively in England or in America. Many descriptive words have now evolved to describe landforms – landform refers to the morphology and character of the land surface resulting from the interaction of physical processes with the surface materials in the land surface environment. Words for particular types or shapes of feature have become adopted as scientific terms so that words such as corrie or the Welsh word cwm for armchair-shaped hollows, became accepted as features of glacially eroded landscapes. Words gradually became adopted in this way so that it is not easy to discern when landform first became basic for the science of geomorphology (see p. 19, Chapter 2).

Scientists in the nineteenth century included Baron Ferdinand von Richthofen (1833–1905), who had trained in geology and geography at Breslau (now Wroclaw) and published a book, arising from his travels, on regional geology and geography in 1886 which may be the first systematic textbook of modern geomorphology (Fairbridge, 1999). Ways in which landforms gradually became assimilated into the scientific literature of studies of the land surface of the Earth are illustrated in Table 1.4, showing the importance of exploration of the American West and the contributions of William Morris Davis which may have

Table 1.4 ILLUSTRATIONS OF THE EMERGENCE OF LANDFORMS AS THE SCIENTIFIC BASIS FOR STUDY (SEE ALSO CHORLEY ET AL., 1964)

Individual contributor	Contribution	Indicative references
Oscar Peschel (1826–75), Professor of Geography, Leipzig	Compared the nature of similar landforms throughout the world, involving classification of surface features and comparison of their morphology.	Peschel, 1870
Ludwig Rutimeyer (1825–95), Professor of Zoology, Basle	In a book on valley and lake formation showed that the largest Alpine valleys had been produced by stream erosion over long periods of geologic time and that different sections of a river course can be marked by distinct types of erosional forms including waterfalls, meanders and floodplains.	Rutimeyer, 1869
J.W. Powell (1834–1902), US Geological Survey	In a study of the Colorado in 1875 identified 3 types of river valleys (antecedent, consequent, superimposed) and referred to landforms.	Powell, 1875
G.K. Gilbert (1843–1918), US Geological Survey	In 1875 discussed formation of alluvial fans. In 1914 produced a masterpiece on fluvial processes.	Gilbert, 1875
Baron Ferdinand von Richthofen (1833–1905)	A guidebook for scientific travellers in 1886 was largely descriptive of landforms, included a classification of mountains.	Von Richthofen, 1886

(Continued)

Table 1.4 *(Continued)*

Individual contributor	Contribution	Indicative references
W.J. McGee (1853–1912), US Geological Survey	Compiled a genetic classification of landforms similar to those used in subsequent textbooks.	McGee, 1888
W.M. Davis (1850–1934), Harvard	Associated landforms with stages in the cycle of erosion and furnished over 150 terms and phrases, some relating to landforms, with probably at least 100 generated by his students.	Davis, 1884, 1900
The position achieved by 2008 is perhaps reflected by Wikipedia which defines a landform as: a geomorphological unit, largely defined by its surface form and location in the landscape, as part of the terrain, and … typically an element of topography.		Wikipedia 2008 provides a list of 169 named landforms in 7 categories

formalized the recognition of the importance of landform (Davis, 1900: 158). Characteristics of the land surface of the Earth and its processes are not only reflected in language but also perceived through cultures, in art and music.

A second way of considering the surface of the Earth involves its contribution to environment or to nature. Einstein is reputed to have described environment as 'everything that isn't me' but usually it is thought of as the set of characteristics or conditions surrounding an individual human being, an organism, a group of organisms or a community. Of the several types of defined environment, the natural environment is a theoretical concept, which includes all living and non-living things occurring naturally on Earth, whereas the physical environment or biophysical environment refers to phenomena excluding humans. Nature comprises the essential qualities of environment: it can be seen as a group of interrelated objects that objectively exist in the world independent from humans, together with the biophysical processes that create and maintain the objects; or as a concept that is socially created and maintained by the human imagination (Soper, 1995: 136; Urban and Rhoads, 2003: 213). Nature, like environment, is therefore seen as a human construct existing only when there are individuals to recognize it, albeit variously perceived by different cultures. Natural history as the study of the physical and biological environment was popular in the nineteenth century but was displaced as environmental sciences became more specialized. Now human impact is so great that one book has been titled *The End of Nature* (McKibben, 1989).

Thirdly, the land surface of the Earth can be thought of as physical places, landscapes or regions. Every country has its well-known places – for example, the Rockies, Himalayas, Sahara Desert, or Amazon basin – each conjuring up a specific picture of the land surface. Place is the word used to refer to that particular part of space occupied by organisms or possessing physical environmental characteristics (Gregory, 2009). Landscape comprises the visible features of an area of land, including physical elements such as landforms, soils, plants and animals, the weather conditions, and also any human components, such as the presence of agriculture or the built environment. Physical places, as enshrined in place names or types of landscape, are not easy to define but progress has been made by recognizing physical or natural regions. Such regions provided a way of dividing up and studying the land surface of the Earth, illustrated by approaches to the regional geomorphology of the USA (Fenneman, 1931, 1938) and North America (Graf, 1987).

Early attempts to describe the land surface of the Earth in terms of physical regions were very detailed but often lacked the necessary understanding of the way in which processes worked; they often relied upon an assumed historical interpretation of physical landscape. One development in the mid-twentieth century recognized the **land system** as a subdivision of a region in which there is a recurring pattern of topography, soils, and vegetation reflecting the underlying rock types and climate of the area and affecting the erosional and depositional surface processes. This idea may have originated in early attempts at land classification associated with national surveys, for example from soil maps, but it was the resource surveys in undeveloped parts of Australia and Papua New Guinea, initiated in 1946 by the Australian Commonwealth Scientific Industrial Research Organization (CSIRO), that initiated the land system approach. By collating information on the geology, climate, geomorphology, soils, vegetation and land use of areas, these surveys designated land systems which could be divided into units or land facets composed of individual slopes or land elements. After the 1960s the advent of Geographic Information Systems (GIS), as the collection, analysis, storage and display of data spatially referenced to the surface of the Earth, enabled identification of patterns and relationships between phenomena and processes (Oguchi and Wasklewicz, 2010). Numerical land classifications based on 1 km grid squares have the advantage of combining many aspects of environmental character without being use-specific.

Fourthly, the surface of the Earth is perceived scientifically in different ways by particular disciplines. A meteorologist may see it in terms of its albedo, pedologists see soil landscapes, whereas ecologists recognize habitats. An overall view of the land surface is as part of Gaia, named after the Greek goddess of the Earth, first explicitly formulated by James Lovelock (see Box 1.1) who suggested that the whole planet acts as a self-regulating, living entity, requiring the interaction of the totality of physical, chemical and biological processes to retain the conditions vital for the survival of all life on Earth, in turn controlling the atmospheric conditions required for the biosphere.

1.3 COMPONENTS OF THE LAND SURFACE

Gaia, the whole planet Earth, is at one extreme, but what are the components that make up the land surface? Analogous to the building blocks of atomic structure has been the search for the building blocks of land surface. It was suggested (Linton, 1951) that the ultimate units of relief are the undivided flat or slope – all land surfaces are composed of a jigsaw of these morphological units. Combinations of these units have been modelled in particular ways (e.g. Figure 4.2, p. 78). There is a hierarchy of combinations of the smallest elements right up to the undivided continent (see Table 1.5), with, in between, physical regions such as the Rockies, the Loess plateau of China or the Deccan plateau of India. All environmental disciplines identify a basic environmental unit: in soil investigations the site was the place at which soil profiles were investigated, and the basic unit for soil study was the vertical section through all the constituent horizons of a soil, recognized as the soil profile since the time of Dokuchaev (see Box 6.1) and other Russian soil scientists in the late nineteenth century. As the profile is two dimensional, the pedon was introduced as the smallest unit or volume of soil that represents, or exemplifies, all the horizons of the soil profile: it is a vertical slice of soil profile of sufficient thickness and width to include all the features that characterize each horizon (Wild, 1993), is usually a horizontal, more or less hexagonal area of 1 m^2 but may be larger (Bates and Jackson, 1980), and is an integral part of many soil survey classification systems. Soil profiles, pedons and sites are then grouped, or classified, into the fundamental soil mapping unit which might be a series, defined as groups of soils with similar profiles formed on lithologically similar parent materials.

In ecology, habitat has come to signify description of where an organism is found, whereas niche is a complete description of how the organism relates to its physical and biological environment. A fundamental niche is that which an individual occupies in the absence of competition with other species; a realized niche is the niche occupied when competition is in progress (Watts,

Table 1.5 HIERARCHICAL CLASSIFICATION OF GEOMORPHOLOGICAL FEATURES (TIME AND SPACE SCALES ARE APPROXIMATE)

Typical units	Spatial scale km²	Time scale years
Continents	10^7	10^8–10^9
Physiographic provinces, mountain ranges	10^6	10^8
Medium and small scale units, domes, volcanoes, troughs	10^2–10^4	10^7–10^8
Erosional/depositional units:		
Large scale, large valleys, deltas, beaches	10–10^2	10^6
Medium scale, floodplains, alluvial fans, cirques, moraines	10^{-1}–10	10^5–10^6
Small scale, offshore bars, sand dunes, terraces	10^{-2}	10^4–10^5
Geomorphic process units:		
Large scale, hillslopes, channel reaches, small drainage basins	10^{-4}	10^3
Medium scale, slope facets, pools, riffles	10^{-6}	10^2
Small scale, sand ripples, pebbles, sand grains, striations	10^{-8}	

Sources: developed from Chorley et al. (1984) and Baker (1986; http://geoinfo.amu.edu.pl/wpk/geos/geo_1/GEO_CHAPTER_1.HTML).

1971). The niche has subsequently been defined as the habitat in which the organism lives, but also the periods of time during which it occurs and is active, and the resources it obtains there. Other terms developed include micro-habitat which is a precise location within a habitat where an individual species is normally found; a biotope is the smallest space occupied by a single life form, as when fungi grow on biotopes found in the hollows of uneven tree trunks. Habitat can also designate the living place of an organism or a com-munity, applying to a range of scales from the microscale, relating to organisms of microscopic or submicroscopic size, through to the macroscale at continen-tal or subcontinental levels.

The hierarchy of building blocks of the Earth's surface (see Table 1.5) sub-sequently developed in at least three ways. First, by involving organisms as well as the environment of the organisms, as reflected in biogeocoenosis, a Russian term equivalent to the Western term ecosystem, and involving both the biocoenosis, a term introduced by Mobius in 1877 for a mixed commu-nity of plants and animals, together with its physical environment (ecotope). Biotope was defined as an area of uniform ecology and organic adaptation, although subsequently thought of as a habitat of a biocoenosis or a micro-habitat within a biocoenosis. Ecological patches forming a mosaic connected by corridors in any scale of landscape can be employed in hydrology, analogous to a patchwork of geomorphological units nested at different scales (Bravard and Gilvear, 1996).

Secondly there has been the need to have an integrated unit for surface form, soil and ecology, and Russian studies recognized the urochischa as a basic physical–geographical unit of landscape with uniform bedrock, hydrological conditions, microclimate, soil and mesorelief (Ye Grishankov, 1973), reminiscent of the land system (Christian and Stewart, 1953).

Thirdly there has been recognition of sequences of units identified for climate (climosequences), relief (toposequences), lithology (lithosequences), ecology (bio-sequences), and time (chronosequences). In geomorphology a nine-unit hypothet-ical landsurface model (Dalrymple et al., 1969) showed how nine particular slope components could occur in landsurface slopes anywhere in the world (see Figure 4.2). Each component was associated with a particular assemblage of processes so that it was possible to predict how slopes could occur under different morphogenetic conditions. A similar approach was applied to pedogeomorphic research (Conacher and Dalrymple, 1977), although a simple 5-unit slope may be sufficient (e.g. Birkeland, 1984) and this scheme can be adapted for particular landscapes such as drylands (see Table 9.4).

The ecosystem is a good example of a functioning unit, a term invented by Tansley (1935) for a community of organisms plus its environment as one unit; therefore embodying the community in a place together with the environmental characteristics, relief, soil, rock type (the habitat) that influence the community. Ecosystems (see Chapter 3, p. 64) can vary in size from 1 to thousands of hectares and could be a pond upstream of a debris dam or a large section of the Russian steppe. The drainage basin, also a functional unit, is the land area

Table 1.6 DRAINAGE BASIN COMPONENTS (DEVELOPED FROM DOWNS AND GREGORY, 2004)

Drainage basin component	Provisional definition
River channel	Linear feature along which surface water may flow, usually clearly differentiated from the adjacent flood plain or valley floor.
River reach	A homogeneous section of a river channel along which the controlling factors do not change significantly.
Channel pattern	Or channel planform, is the plan of the river channel from the air; may be either single thread or multithread, varying according to discharge.
Floodplain	Valley floor area adjacent to the river channel.
River corridor	Linear features of the landscape bordering the river channel.
Drainage network	Network of stream and river channels within a specific basin; may be perennial, intermittent or ephemeral.
Drainage basin or catchment	Delimited by a topographic divide or watershed as the land area which collects all the surface runoff flowing in a network of channels to exit at a particular point on a river.

delimited by a topographic divide or watershed which collects all the surface runoff flowing in a network of channels to exit at a particular point along a river. The subsurface phreatic divide (the underground watershed defined by the water table) may not correspond exactly to the topographic divide on the surface. In the USA the term 'watershed' is often applied to small and medium-sized drainage basins, 'river basin' is used for larger areas, and in some countries the term catchment is used to describe small drainage basins. The drainage basin is dynamic because the water and sediment-producing areas expand and contract depending upon the catchment characteristics, the antecedent conditions prior to any storm event, and the character of the storm input. The components of the drainage basin nested within each other (see Table 1.6) are all affected by the dynamic character of the basin. The functional significance of the drainage basin is the reason it has been used as a basis for collecting hydrological information, for the **modelling** of flows such as flood forecasting, and for the management of physical resources.

More than 200 river basins, accounting for about 60% of the Earth's land area, extend over two or more countries. Fragmented planning and development of the associated trans-boundary river, lake and coastal basins are the rule rather than the exception. Although more than 300 treaties have been signed by countries to deal with specific concerns about international water resources and more than 2000 treaties have provisions related to water, co-ordinated management of international river basins is still rare, resulting in economic losses, environmental degradation and international conflict (World Bank, 1993). Many disciplines have an interest in the land surface, as considered in the next chapter.

Box 1.1

PROFESSOR JAMES LOVELOCK

Professor James Lovelock (1919–), trained in Chemistry (BSc Manchester, 1941), and Medicine (PhD 1948, London), is a pioneer in the development of outstanding environmental concepts, having been described as one of the great thinkers of our time. His ideas have created a context which can be considered when understanding the land surface of the Earth. Working as an independent scientist since 1961, he has been associated with universities in the UK and the USA, was made FRS in 1974, and a Companion of Honour in 2005. He is renowned for the Gaia hypothesis, formulated in the 1960s, and although it was largely ignored until the 1970s, it has subsequently become generally accepted as the Gaia theory. His 1979 book launched the idea into both scientific and humanistic communities and was at first much derided by the former, although subsequent books (1988, 2006, 2009) have been taken much more seriously. In 2006 he argued that change is now so substantial that Gaia may no longer be able to adjust, drawing the analogy of an old lady sharing her house with a growing, destructive group of teenagers – Gaia grows angry and if they do not mend their ways she will evict them (2006: 47). Referring to the deadly 3Cs (combustion, cattle, and chainsaw – Lovelock, 2006: 132), he suggests that we could be past the tipping point – the threshold at which the characteristics of the land surface change dramatically, identifying the potential fragility of the land surface of the Earth. His most recent book (Lovelock, 2009) argues that, although the planet will look after itself, humans need to be saved – and soon. Some hope is offered arising from geo-engineering.

Professor Lovelock is a scientist whose independent thought provides a context for analysing the land surface of the Earth, giving a challenging framework within which to think holistically about the Earth – present and future. His *Homage to GAIA – The Life of an Independent Scientist* was published in 2000. How should study of the land surface be conditioned by thinking of Gaia and the possibility that global change is inevitable (see Chapter 12, p. 281) and requires action?

FURTHER READING

Many books provide an introduction to the surface of the Earth including the oceans, such as:

Skinner, B.J., Porter, S.C. and Botkin, D.B. (1999) *The Blue Planet: An Introduction to Earth System Science.* Wiley, New York, 2nd edn.

and with numerous illustrations:

Janson-Smith, D., Cressey, G. and Fleet, A. (2008) *Earth's Restless Surface*. Natural History Museum, London.

An excellent summary covering the context for spatial units is:

Kent, M. (2009) Space: Making room for space in physical geography. In N. Clifford, S.L. Holloway, S.P. Rice and G. Valentine, *Key Concepts in Geography*. Sage, London, 2nd edn, pp. 97–118.

Throughout the following chapters it could be useful to refer to a dictionary such as:

Thomas, D. and Goudie, A. (eds) (2000) *The Dictionary of Physical Geography*. Blackwell, Oxford.

or a Companion:

Gregory, K.J., Simmons, I.G., Brazel, A.J., Day, J.W., Keller, E.A., Sylvester, A.G. and Yanez-Arancibia, Y. (2009) *Environmental Sciences. A Student's Companion*. Sage, London.

A stimulating read is:

Tooth, S. (2009) Invisible geomorphology? *Earth Surface Processes and Landforms* 34: 752–754.

TOPICS

1 As noted in the Preface all the individuals described in this and subsequent chapters are male. Many very significant contributions have been made by females in the subsequent generations – identify individuals and research their contributions (suggestions to consider: Cuchlaine A.M. King, Marie E. Morisawa, Angela M. Gurnell, Marjorie Sweeting).
2 Find values from the internet and other sources for a particular dimension (e.g. mean height, 10 highest mountains, 4 longest rivers) of the land surface, see how much variation occurs and suggest why.
3 For a specific area of the land surface consider how several time scales (see Table 1.3) might be applied – how easy is it to relate studies undertaken at different time scales?
4 Is nature possible without human beings?

2

STUDY OF THE
LAND SURFACE

In the last chapter it was suggested that study of the surface of the Earth may have been associated with disciplines whose prime purpose often lies elsewhere.

2.1 DISCIPLINES FOR THE LAND SURFACE

A discipline is defined by the *Oxford English Dictionary* as 'a branch of learning or scholarly instruction'. Instruction is provided in programmes of study which effectively create the academic world inhabited by scholars. Academic disciplines are often regarded as branches of knowledge taught and researched at higher education level, recognized by the academic journals in which research is published and by the learned societies and university departments to which practitioners belong. Although it is useful for academics to distinguish between disciplines, between geology and geography for example, one individual may not understand why the distinction is necessary or why there are differences in approach. Differences between disciplines can be reinforced by syllabi in schools and universities, or by the content of journals and books, and it has been suggested (Martin, 1998) that the way in which knowledge is organized and divided can be the subject of a power struggle so that confrontations almost like tribal wars may develop between disciplines! Against this background it is understandable why Rhodes (Rhodes and Stone, 1981; Rhodes et al., 2008: xii) contended that 'One of the problems with our conventional styles of teaching and conventional patterns of learning at the introductory undergraduate level is that the "subject" – whatever it may be – all too easily emerges as given, frozen, complete, canned'.

Whereas universities in mediaeval Europe had just four faculties (Theology, Medicine, Jurisprudence and Arts), university development in the middle and late nineteenth century saw the expansion of the curriculum to include non-classical languages and literature, science and technology. Each discipline developed its own epistemology or theory of knowledge, so that the philosophy of any one discipline evolved particular methods and concepts. Epistemology also developed as a core area of the study of philosophy concerned with the nature, origins and limits of knowledge. The size of any discipline is limited by what is termed closure, and it has been argued that there may be a spectrum of disciplines: some basic and very detailed, including physics and chemistry, others composite such as geology and geography (Osterkamp and Hupp, 1996). Any discipline has a set of defining practices or **paradigms** according to Kuhn who suggested (Kuhn, 1970: 12) that 'Successive transition from one paradigm to another via revolution is the usual developmental pattern of mature science'. Although some people argue that Kuhn's original paradigm idea is now too limited, it is valuable not least for the way in which it recognizes that a 'paradigm shift' can occur, although paradigms are shaped by both cultural background and historical context.

It is probably inevitable that disciplines concerned with the surface of the Earth are composite rather than basic. The first disciplines involving aspects of the land surface of the Earth were probably geology and biology. Although it has been argued that several disciplines, including geology, had their origins more than 1000 years ago, geology was really conceived in 1785 when James Hutton, often viewed as the first modern geologist, reflected his belief that the evolution of the Earth required millions of years when he presented a paper entitled *Theory of the Earth* to the Royal Society of Edinburgh. However Hutton's ideas were not extensively promulgated until nearly 50 years later, when they were included in a publication by Sir Charles Lyell in 1830. Geology is now defined in the AGI glossary (see Bates and Jackson, 1987) as 'The study of the planet Earth – the materials of which it is made, the processes that act on these materials, the products formed, and the history of the planet and its life forms since its origin. Geology considers the physical forces that act on the Earth, the chemistry of its constituent materials, and the biology of its past inhabitants as revealed by fossils' (Gregory et al., 2008).

Biology as the scientific study of life, examining the structure, function, growth, origin, evolution and distribution of living things, arose also as a single coherent field in the nineteenth century, when individual scholars were very influential: Hutton and Lyell in geology and Charles Darwin (1809–1882) in biology through development of the theory of evolution (see Box 3.1, p. 68). In the previous century Carl Linnaeus (1707–1778) had published the first edition of his classification of living things in 1735, the *Systema Naturae* which introduced a classification scheme employing Latin names for living organisms, paving the way for modern approaches to classification of plants and animals in taxonomy. Some individuals, like the explorer naturalist Alexander von Humboldt (1769–1859), influenced the development of studies of the surface of the Earth by the information gained from their travels, and Humboldt is often credited with a formative influence on the development of biogeography.

Whereas geology was most concerned with rocks and the evolution of the Earth and biology most concerned with life upon it, the discipline primarily concerned with the land surface of the Earth is geomorphology, which literally means write about (Greek *logos*) the shape or form (*morphe*) of the Earth (*ge*). The name first appeared in 1858 in the German literature but came into general use, including by the US Geological Survey, after about 1890. For some time the term physiography persisted in North America, eventually being used for regional geomorphology. Although originating in geology, geomorphology became more geographically based with the contributions of W.M. Davis (1850–1934) who developed a normal cycle of erosion, wrote more than 500 papers and books and came to have an extremely influential role on the development of understanding of the surface of the Earth – an influence which lasted until at least the 1960s (Gregory, 2000).

Early progress in the study of the land surface of the Earth was greatly influenced by individuals in particular disciplines. There were also differences between countries. In the twentieth century geology in the UK did not give much attention to the land surface *per se*, with a few notable exceptions. These included Arthur Holmes (1890–1965) who in 1913 proposed the first geological time scale, and wrote his textbook *Principles of Physical Geology*, first published in 1944; and J.K. Charlesworth (Finnegan, 2004) who was one of the most important individuals in the development of Quaternary science in Britain, publishing *The Quaternary Era* in 1957. The geological column was so extensively represented in the UK that, at a time of expansion of the range of sub-branches of geology, there was so much for geologists to investigate that geomorphology tended to figure within academic geography. In the USA a different situation arose, despite W.M. Davis as a Professor of Geography, because for much of the first three-quarters of the twentieth century geomorphology featured in geology departments, with notable contributors being Professors of Geology. Only in recent decades has there been a growth of geomorphology in US geography departments.

Wherever geomorphology was located, the disciplines concerned with the land surface of the Earth became progressively more specialized throughout the twentieth century, just as the disciplines of biology and geology recognized a series of separate branches. As specialized branches developed, the key role of particular individuals still prevailed but in addition there was a tendency for specialisms to develop in particular places. Thus W.M. Davis, who spent most of his working life in the stable environment of the east coast of the USA, commented, when he retired to the west coast, how he found a very different environment and concluded that 'the scale on which deposition, deformation and denudation have gone on by thousands and thousands of feet in this new-made country is 10 or 20 fold greater than that of corresponding processes on my old tramping ground' (Chorley et al., 1973: 647). It is interesting to speculate how geomorphology may have evolved differently if Davis or his equivalent had spent their life on the west coast of the USA. According to Tinkler (1985: 12) Charles Lyell in 1833 stated 'I occasionally amused myself with speculating on the different rate of progress which Geology might have made, had it been first cultivated with success in Catania where ... the changes produced in the historical era by the Calabrian earthquakes would have been familiarly known'. Such examples show how geomorphologists are affected by place and their environment. Someone living in close proximity to the San Andreas fault cannot fail to be influenced by the possibility of tectonic influence on the landscape, whereas earthquake activity is much less significant in Boston and on the east coast of the USA.

Environment is one reason why fashions have been important in the growth of geomorphology but in addition there is the influence of other scientific developments and of the intellectual climate of the times. Thus Sherman (1996) adopted the idea of fashion change (Sperber, 1990), that progress in the goals, subjects, methods and philosophies of science can often be attributed to the emergence of an opinion created by a new fashion leader. Fashion dudes make significant advances in their disciplines and Sherman (1996) instanced Davis, Gilbert, Strahler and Chorley as influencing the course of development of geomorphology, one of whom is introduced in Box 2.1.

One consequence of the growth of academic subdivisions and increased specialization, for geomorphology and for other subjects, was the realization that we should not lose sight of interrelationships affecting the land surface of the Earth. To ensure that land surface form and process was not isolated from study of soils, rocks and ecology, a multidisciplinary trend developed, reflected in the growth of environmental sciences and environmental studies. There are at least two environmental sciences: first a single-science, multidisciplinary field that began to develop in the 1960s and 1970s; and second environmental sciences as a generic term for all those disciplines which contribute to, and illuminate investigation of, the environment (Gregory et al., 2008). Environmental sciences are concerned with organisms and where they live, thus embracing the living (biotic) and inanimate (abiotic) components of the Earth's surface concentrated in the envelope within 50 km above the surface and a few hundred metres below it. One definition of environmental science is 'the sciences concerned with investigating the state and condition of the Earth' (ERFF, 2003).

A more recent development has been earth system science as the study of the Earth in terms of its various component **systems**, including the atmosphere, hydrosphere, biosphere, and lithosphere, embracing global cycling of important nutrients and other elements that maintain ecosystems on a planetary scale. However Clifford and Richards (2005) concluded that earth system science (ESS) constitutes an oxymoron, that neither should it be seen as an alternative to the traditional scientific disciplines, or to environmental science itself, nor regarded as a wholesale replacement for a traditional vision of environmental science, but rather as an adjunct approach. Subsequently it was suggested (Richards and Clifford, 2008) that LESS (local environmental systems science) would be a more appropriate focus for geomorphology.

We tend to interpret the land surface according to the way in which it has been studied and at least three alternatives are now perceived: geographical, interpreting the morphology and processes; geophysical, concentrating upon the broad structural outlines (see Church, 2005; Summerfield, 2005a, 2005b); and chronological, focused on the history of change (see **Quaternary Chronology**, Glossary p. 306). A further perspective could be added in planetary terms because it has been suggested (Baker, 2008b) that to be a complete science of landforms and landscapes geomorphology should not be restricted to the terrestrial portions of the Earth's surface but could include the landforms of the ocean floors and our neighbouring planets.

Studies of the land surface of the Earth developed with successive paradigms against a background where certain developments in understanding gradually became established (see Table 2.1). Although now often taken for granted, many represented great advances in their time, some relying upon a contribution by one individual, others emerging gradually over a number of years. It now seems difficult to believe that the views of Bishop Usher in 1654, that creation occurred on 23 October 4004BC and that the great Noachian flood occurred from 7 December 2349BC to 6 May 2348BC, held sway for so long and required sustained arguments to dislodge. Looking at Niagara Falls has prompted very different reactions (see Figure 2.1).

Table 2.1 SOME FOUNDATION MILESTONES FOR STUDYING THE LAND SURFACE OF THE EARTH

Advance	Particular individuals and dates	Significance
Surface processes	Leonardo da Vinci (1452–1519)	Notebooks show that he may have marked the transition from theoretical to observational and deductive methods but he was succeeded by others including Palissy (1510–1590), Bauer or Agricola (1494–1555).
Hydrological cycle	Pierre Perrault (1611–1680)	Showed for the Seine basin that precipitation was sufficient to sustain the flow of rivers in contrast to the long-held belief that subterranean condensation or return flow of seawater explained the discharge of water in springs and rivers. He probably provided the

(Continued)

Table 2.1   *(Continued)*

Advance	Particular individuals and dates	Significance
		foundation for our understanding of the hydrological cycle.
Natural history	Gilbert White (1720–1793)	Published the *Natural History of Selborne* in 1789, which transformed the way we look at the natural world, by focusing on natural history, so that he was recognized as one of the fathers of ecology.
Superposition	William Smith (1769–1839)	Credited with creating the first nationwide geological map in 1815, embracing the principle of superposition so that he became known as 'Strata Smith'.
Uniformitarianism	Charles Lyell (1795–1875)	Published *Principles of Geology* in 1830 with a subtitle *An attempt to explain the former changes of the earth's surface by reference to causes now in operation*. Uniformitarianism has been thought of as 'the present is the key to the past'. Includes actualism (effects of present processes) and gradualism (surface changes require long periods of time).
Glacial erosion	Louis Agassiz (1807–1873)	Credited with the idea in 1840 that glaciers erode and are responsible for many features in areas not now occupied by glaciers.
Evolution	Charles Darwin (1809–1882) – see Box 3.1	Published *On the Origin of Species* in 1859 proposing that progressive changes in populations occurred through sequential generations by the process of natural selection. This influenced thinking about aspects of the earth's surface including the cycle of erosion.
Human activity	G.P. Marsh (1801–1882)	In 1864 published *Man and Nature* which illustrated that man is 'a power of a higher

Table 2.1 *(Continued)*

Advance	Particular individuals and dates	Significance
		order than any of the other forms of animated life' and initiated conservation movement. 'Anthropogenic', referring to activities of humans, used in Russian literature from 1922.
Cyclic change	W.M. Davis (1850–1934)	Proposed that landscape can be understood in terms of structure, process and stage and that there are cycles of erosion whereby the land surface proceeds through stages of youth, maturity and old age.
Continental drift	Alfred Wegener (1880–1930)	His suggestion in 1915 of continental drift was later feasible with the advent of plate tectonics.
Systems	R.J. Chorley (1927–2002) – see Box 2.1	Introduced the systems approach to the study of the land surface of the Earth in accord with general systems theory as suggested by Von Bertalanffy in 1962.
Glacial chronology and oxygen-isotope stages	Cesare Emiliani (1922–1995), Harold Urey (1893–1981), Sir Nicholas Shackleton (1937–2006) – see Box 7.1 – and Neil Opdyke (1933–)	Relationships established between stable isotopes and environmental variables, following work by Urey and his students including Emiliani involved studies of the relation between oxygen isotopes and temperature in recent molluscs, and its application to determination of paleotemperatures. Oxygen isotope analysis of calcareous foraminifera within deep-sea cores has been one of the main techniques used for correlation and climatic reconstruction during the past 40 years. Shackleton and Opdyke (1973) identified 22 stages, interpreted the record in terms of continental ice-volume changes and assigned ages to each stage boundary,

(Continued)

Table 2.1   *(Continued)*

Advance	Particular individuals and dates	Significance
		providing a template widely used for correlation and for interpreting the terrestrial record.
Plate tectonics	Harry Hess (1906–69)	Emerged at a symposium in Tasmania in 1956 but had a number of earlier contributing elements including continental drift proposed by Wegener. Later work on sea floor spreading and magnetic field reversals by Hess and Mason was important in leading to construction of the theory in 1961.
Time scales	S.A. Schumm and R.W. Lichty	In 1965 recognized steady, graded and cyclic time scales (see Table 1.3). The geologic time scale had been developed over the period 1800–1850 but this paper showed how it was possible to link time scales to understand the land surface.

2.2 METHODS FOR MEASUREMENT AND ANALYSIS

Careful analysis by David Alexander (1982) showed that Leonardo da Vinci (1452–1519) progressed the move to more observational and deductive methods (see Table 2.1). However it took nearly 400 years before real understanding was achieved utilizing methods of investigation which involved basic data collection, analysis techniques, and scenarios for conclusions to be reached.

One of the most important requirements was the availability of maps, not only to locate places but also to give information about the shape and character of the land surface. In the UK the foundation of the Ordnance Survey in Britain in 1795 and the Geological Survey in 1801 were the beginnings of surveys providing basic information. Topographic maps often used contours to depict the land surface so that the spacing, shapes and patterns of those contours had to be interpreted to 'read' the shape of the land. More directly relevant were slope maps, derived by showing that areas of slope of particular angles could relate directly to land use practices, because slope categories could be directly related to angles at which agricultural implements can operate,

Figure 2.1 Niagara Falls

Reactions range from 'there's nothing to stop it', to Mahler 'Fortissimo at last', to scientific investigations (Tinkler, 1985: 96–8) that include recession rates and flow abstraction (Tinkler, 1993):

Period	Percentage of total flow after water abstraction for power generation	Recession rate m.yr^{-1}
1842–1905	100	1.28–1.52
1905/6–1927	72	0.98
1927–1950	60	0.67
Post 1950	34	0.10

Hayakawa and Matsukura (2009) suggest that the decreased recession rate of Horseshoe Falls is related to both artificial reduction in river discharge and natural increase in waterfall lip length, whereas that of American Falls is solely due to the reduction in flow volume.

or to the slopes angles at which mass movements occur. Morphological maps showing the distribution of slopes, were succeeded by geomorphological maps which included specific landforms. Such maps ideally characterize the surface morphology, indicate landform origin, date each section of the land surface, and indicate the rock types, sediments and soils beneath the surface. These requirements could not all be achieved in any one map series so that geomorphological maps produced in several countries each had their particular emphases. In one of the most successful schemes in Poland many physical geographers contributed to a national scheme producing maps at the scale of 1:50,000. Mapping schemes capable of international application were evolved and an approach summarized by Cooke and Doornkamp (1974, 1990) was adapted to conditions in several areas (Cooke et al., 1982), so that for many applied projects, geomorphological mapping was the central technique.

Geomorphological maps of Germany are available online at http://gidimap. giub.uni-bonn.de/gmk.digital/home_en.htm. More than 30 maps are available at two different scales. They were surveyed and generated during the priority research programme *Geomorphologische Detailkartierung in der Bundesrepublik Deutschland* (Detailed Geomorphological Mapping in the Federal Republic of Germany). The programme was funded by the German Research Foundation (DFG) and completed in 1986. The cartographic map consists of 27 maps at 1:25,000 and 8 maps at 1:100.000 scale as well as complementary booklets with annotations.

Enthusiasm for geomorphological maps was limited, because their production, certainly for whole countries, was prohibitively expensive. Therefore information already routinely collected by national mapping agencies was used wherever possible. However, a difficulty with many mapping series is that they require many years to complete and then need constant revision and updating. National topographic agencies undertaking this task include the United States Geological Survey (USGS) established in 1879 to collect data on land, water, energy and mineral resources. Information on other aspects of the character of the land surface included soil data – the national soil survey in the UK dated from 1949, and approximately half the area of continental USA was covered by soil maps 1899–1935 (Barnes, 1954). More integrated interpretative approaches focusing on integration of several aspects of the physical landscape include the land systems method (e.g. Cooke and Doornkamp, 1990; Mitchell, 1991; Verstappen, 1983), which can now be effected by databases and GIS, as remote sensing has greatly enhanced the availability of information about the land surface of the Earth (Lillesand et al., 2004).

Process data are also required for the land surface, including hydrological measurements, glacier surveys and coastal information. The first records made of river stage included those for the Elbe at Magdeburg 1727–1869 (Biswas, 1970); continuous river discharge measurements have been made on the river Thames since 1883; and there were c.1200 river gauging stations throughout Britain by 1975. Continuous measurements of streamflow in the USA began in 1900, with the basic network of gauging stations established during the period 1910–1940, so that by 1950 observation occurred regularly at about 6000 points. Permanent records have been made by tide gauges at coastal sites in the UK since 1860. Such growth in environmental monitoring emphasizes how recent the acquisition of data on physical environment has been, often for much less than 100 years. Many data sets and satellite imagery can now be downloaded (for example, The Global Runoff Data Center, UNEP, 2006).

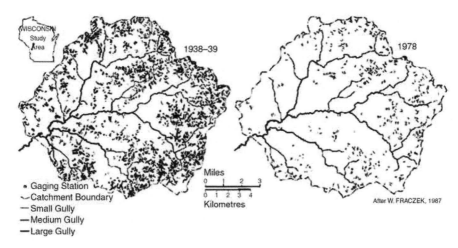

Figure 2.2 Decrease of drainage density 1938–78 in the upper Coon
Creek basin (adapted from Fraczek,1987 by Trimble, 2008)

Analysis of drainage density and other data suggests that present flood peaks
are a fraction of those of the late 1930s (see Figure 8.4).

Reprinted with permission.

Information on change can sometimes be obtained by comparing topographic
maps of different dates. However more frequent surveys are often required in the
case of glaciers for example. When using maps produced by national agencies it
is important to understand the conventions used for the description of land
surface features shown on particular map scales and editions for any one map
series. Thus in the case of the drainage network some maps show rivers and streams
that flow at 'normal winter level', others may show the extent of stream channels regardless of whether water is flowing in them or not, and there are
differences in conventions used not only from one map series to another but also
from one scale to another. The range of historical sources available for compar-
isons at different dates (Hooke and Kain, 1982) and data and artifacts for
dating geomorphological processes over the past century or more (Trimble, 2008)
can be valuable (see, for example, Figure 2.2).

All of the data needed for an investigation are seldom available from previously collected records, maps or surveys. Field techniques are therefore necessary
to acquire information, often quantitative, for analysis, and are for two major
purposes – to collect information on the shape and composition of the land surface and to measure processes. A great range of measurement techniques is now
available, some very innovative (see Table 2.2).

In addition to mapping as noted above, detailed surveying may be required
using techniques to map landscape materials; to obtain details of depth and
thickness of superficial deposits by coring or geophysical techniques; to
describe the character of landscape-forming materials including rock, sediment and soils; and to date changes (Dackcombe and Gardiner, 1983).
Secondly, measurements of process effectively provide information on the

Table 2.2 SOME RECENT TECHNIQUES THAT HAVE TRANSFORMED INVESTIGATIONS OF THE LAND SURFACE (SEE GLOSSARY, P. 303)

Purpose	Technique	Example of application
Site characterization and location	**Electronic distance measurement** (EDM)	Rapid field survey
	Close-range digital work station	Improved field monitoring
	Global positioning systems (GPS)	Rapid field survey
	Digital elevation models (DEMs)	Compute short-term spatial changes
	Digital camera, digital mapping	Coastal landform change
	Terrestrial photogrammetry	River bank erosion
	Ground penetrating radar	Sediment variations
	Airborne radar and radio echo-sounding	Basal ice conditions, water volumes in lakes in ice sheets
Process measurements	Digital loggers	Water quality monitoring, turbidity monitoring
	Continuous monitoring	Sand traps for aeolian events, bedload monitoring
	Acoustic Doppler velocimetry	3D velocities in rivers
	Magnetic techniques	Beach sediment sources
	Magnetic resonance imaging	Infiltration into soils
	O^{18} Deuterium	Hydrograph separation
Laboratory analytical techniques	Automated analysis	Greater number of samples can be processed and more properties analysed
	Scanning electron microscopy	Grain characteristics in sediments to indicate transport conditions
Modelling	Generalized linear modelling	Glacier surging, landslide susceptibility

transfer or flux of energy or mass within the physical environment of the land surface. However, because there is an infinite number of points in space and time with several process elements that could be measured at each point, a sampling strategy has to be used to select from what is an infinite population. Process monitoring may require empirical measurements over several years because existing records may not be available with spatial and temporal frequencies sufficient to meet the requirements of research programmes. Additional measurements can be obtained from small experimental areas but the enthusiasm for small instrumented areas encountered a number of problems including lack of control, replicability of measurements, insufficient representativeness, accuracy of data, and problems of finding suitable methods of analysis for the large amounts of data collected. However the development of process measurements often required where there are no regular national monitoring frameworks for process measurements can provide the basis for significant research as shown by Professor Des Walling (see Box 8.1).

In view of the difficulties of establishing and maintaining process measurements alternative methods of obtaining data are available. *Experimental investigations* embrace a range of approaches which include field plot experiments; laboratory hardware models which attempt to use scaled down versions of the real world; and analogue models which employ a different medium for investigation. Measurements by rainfall simulators, flumes, wave tanks and wind tunnels are examples, and their potential has probably not been fully explored (Mosley and Zimpfer, 1978) but the difficulty of overcoming the scale problem and of relating the observations to geophysical event sequences have been most evident. *Theoretical approaches* do not require the establishment of a monitoring framework, or long periods of time for empirical data collection, with theoretical models capable of application to much larger areas or systems than those that can be monitored in field experiments. However they require basic conservation equations, including energy and water balance equations, and they necessarily depend upon stipulating simplifying assumptions. They have been used most effectively for relatively simple morphological sequences where there is a clear relationship between the existing system and the processes operating upon it, as on hill slopes.

Methods for investigation of the land surface have recently benefited from great strides in techniques of remote sensing and GPS, in GIS and real time computation, and in dating techniques, for which some of the great advances are listed in Table 2.3. Remote sensing has not only revealed details previously only dreamed of, but can also easily provide frequently repeated imagery and access aspects of the land surface not previously possible. LiDAR (Light detection and ranging) uses laser pulses, has many applications including detection of faults, measurement of land uplift, monitoring glacier changes as well as measurement of forest canopy characteristics and can be used very effectively in conjunction with GIS. Global positioning systems (GPS) use a constellation of 24 satellites placed into orbit by the US Department of Defense which works anywhere in the world, in all weather conditions, 24 hours a day. Geographical information systems refer to the collection, analysis, storage and display of data which are spatially referenced to the surface of the Earth by a set of geographic co-ordinates (Heywood et al., 1998).

2.3 CONCEPTUAL IDEAS

Data about the form of the land surface or the processes operating upon it must be collected according to some a priori hypothesis. Basic theory was not always explicit, so that a well-known quotation opening a chapter on bases for theory in geomorphology (Chorley, 1978: 1) was 'Whenever anyone mentions theory to a geomorphologist, he instinctively reaches for his soil auger' – a comment made because geomorphological studies of the land surface had concentrated upon empirical observations and field investigations. A change

Table 2.3 SOME DATING METHODS FOR QUATERNARY DEPOSITS (ADAPTED FROM SOWERS ET AL., 2000, STOKES AND WALLING, 2003 AND GREGORY AND DOWNS, 2008)

Type of method	Method	Approximate age range	Basis of method
Siderial	Dendrochronology	10–4000	Ageing of living tree or correlation to chronologies for other trees
	Varve chronology	10–9000	Counting seasonal sediment layers back from present
	Sclerochronology	10–600	Counting annual growth bands in molluscs and corals
Isotopic	Radiocarbon	100–30,000	Radioactive decay of ^{14}C to ^{14}N in organic fissures tissues and carbonates
	Cosmogenic nuclides ^{10}Be, ^{26}Al, ^{36}Cl, ^{3}He, ^{14}C	400–10,000,000	Formation and decay of nuclides in rocks exposed to cosmic rays
	Potassium-argon (K-Ar), argon-argon (Ar-Ar)	10,000–20,000,000	Radioactive decay of ^{40}K in K-bearing silicate minerals
	Uranium series (^{234}U-^{230}Th, ^{235}U-^{231}Pa)	10–400,000	Radioactive decay of uranium and protégés in sedimentary minerals
	Short-lived radionuclides, lead-210 (^{210}Pb)	10–70	Radioactive decay of ^{210}Pb to ^{206}Pb
	Short-lived radionuclides, caesium-137 (^{137}Cs)	10–100	Radioactive decay of ^{137}Cs to ^{137}Ba
	Uranium-lead (U-Pb), thorium-lead (Th-Pb)	10,000–20,000,000	Measurement of Pb enrichment from decay of radiogenic Th and U
Radiogenic	Fission track	2000–20,000,000	Accumulation of damage trails from natural fission decay of ^{238}U
	Luminescence (TL, OSL, IRSL)	10–1,000,000	Accumulation of electrons in crystal defects due to radiation
	Electron-spin resonance	1000–1,000,000	Accumulation of electrons in crystal defects due to radiation

Table 2.3 (Continued)

Type of method	Method	Approximate age range	Basis of method
Chemical and biological	Amino-acid racemization (AAR)	200–2,000,000	Racemization of L-amino acid to D-amino acid in organic material
	Obsidian hydration	10–1,000,000	Increase in thickness of hydration rind on obsidian surface
	Lichenometry	10–10,000	Growth of lichens on freshly exposed rock surfaces
Geomorphic	Soil profile development	3000–100,000	Systematic changes in soil properties due to soil processes
	Rock and mineral weathering	10–100,000	Systematic alteration of rocks and minerals due to weathering
	Scarp morphology	2000–30,000	Progressive changes in scarp profiles due to surface processes
Correlation	Palaeomagnetism, secular variation	10–6000	Secular variations in the Earth's magnetic field
	Pamaeomagnetism, geomagnetic reversal stratigraphy	400,000–2,000,000	Reversal of the Earth's magnetic field recorded in magnetic minerals
	Tephrochronology	10–2,000,000	Recognition and correlation of tephra layers via unique properties
	Palaeontology	50,000–500,000	Progressive evolution
	Climatic correlations	1000–1,000,000	Correlation of landforms and deposits to known global climate changes

occurred, with greater awareness of general philosophical thinking, of a more scientific foundation, together with the advent of statistical and mathematical methods and with recognition of the types of model available to assist understanding.

Greater awareness of philosophical thinking meant that paradigms were recognized and thought was given to ways in which data were collected and analysed and how conclusions were reached. Distinction between inductive and deductive methods coincided with investigations seeking general models rather than those based exclusively upon detailed field investigations of specific areas. Positivists dominated until the late 1950s (Brown, 1996) maintaining that scientific theories should be evaluated solely on the basis of observational data in accordance with a set of formal rules, but this approach was flawed because of the absence of sound principles of verification or induction and because observations are actually theory-dependent (Haines-Young and Petch, 1986). A move towards a critical rationalist view, whereby a rational basis for scientific knowledge is provided by deducing the consequences of theories and then attempting to expose their falsity by critical testing, appreciated that facts are not objective because they are observations perceived in a particular way, according to the technology available for measurement and observation. The variables perceived to be important, and selected for measurement, are chosen in the light of some preconceived theory. In this post-positivist state it is possible to have a multi-paradigm state; it takes years to develop a new paradigm, heretical thinking must go on for a long time before paradigm change, so that pluralism is required (Slaymaker, 1997).

One of the prevailing approaches is described as scientific realism, deriving from the contributions of C.S. Peirce at the end of the nineteenth century and of Karl Popper since the 1930s. Critical realism adopts the perspective that the aim of science is *to seek the truth* not merely to solve problems, readily acknowledging that all aspects of scientific enquiry are theory laden, that current theories are approximately true and are the foundation for scientific progress (Baker, 1996).

Greater awareness of the philosophy of science paralleled a greater scientific foundation. Yatsu (1966: 13) expressed the idea that 'Geomorphologists have been trying to answer the *what, where* and *when* of things, but they have seldom tried to ask *how*. And they have never asked *why*. It is a great mystery why they have never asked *why*'. Asking the why question required awareness of basic scientific principles. As geomorphology has been characterized as a mesoscale science, then, as in other earth sciences, the extent to which it is realistic to extend geomorphological processes to the micro scale is debatable. Geomorphology needs to identify the physical principles underlying landscape processes because this will not necessarily prejudice a mesoscale approach and has been achieved by sedimentology (for example, Allen, 1970) and by soil science.

A model is any abstraction or simplification of reality providing a major tool in addressing the limitations of laboratory experimentation and environmental records in extending the bounds of space and time in environmental understanding (Lane, 2003). Several stages of modelling can be employed as a basis for investigations of the land surface of the Earth in a problem-solving context including (Huggett, 1980):

- *a lexical phase* identifying the components investigated in a particular problem
- the *parsing phase* establishing relationships between the components

- the *modelling phase* expressing relationships in a type of model (conceptual, mathematical empirical or mathematical deterministic) followed by calibration of the model
- the *analysis phase* attempting to solve the system model

If not successful the procedure is repeated with a modified model.

The enormous growth in design, creation and use of databases, together with the rapid decrease in the cost of computing power with micro computers, means that separate categories of statistical, mathematical and databased modelling are already becoming redundant (Macmillan, 1989: 310).

2.4 DEBATES AND PARADIGM SHIFTS

Communication of ideas requires societies, journals and books as well as the internet. Although geomorphology research was originally published as part of geographical (earliest founded in 1821 in Paris) or geological societies (London founded in 1810), geomorphological societies were established later, including the British Geomorphology Research Group (now the British Society for Geomorphology) founded in 1959/60. Many societies publish journals but others were independently created, some developed for subfields (see Table 2.4). Most journals have expanded enormously so that *Earth Surface Processes* which began in 1976 with 4 issues (totalling 395 pages) had 14 issues (2306 pages) in 2008.

Increased knowledge about the surface of the Earth, with consequent increases in societies, journals, books and students, inevitably encouraged major debates – or paradigm shifts. Many occurred during the history of geomorphology, impressively dealt with by R.J. Chorley (see Box 2.1) in three scholarly volumes (Chorley et al., 1964, 1973; Beckinsale and Chorley, 1991) now complemented by a fourth (Burt et al., 2008). A major debate concerned the contributions by W.M. Davis – with most of one 874-page volume devoted to his contribution (Chorley et al., 1973) and more than 20% in Part II of the subsequent volume (Beckinsale and Chorley, 1991). His contributions are noted in Table 2.5 with a selection of other paradigms. Other topics that might have been included: whether **uniformitarianism/catastrophism** is necessary for our understanding of the land surface; is there certainty in explanations or are they ruled by chaos; and should we examine cultural differences and ethical considerations. However Table 2.5 demonstrates how study of the land surface has to be undertaken in the context of a prevailing idea or conceptual hypothesis, with any investigation subject to ideas prevailing at the time. As suggested above, pluralism is necessary and Slaymaker (1997) argued that there is no recognizable central concept in geomorphology and no problem focus. That could be a good thing – we should approach the land surface of the Earth with an open mind but with a range of ideas to test. A recent debate concerns present global change – are those who study the land surface of the Earth doing enough to explore the possible impact of present trends in global climate change?

Table 2.4   EXAMPLES OF JOURNALS PUBLISHING PAPERS ON THE LAND SURFACE OF THE EARTH

Geomorphological journals are given in bold, followed by examples of other categories. Many geographical and geological journals such as *Geographical Journal* (1831–), *Bulletin Geological Society of America* (1890–), also contain important geomorphological papers.

Year initiated	Journal	Comments
	Zeitschrift fur Geomorphologie	Publishes papers from the entire field of geomorphological research, both applied and theoretical. Since 1960 has published 153 Supplementbände (Supplementary volumes) for specific topics.
1950	***Revue de Geomorphologie Dynamique***	Edited and inspired by Professor Jean Tricart.
1960	***Geomorphological Abstracts***	At first published abstracts of papers in geomorphology but later expanded to Geo Abstracts covering many related disciplines.
1977	***Earth Surface Processes and Landforms***	From 1977–1979 was *Earth Surface Processes* but then expanded its name. Described as an International Journal of Geomorphology publishing on all aspects of Earth Surface Science.
1989	***Geomorphology***	Publishes peer-reviewed works across the full spectrum of the discipline from fundamental theory and science to applied research of relevance to sustainable management of the environment.
Hydrological	*1963 Journal of Hydrology* *1970 Nordic Hydrology* *1971 Water, Air and Soil Pollution* *1984 Regulated Rivers* *1987 Hydrological Processes*	
Glacial	*1947 Journal of Glaciology* *1977 Polar Geography and Geology* *1980 Annals of Glaciology* *1990 Permafrost and Periglacial Processes* *1990 Polar and Glaciological Abstracts*	
Coastal	*1973 Coastal Zone Management* *1984 Journal of Coastal Research*	
Arid	*1981 Journal of Arid Environments* *2009 Aeolian Research*	
Quaternary	*1970 Quaternary Research, Quaternary Newsletter* *1972 Boreas* *1982 Quaternary Science Reviews* *1985 Journal of Quaternary Science* *1990 Quaternary Perspectives, Quaternary International* *1991 The Holocene*	
Physical geology	*1973 Geology* *1975 Environmental Geology*	

Table 2.4   *(Continued)*

Year initiated	Journal	Comments
Physical geography	1977 *Progress in Physical Geography* 1980 *Physical Geography*	
Environment	1972 *Science of the Total Environment* 1973 *Catena* 1976 *Geo Journal, Environmental Management* 1990 *Global Environmental Change* 1997 *Global Environmental Outlook*	

Table 2.5   SOME DEBATES OR PARADIGM SHIFTS

Subject/issue	Established position	Alternative view
Davisian cycle of erosion	Landforms are a function of structure, process and time, and evolve through stages of youth, maturity and old age. This conceptual model was devised for a normal cycle of erosion, applied to temperate landscapes, but alternatives of arid and marine cycles also proposed, and in the course of landscape evolution, there could be accidents, either glacial or volcanic. Land surface was interpreted in terms of the stage reached in the cycle of erosion and came to be dominated by a historical interpretation concentrating upon the way in which landscapes had been shaped during progression through stages in a particular cycle, towards peneplanation – an approach termed denudation chronology (see Gregory, 2000: 38–42).	Approach was essentially qualitative and did not have a sound scientific foundation; it appealed to persons with little training in basic physical sciences but who like scenery and outdoor life, and focused on parts of the land surface and ignored others. It was partial in that it focused exclusively upon the historical development of the land surface of the earth without giving sufficient attention to the formative processes operating (see Chorley et al., 1973).
Developments of cycle	Debates about origin of planation surfaces as of subaerial or marine in origin.	Attracted similar objections to those to the cycle approach.
Alternative models: slope-based	Walther Penck in 1924 proposed parallel recession of slopes rather than progressive decline as suggested in the Davisian cycle.	Subject to similar limitations as Davisian cycle and some slopes shown to decline (see Tinkler, 1985: 166–9).
Alternative models: pediplanation	Series of papers culminated in book (King, 1962) which argued that pediplanation	Subject to some of the criticisms levelled at the Davisian approach, overtaken

(Continued)

Table 2.5 *(Continued)*

Subject/issue	Established position	Alternative view
	was the norm, that lower latitudes were the norm rather than temperate areas and correlated surfaces from Africa to South America, Australia and other parts of the world's plainlands. Embraced earth movements in terms of cymatogenic arching (see Gregory, 2000: 142–3). Formalized approach in 50 canons of landscape evolution (King, 1953: 747–50).	by the advent of plate tectonics, and overshadowed by other approaches including climatic geomorphology in Europe including interpretation of landscape in terms of double surface of levelling (see Ollier, 1995).
Emphasis on earth surface processes	Lead given by Gilbert (1914) was largely ignored until 1960s when more attention given to stresses acting on materials, aided by mathematical and statistical methods and by development of new models. The book by Leopold et al. (1964) was particularly influential (see Box 4.1).	Towards the end of the twentieth century some geomorphologists felt that the study of the energetics of the land surface had 'perhaps robbed the subject of some of its scope and depth' (Thomas, 1980), that the original intention of process research, to explain landforms, had been forgotten (Conacher, 1988: 161), and that evolutionary geomorphology (Ollier, 1979, 1981) is more appropriate to some areas of the world such as Australia.
Quaternary science and interdisciplinary research	International Union for Quaternary Research (INQUA) established in 1928 and one example of need to involve range of earth and environmental scientists including archaeologists, biologists, oceanographers, limnologists. Catalysed by developments in pollen analysis, radiocarbon dating and subsequent dating methods, and by refinement of the Quaternary time scale.	Quaternary science community focused on chronology and stages of development tended to become separate from that concerned with earth surface processes and with landforms. At the end of the twentieth century the two communities have interacted much more profitably.
Landscape reenchantment of geomorphology	Baker and Twidale (1991) perceived disenchantment to have arisen from denigration of the study of landform, the infatuation with theory, the dominance of models, and the emphasis upon applications, so that they proposed	That studies of process continued to be necessary together with theory and modelling to provide the necessary foundation for understanding how the land surface works. Some of the process studies necessarily began at small detailed scales

Table 2.5 *(Continued)*

Subject/issue	Established position	Alternative view
	reenchantment where 'the landscape must be viewed with awe and wonder, that is, as something far superior to the idealizations that we seek to impose upon it'.	but could later be extended to regional or continental scales.
Macro geomorphology	Proposed to have a more secure basis of geophysical, sedimentological and geochronometric data (Summerfield, 1981) and led to the first textbook to fully integrate global tectonics into the study of landforms (Summerfield, 1991).	Distinction drawn by Church (2005) between the diminishing role of 'geographical geomorphologists' and the growing role of geophysicists whereas Summerfield (2005a, 2005b) counters that there is enormous scope to advance geomorphology as a whole, probably at its most exciting time since it emerged as a discipline.
Human activity and applications	Influence of human activity had been ignored in research until mid-twentieth century and potentially very influential on processes.	Alternative group tended to ignore human activity and focus upon land systems relatively little affected by human impact.

BOX 2.1

PROFESSOR DICK CHORLEY

Professor Dick Chorley (1927–2002) was inspirational in evolving geomorphology from the first to the second half of the twentieth century. 'A reformer with a cause', the very apt title of the first chapter (Beckinsale, 1997) of a book compiled in his honour (Stoddart, 1997), was so appropriate for someone so extremely pleasant, unassuming yet ebullient, with such warm good humour and gentle self-effacement, but responsible for bringing many ideas critical to the development of geomorphology – including general systems theory and quantification. After his school career in Somerset, including Minehead Grammar, he was a lieutenant in the Royal Engineers (1946–48), studied at the University of Oxford and graduated with a BA in Geography in 1951. Subsequently he spent several years in the USA in Geology departments where he interacted with A.N. Strahler and his students including Stanley Schumm and Mark Melton, and was inspired by the contributions of Luna Leopold (see Box 4.1, p. 120). In 1958 he was appointed to

(Continued)

(Continued)

the School of Geography, University of Cambridge where he was subsequently Lecturer (1962), Reader (1970) and Professor (1974). Interaction with US geomorphologists early in his career was a fundamental influence conditioning his approach to geomorphology as the study of landforms, and his conviction that the discipline should be scientific, quantitative, process-based and rational. He sought to replace the prevailing paradigm of the Davisian cycles of erosion with a quantitative model-based paradigm which emphasized General Systems Theory and numerical modelling. Robert Beckinsale (1997: 5) reveals that in his first degree examinations Chorley obtained two of his lowest marks in the landform papers, thus instigating his move to the USA, so vital for the development of his influential ideas – and for the way in which geomorphology was to develop.

In addition to authoring many extremely influential papers, his book with Barbara Kennedy in 1971 on *Physical Geography: A Systems Approach* provided a breath of fresh air. Other strands to his bow were the magisterial *History of the Study of Landforms* (3 volumes totalling 2048 pages published 1964, 1973 and Beckinsale and Chorley, 1991), and *Geomorphology* with S.A. Schumm and D.E. Sugden in 1984. His contributions on models and other key developments were influential not only in geomorphology but throughout geography, including his 1978 book with R.J. Bennett on *Environmental Systems*, and his 1968 text co-authored with Roger Barry on *Atmosphere, Weather and Climate* which continued for eight editions. Many of his written and edited contributions involved collaboration with others, including the influential *Water, Earth and Man* (1969) with its subtitle *A Synthesis of Hydrology, Geomorphology and Socio-economic Geography*. His many awards include the Patron's medal of the Royal Geographical Society (1987), his many contributions include originating *Progress in Geography* which evolved to become *Progress in Physical Geography* and *Progress in Human Geography*, and many of his fruitful collaborations were with Professor Peter Haggett.

This range of contributions shows what is needed to fundamentally change the direction of landform studies, which Professor Dick Chorley achieved, but to fully appreciate the 'climate' of those times, the transforming changes to which he contributed, and some of the anecdotal context, read:

Beckinsale, R.P. (1997) Richard J. Chorley: A reformer with a cause. In D.R. Stoddart (ed.), *Process and Form in Geomorphology*. Routledge, London and New York, pp. 3–12.

Stoddart, D.R. (1997) Richard J. Chorley and modern geomorphology. In D.R. Stoddart (ed.), *Process and Form in Geomorphology*. Routledge, London and New York, pp. 383–399.

FURTHER READING

Further details on the development of geomorphology are included in:

Gregory, K.J. (2000) *The Nature of Physical Geography.* Arnold, London, esp. pp. 63–66, 118–124.

An indication of how a range of disciplines contribute to our cumulative understanding is:

Rhodes, F.H.T., Stone, R.O. and Malamud, B.D. (2008) *Language of the Earth.* Blackwell, Oxford.

A stimulating read is provided by:

Richards, K.S. and Clifford, N. (2008) Science, systems and geomorphologies: why LESS may be more. *Earth Surface Processes and Landforms* 33: 1323–1340.

A comprehensive coverage of geomorphology is provided in:

Summerfield, M.A. (1991) *Global Geomorphology.* Longman, Harlow.

TOPICS

1 Access a model on the internet and explore the limitations and applications (see Brooks, 2003 in Rogers and Viles, 2003 for website addresses).
2 What disciplines are involved in research investigations of the surface of the Earth?
3 Should geomorphology be thought of as the science of the study of landforms or as the study of the processes and form of the land surface of the Earth? Should it include other planets? (See Baker, 2008b.)
4 For an area/landscape that you know well, envisage how an investigation could employ different geomorphological approaches.
5 Could you conceive of a geomorphology without contributions from W.M. Davis? Do you agree that without Davisian geomorphology the discipline would have not been as coherent as it is?

PART II
DYNAMICS OF THE LAND SURFACE

3

CONTROLS AFFECTING THE LAND SURFACE

Rocks, climate, soil, vegetation, land use and human activity all exercise controls upon the land surface, often by influencing Earth surface processes. A context (see section 3.1) is necessary for the controls upon the surface (section 3.2) and the processes operating (Chapter 4).

3.1 SPHERES, CYCLES AND SYSTEMS

The Earth and its environment, in the envelope from about 200 km above the Earth's surface to the centre of the Earth, can be thought of as comprising a series of spheres, rather like concentric shells, although some are not discrete and overlap with others. The land surface is one such sphere. Such division of the Earth's environments is a useful basis for positioning study by different disciplines. Atmosphere had been used since the late seventeenth century but because land, water, living things and air are the main ingredients of environment, it was the four spheres of lithosphere, hydrosphere, biosphere and atmosphere which were identified by the Austrian geologist Suess in 1875. Other spheres were subsequently identified (see Table 3.1), producing so many that earth and life scientists may have become sphere crazy (Huggett, 1997: 3).

Because any academic discipline is limited in size by **closure** (Glossary, p. 304), no single discipline can study all the major spheres now identified. However spheres can help in defining disciplines and research investigations in three ways. First, some spheres are largely the province of one discipline, in the way that the biosphere is dominantly the realm of the biological sciences, meteorology is concerned with the atmosphere, and the lithosphere is dominantly the domain of the geologist. Secondly some disciplines transcend several

Table 3.1   SOME EARTH SPHERES THAT HAVE BEEN SUGGESTED
(THOSE DIRECTLY RELEVANT TO THE LAND SURFACE ARE SHOWN)

Sphere	Definition
Stratosphere	Stable layer about 10–50 km above the troposphere in which temperature is largely independent of altitude and averages –60°C.
Troposphere	Part of the atmosphere between the earth's surface and the tropopause (8–16 km above earth's surface) which marks a sharp change in the lapse rate of temperature change with altitude above the surface of the earth.
Atmosphere	Gaseous envelope of air surrounding the earth and maintained by gravitational attraction, without a clear upper limit but about 200 km maximum.
Heterosphere	Upper portion of a two-part division of the atmosphere (the lower portion is the homosphere) where molecular diffusion dominates and the chemical composition of the atmosphere varies according to chemical species. No convective heating at this height.
Homosphere	Portion of the earth's atmosphere, up to an altitude of about 80 km above sea level, in which there is continuous turbulent mixing, and hence the composition of the atmosphere is relatively constant. (Homosphere sometimes used for the biosphere as modified by human activity.)
Hydrosphere	Water body of the earth in liquid, solid or gaseous state and occurring in fresh and saline forms.
Geosphere	Usually several spheres (e.g. lithosphere + hydrosphere + atmosphere, or core, mantle and all layers of crust) or for the zone of interaction on, or near, the Earth's surface involving the atmosphere, hydrosphere, biosphere, lithosphere, pedosphere and noosphere. Sometimes used for the lithosphere.
Relief sphere	Totality of the earth's topography.
Toposphere	Representing the interface of the pedosphere, atmosphere and hydrosphere.
Landscape sphere	Zone of interaction at or near the Earth's surface involving atmosphere, hydrosphere, biosphere, lithosphere, pedosphere and noosphere.
Cryosphere	That part of the Earth's surface where water is in a solid form, as snow or ice, including glaciers, ice shelves, snow, icebergs and snow fields.
Noosphere	The realm of human consciousness in nature or the 'thinking' layer arising from the transformation of the biosphere under the influence of human activity. May be regarded as synonymous with the anthroposphere.
Anthroposphere	Including human activity and constructions by the human population, including cities, bridges, dams and roads.
Geoecosphere	Where other spheres (biosphere, troposphere, atmosphere, pedosphere and hydrosphere) interact.
Biosphere	Where living organisms occur on earth thus overlapping with the hydrosphere, lithosphere and atmosphere.
Ecosphere	Living and non-living components of the biosphere.
Pedosphere	Outermost layer of the Earth composed of soil and subject to soil-forming processes.
Lithosphere	Rocks of the earth's crust and a portion of the upper mantle, up to 300 km thick and more rigid than the asthenosphere beneath.

Table 3.1   *(Continued)*

Sphere	Definition
Asthenosphere	Deformable zone within the upper mantle of the Earth extending from 50 to 300 km below the surface of the Earth sometimes to a depth of 700 km. Characterized by low-density, partially molten rock material chemically similar to the overlying lithosphere. Upper part of the asthenosphere believed to be the zone involved in plate movements and isostatic adjustments.

Source: developed from Gregory et al., 2008: 103–105, where other spheres are defined.

spheres, being concerned with the interactions between them in the way that physical geography is concerned with the surface of the Earth and with the interactions between several spheres, particularly between the atmosphere, lithosphere and biosphere. Focused on the hydrosphere, hydrology developed in several disciplines, including civil engineering, geology, physical geography, geomorphology and ecology, but eventually emerged as a discrete discipline in view of its importance in relation to flow measurement, water resources and flood management. Similarly pedology or soil science subsequently became a separate discipline in view of its importance for soil survey and agricultural land planning. Thirdly the framework provides a convenient way of identifying the controls upon a particular sphere and the associated disciplines, showing how the land surface of the Earth is positioned (see Table 3.1).

At first it may seem unsatisfactory that no one single sphere relates primarily to the land surface but this is because many of the dynamic interactions between the spheres have been expressed by cycles and systems. Cycles express the way in which matter and energy is transferred between the spheres because, as material on the earth's surface cannot be created or destroyed, then materials must cycle. Understanding of particular cycles emerged gradually. For example the hydrological or water cycle expresses the way in which water in the hydrosphere (the water of the earth in liquid, solid or gaseous state and occurring in fresh and saline forms) moves between stores in the oceans (c.94%); in groundwater (>4%); in ice sheets and glaciers (c.2%); with smaller amounts in the atmosphere and on land – in lakes and seas, rivers and in the soil. Solar energy drives the hydrological cycle, providing the lift against gravity necessary as water vapour is transferred by evaporation to form atmospheric moisture and then precipitation. The energy required to drive the global hydrological cycle is equivalent to the output of about 40 million major (1000 MW) power stations (Walling, 1977), and about one-third of the solar energy reaching earth is used evaporating water with about 400,000 km^3 evaporated each year; the entire contents of the oceans would take about 1 million years to pass through the water cycle. Water (H_2O) that is constantly moving between the stores changes in phase from gaseous to liquid to solid in different parts of the cycle; such changes or fluxes are extremely important and complex but some, such as precipitation, occur for a relatively small proportion of the time.

This example indicates the questions we can ask, including what transfers (water in this example), where the stores (reservoirs) are, what transfers or fluxes (amount of material transferred from one reservoir to another per unit time) occur, and what cycles (movements including at least two or more connected reservoirs) exist. Some answers to these four questions are shown in Table 3.2, and provide the basis for the scientific study of energy flows, known as energetics. This way of describing the Earth's engine shows how the land surface of the Earth plays a part. Cycles detailed in Table 3.2(A) illustrate how it is possible to envisage the Earth's environment in terms of the energy supplied to drive it and how energy flows relate to world cycles for which laws have been stated to express the controls upon energy distribution (Table 3.2(B)) since the time of Isaac Newton (1643–1727).

Although some cycles have been known since the late nineteenth century, it was in the twentieth century that systems were clearly identified. *General systems theory* was proposed at a philosophical seminar in Chicago in 1937 by the biologist Ludwig van Bertalanffy as an analytical framework and procedure for all

Table 3.2   CYCLING AND ENERGETICS

(A) Cycle	Stores or reservoirs	Aspects of transfers or flux and budget
Hydrological	Oceans (c.94%), groundwater (>4%), in ice sheets and glaciers (c.2%), with smaller amounts in the atmosphere and on land – in lakes and seas and rivers and in the soil.	Evaporation, precipitation, runoff. Solar energy – about one-third of that reaching the Earth is used evaporating water with about 400,000 km^3 evaporated each year. Entire contents of the oceans would take about 1 million years to pass through the water cycle.
Geological: the creation and destruction of rocks, usually measured in hundreds of millions to billions of years	*Rock*: Igneous, sedimentary and metamorphic rocks.	Involves the processes that produce, change and weather rocks and soils, involving weathering, transport and deposition of sediment.
	Tectonic: involves movement of large plates in the lithosphere.	Driven by geothermal energy and can provide pressure and thermal conditions to metamorphose rocks (see Table 3.4).
Biogeochemical: cycling of elements, minerals and compounds	The whole or part of the atmosphere, ocean, sediments and living organisms.	The 'big six' elements – carbon, nitrogen, phosphorous, hydrogen, oxygen and sulphur – are the building blocks of life. Variations in residence times
	Carbon: Over 99% in carbonate rocks and organic deposits.	Photosynthesis fixes carbon dioxide from air and water. Plants and animal respiration, decomposition.

Table 3.2 *(Continued)*

(A) Cycle	Stores or reservoirs	Aspects of transfers or flux and budget
	Nitrogen: The nitrogen cycle is one of the most complicated material cycles in nature. The atmosphere (80% N_2) contains most of the nitrogen but there are also significant amounts in living organisms and in the soil.	Nitrogen fixation is the process of converting N_2 to nitrate or ammonia which can then be used. Major processes in the global nitrogen cycle include nitrogen fixation (by both biological and industrial processes), uptake and release by organisms, denitrification (the conversion of nitrate back to N_2), soil erosion, runoff, and flux in rivers, and burial in marine sediments. Eutrophication of lakes and rivers. Five main reactions in nitrogen cycling are: fixation, assimilation of nitrate to organic N, mineralization, nitrification oxidation process, and denitrification.
	Sulphur: Mainly in lithosphere, also in hydrosphere and vital in proteins.	Second to bicarbonate as the most abundant anion in rivers, major cause of acidity in natural and polluted rain water, a key ingredient in rock weathering. Sulphur cycle one of the most affected by human activity.
(B) 'Laws' suggested for ecology (Commoner, 1972)	Indicate principles of energy flows: • everything is connected to everything else • everything must go somewhere • Nature knows best • There's no such thing as a free lunch, because somebody somewhere must foot the bill	
Principles or laws (see Odum and Odum, 1976; Phillips, 1999)	Conservation of energy	Energy cannot be created or destroyed but can change form.
	Degradation of energy	Introduces entropy as a measure of technical disorder to signify the extent to which energy is unable to do work.
	Systems which use energy best survive	Maximum power principle or minimum energy expenditure principle.

sciences. However it was not widely accepted in geomorphology until a paper by Chorley (1962, see Box 2.1). The term ecosystem had already been proposed as a general term for both the biome which was 'the whole complex of organisms – both

Table 3.3   SYSTEM TYPES AND PRINCIPLES

System types (after Chorley and Kennedy, 1971)	Definition
Morphological systems	Comprise morphological or formal instantaneous properties integrated to form recognizable operational parts of physical reality, with the strength and direction of connectivity revealed by correlation analysis.
Cascading systems	Composed of chains of subsystems which are dynamically linked by a cascade of mass or energy so that output from one subsystem becomes the input for the adjacent subsystem.
Process-response systems	Formed by the intersection of morphological and cascading systems and emphasizing processes and the resulting forms.
Control systems	Where intelligence can intervene to produce operational changes in the distribution of energy and mass.

Eleven principles of earth surface systems (Phillips, 1999) (See Glossary p. 307)
- Earth surface systems are inherently unstable, chaotic and self-organizing
- Earth surface systems are inherently orderly
- Order and complexity are emergent properties of earth surface systems
- Earth surface systems have both self-organizing and non-self organizing modes
- Both unstable/chaotic and stable/nonchaotic features may coexist in the same landscape at the same time
- Simultaneous order and irregularity may be explained by a view of earth systems as complex nonlinear dynamical systems
- The tendency of small perturbations to persist and grow over times and spaces is an inevitable outcome of earth surface systems dynamics
- Earth surface systems do not necessarily evolve toward increasing complexity
- Neither stable, non self-organizing nor unstable, self-organizing evolutionary pathways can continue indefinitely in earth surface systems
- Environmental processes and controls operating at distinctly different spatial and temporal scales are independent
- Scale independence is a function of the relative rates, frequencies and durations of earth surface phenomena

animals and plants – naturally living together as a sociological unit' and for its habitat, with all the parts of the ecosystem 'regarded as interacting factors which, in a mature ecosystem, are in approximate equilibrium: it is through their interactions that the whole system is maintained' (Tansley, 1946: 207). Applied to the land surface, four types of system were identified (see Table 3.3), with a system defined as a set of elements with variable characteristics, the relationships between the characteristics of the elements, and the relationships between the environment and the characteristics of the elements (Chorley and Kennedy, 1971).

The systems approach was not immediately accepted by all scientists concerned with the land surface because of conflict with the Davisian approach: the systems approach focused on the entire landscape and its dynamics whereas the Davisian approach had concentrated upon landscape evolution. Discussion was not always serious as illustrated by Huggett (1985) who quoted a report (Van Dyne, 1980: 889) of a facetious comment overheard, 'In instances where there are from one to two variables in a study you have a science, where there are from four to seven variables you have an art, and where there are more than seven variables you have a system'. However, many advantages of the approach include the way in which it comprehensively focuses on all aspects of a problem, requiring identification and analysis of all the multivariate components, concentrating on the general (nomothetic) rather than the particular (idiographic), offering an orderly, rational and comprehensible structure enabling focus on the links between the structures, and representing the function and the throughput of matter and energy. It provides a dynamic approach to complement the historical one, and focuses on the whole landscape assemblage and upon adjustment of landform (p. 50) to processes. It directs attention to the holism of the Earth's environment and to its dynamics so that systems diagrams can be constructed to demonstrate all the variables and interactions involved. An energetic approach can use power (the rate of doing work) to integrate energy sources, energy circulation and transfers or fluxes, energy budget, energy related to morphology, as well as changes of energy distribution, requiring standardization in terms of units and measures (Gregory, 1987) in consistent units (see Figure 3.1).

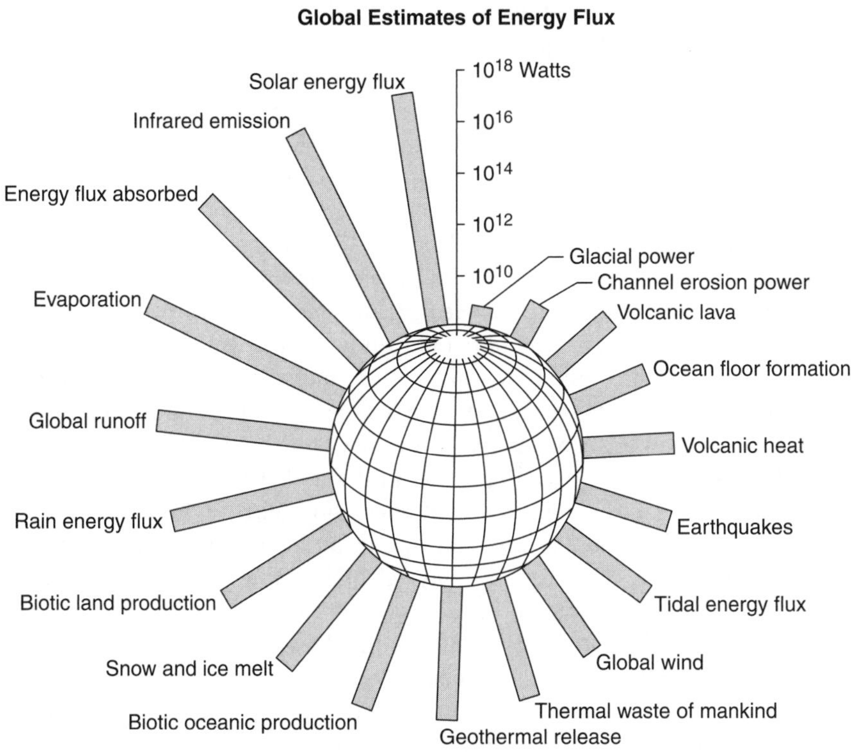

Global Estimates of Energy Flux

Figure 3.1   World energy flux

A comparison of major fluxes in the Earth's physical environment. The scale is logarithmic and the figures represent values of power over the Earth's surface (after Gregory, 1987).

3.2 CONTROLS UPON THE LAND SURFACE SYSTEM

The land surface of the Earth interacts with the lithosphere, atmosphere, biosphere and hydrosphere in relationships expressed in systems analysis.

3.2.1. LITHOSPHERE

The lithosphere has an important controlling influence on the land surface through tectonic effects, lithological influences, and the distinctive character of certain rock types.

Tectonic effects, arising from endogenetic processes of faulting and vulcanicity (Chapter 4, p. 109), account for the broad outlines of the Earth's surface. In earth structure the crust floats on the underlying mantle, in turn underlain by the outer core made up of molten iron surrounding the inner core which is more rigid as a result of the high pressures. The lithosphere is 100–300 km thick and composed of the crust and the upper part of the denser mantle; it consists of a series of 9 major 'plates' and many lesser ones that overlie the mantle 'floating' on the aesthenosphere below (see Figure 3.2, Colour plate 1), thus making the earth's surface very dynamic. Convective flow of the underlying mantle propels independent movement of the plates relative to one another, and is reflected in the youngest rocks in ridges or rifts occurring in the centre of oceans. Rates of movement vary: each side of the Atlantic ocean is moving away from the centre at about 1 cm per year so that the USA has moved nearly 5 m away from the UK since the Declaration of Independence in 1776 and gets 2 cm further away every year (Bowler, 2002). The East Pacific rise moves more rapidly at 10 cm per year, and the eastern Pacific ocean is moving north relative to California at about 6 cm per year. These horizontal movements, usually in the range of 1–14 cm per year, of torsionally rigid bodies, mean that seismic and tectonic activity occur at the plate boundaries. Three types of plate boundary exist: if the two adjacent plates are moving apart in a divergent boundary magma rises to the surface to form new ocean crust, as in the mid Atlantic ridge; where two plates slide past each other in a transform boundary a fault line occurs, exemplified by the San Andreas fault in California; and when an oceanic and a continental plate converge, in a convergent boundary, the denser oceanic plate plunges into the mantle beneath the continental plate in a subduction zone, a situation illustrated by the Andes, the Alps and the Himalayas. These three types of plate boundary are associated with crustal deformation, leading to mountain building or creation of ocean trenches, by long, narrow zones of earthquakes and active volcanoes (see Figure 3.2 A–C, Colour plate 1).

Plate tectonics revolutionized thinking about the surface of the earth (see Table 2.1), imaginatively integrating earlier ideas such as continental drift that were not generally accepted when first proposed. One scientist regarded the concept as so audacious that he described his first account of it as geopoetry (Rhodes et al., 2008: 151). The theory is core to geology but has significant implications for understanding the land surface of the Earth (see Table 3.4).

Table 3.4 PLATE TECTONICS

Suggested to be the first theory providing a unified explanation for the Earth's major surface features (Kearey and Vine, 1990) linking many aspects previously considered independent and unrelated, such as past distribution of flora and fauna, spatial relationships of volcanic rock suites at plate margins, distribution in space and time of different metamorphic facies, and the scheme of deformation in mountain belts. Although of great significance for geology, many implications for the land surface (Summerfield, 2000) include:

Implication for land surface	Significance
Major features of continents (see Table 2.5)	Account for distribution of mountain ranges, folded and faulted features, alignment of some major rivers along lines of structural weakness. World's largest rivers drain to passive continental margins with 25 of the largest deltas.
Relative roles of tectonics and climate	In the past changes in continental distribution and relief modified atmospheric circulation patterns and climatic regimes (Burbank et al., 2003).
Distribution of earthquakes and volcanic activity	Provides explanation for their distribution. Volcanoes around the Pacific perimeter known as the circum-Pacific ring of fire.
Collision zones	Extensive linear feature where two continental plates collide, characterized by young fold mountains and earthquakes. Ancient collision zones are eroded but can be reconstructed.
Volcanoes	Explains why certain types of volcanic activity occur – rift valley volcanoes have basic, fast-flowing lava, and may be lava floes, flowing typically at 60 km per hour; subduction zone volcanoes occur because the mantle has melted as a result of subduction, as in the Andes; in volcanic islands, such as the Aleutians, subduction occurs where one oceanic plate descends beneath another ocean plate; hot spot volcanoes, as in Hawaii, occur well away from plate boundaries but where a plume of hotter rock rises from deep inside the mantle.
Earthquakes	Earthquake types relate to the types of plate boundary: in mid-ocean ridges, newly formed crust fractures to enable more molten rock to reach the surface and so cracks as it cools and causes earthquakes; transform faults have earthquakes that arise from largely horizontal movement; subduction zones also have shallow earthquakes in the overlying rocks as the oceanic plate moves towards the continent.
Models of long-term landscape evolution	Rely on inter-continental correlations and so require knowledge of plate movement and associated tectonic processes. Integrates rock uplift, land uplift and surface processes (see Bishop, 2007).
Relation of landform elements to landscape development	Can affect scarp retreat, together with hillslope-channel linkage.

Figure 3.3   Grand Canyon from Powell Memorial

A second way in which the lithosphere affects the land surface is through the influence of rock types on landforms: we can all think of examples, such as the Grand Canyon (see Figure 3.3), where rocks seem to control landforms and scenery. Although soil and rock are not always distinguished in the same way, soil is an organic-rich material produced by physical, chemical and biological processes acting on parent material which could be rock, superficial deposits or earlier soils. Rock is a coherent, consolidated and compact mass of mineral material which affects scenery as a result of its lithology and structure. Lithology refers to the rock characteristics including the grain size, particle size and their physical and chemical characteristics. The long standing threefold division of rocks into igneous (formed from the crystallization and solidification of magma), sedimentary (formed by accumulation of material derived from pre-existing rocks or from organic sources), and metamorphic (changed from their original state by metamorphic processes usually as a result of mountain building or the intrusion and extrusion of magma) may be complemented by a fourth category of 'artificial man-made ground and natural superficial deposits' (BGS, 2009).

The lithological characteristics of these four categories of rock account for some landform characteristics – observing certain landscapes it is possible to deduce what sort of rock exists beneath the land surface. Igneous rocks underlie extensive areas with eroded granite plutons outcropping over more than 10% of the present land surface, often presenting domed surfaces or landscapes punctuated by tors. Sedimentary rocks include one category of clastic sediments, conglomerates, sandstones and a second category of chemical and biochemical rocks which include carbonates. Metamorphic rocks may have been created by the contact metamorphism of rocks adjacent to igneous intrusions, by dislocation associated with movement along fault planes, or regionally in areas of mountain building (orogenic uplift). Such metamorphic processes often increase rock resistance, as calcite or dolomite is recrystallized to become marble, granitic rocks are metamorphosed to gneiss, slate is derived from mudstones or shales, and schist from argillaceous rocks – all changes to more resistant rocks. Rock types affect landforms at macro and micro levels through lithology, structure and water relationships according to their characteristic properties (see Table 3.5), many of which can be described, using techniques of soil mechanics employed by civil engineers to describe properties of materials for building projects. It is difficult to obtain an overall appreciation of the effect of a particular rock (see Table 3.5) but ratings according to geomorphic mass strength classification have been developed (Moon, 1984; Selby, 1980). Such overall characteristics also include rock strength and resistance, hardness and rock colours all of which may relate to landform character.

Table 3.5 ROCK CHARACTERISTICS RELATED TO LANDFORM AND LANDSCAPE

Features	Rock characteristics	Influence on landform
Lithology	Grain size and particle size, sorting, particle shape, fabric (orientation and packing). Physical and chemical composition.	Arenaceous (largely sand), argillaceous (largely clay) and calcareous landforms, igneous and metamorphic landforms.
Structure	Internal characteristics – joints, bedding planes. Folding, faulting, dip, fractures.	Detailed shape of slopes and landforms, tors. Faulted landforms, fault scarps, fault-line scarps, horst and graben, cuestas, dome structures.
Water relationships	Porosity (proportion of voids in the rock, their size distribution, arrangement and degree of compaction); microporosity (proportion of pores <0.005 mm). Water-related /ease of water absorption and transmission. Permeability (measure of ability of water to pass through), reflecting pore spaces and joints, fissures etc. Water absorption (amount of water absorbed in unit time e.g. 24 hours). Saturation coefficient (measure of the amount of water absorbed in a unit of time expressed as a fraction of the pore space).	Frequency of fluvial landforms such as drainage density (length of stream channel divided by basin area), karst landforms (see p. 54).

Table 3.5 *(Continued)*

Features	Rock characteristics	Influence on landform
Overall characteristics	Rock mass strength measurement system of five categories based on intact strength rating, weathering, joint characteristics, ground-water (Moon, 1984; Selby, 1980).	Categories relate to angles of rock slopes.
Rock strength and resistance	Properties which relate to the strength (measured by compressive strength, tensile strength, shear strength). Can reflect engineering properties: tensile strength, Atterberg limits, compressive and shear strengths, particle size, infiltration capacity, erodibility, bulk density, consolidation, bearing capacity.	Resistance to erosion, volcanic plugs.
Hardness	Properties measuring the hardness (resistance to abrasion) or toughness of rocks (resistance to crushing or impact). Mohs' scale discriminates gypsum, muscovite (up to 2.5), apatite (5) and quartz (7).	Abrasion of rocks on beaches can give differential erosion.
Rock colour	Reflects petrological content and effects of weathering.	Gives colour to landforms and landscapes.

Of all the geological influences upon the surface of the Earth it is perhaps carbonate rocks that provide the most distinctive assemblage of features. Described by the word karst, (a German form of the Slovene word *kras*, used since the thirteenth century because classic karst scenery is found in Slovenia in the Balkans), the landscape near Kweilin, China (see Figure 3.4) demonstrates how dramatically limestone can affect scenery. Distinctive features of karst scenery arise because limestone is affected by solution. This occurs because rainwater is slightly acidic, having dissolved some carbon dioxide (CO_2) in the atmosphere. The resulting weak carbonic acid (H_2CO_3) reacts with the limestone ($CaCO_3$) to produce soluble calcium bicarbonate ($CaHCO_3$), a solution process which may be associated with other chemical reactions. If the limestone is covered by a sufficient thickness of superficial deposits, glacial deposits for example, then karst features may be absent from the surface.

The great range of different limestones usually have at least 50% carbonate minerals of which calcite ($CaCO_3$) is the most frequent, but dolomite ($CaMg(CO_3)_2$) also occurs and in some areas gypsum presents distinctive features. The amount of $CaHCO_3$ that can be dissolved in the water flowing through karst landscapes tends to be greater at lower temperatures (solubility of CO_2 in water at 0°C is twice that at 20°C) and water may often become saturated as soon as it has passed

Figure 3.4   Tower karst near Kweilin, China

through the soil profile or flowed over the land surface. Variations in landforms occur according to the type of limestone, so that in the UK where there are three principal types of limestone, Carboniferous, Jurassic and Cretaceous in age, the first exhibits classic features including small scale clints and grikes and limestone pavement, whereas such pavement does not occur on the porous chalk outcrops. Although there are differences according to the type of limestone the landforms produced can be envisaged as a sequence. Water infiltrates along joints, widening them by solution to form grikes (a term used in England, but lapis in France and karren in Germany), with blocks called clints between – the two constituting limestone pavement. Streams can dissolve their way into the rock, usually at swallow holes or dolines which can be up to 100 m deep and up to 1000 m in diameter. These may occur in groups with as many as 2460 per km^2 (Chorley et al., 1984: 187), and have been suggested to be the characteristic features of karst. In addition to solution, collapse of a cave, or subsidence of deposits overlying limestone, may contribute to the formation and characteristics of the doline. Larger depressions on the surface, called polje, up to 50 km long and 6 km wide, may have developed either where there are faults or weaknesses, which are exploited by solution, or where rocks other than limestone limit downward solution. They are often seasonally flooded. Variations in surface scenery occur according to climate so that, in the tropics, tower karst (see Figure 3.4) can be composed of rock towers rising up to 200 m above valley floors or alluvial plains. Although there is comparatively little continuous water flow on the surface, some areas do have dramatic gorges, with exposures of limestone in their steep side slopes, and dry valleys may occur although these are not necessarily restricted to limestone outcrops.

Water progressing into the limestone can technically continue until it reaches the water table, the phreatic zone, in which all spaces are filled with water. In some limestones the numerous joints and fissures can be widened by solution so that the surface doline may connect to an uvula and then to a cavern below the surface. Caverns, often with many features of redeposited limestone precipitated from the karst waters as stalactites or stalagmites, or as a covering on the rock walls of the cavern, can be up to 100 km long and over 1000 m deep, and can be part of systems with passages connecting caves, tunnels and shafts. The great range of dramatic features, evident to tourists who visit cave systems, are not all the result of limestone solution because mechanical abrasion by pebbles and stones carried in underground streams, and subsidence when caves enlarge so that the roof may collapse, contribute to the diversity of cave forms that exist. Underground stream systems emerge on the surface again at springs or resurgences and in limestones such as chalk, which do not have underground stream systems, springs may occur where the water table intersects the surface – often in lines which may have been important for location of settlements.

Landforms in areas of limestone outcrop fascinate many tourists who visit them, but they are equally fascinating to the geomorphologist. Such areas demonstrate how many local names from different languages have been adopted to build up the framework of names for landforms, and how it has been possible to deduce how rapidly an area changes and erodes. Estimates first deduced for limestone areas and then elsewhere (see Table 3.6) may encourage the perception that such areas develop slowly, but they can be very dynamic if a sudden rainstorm causes high flows which completely fill an underground tunnel, or cave system.

3.2.2. ATMOSPHERE

We often associate the atmosphere with the land surface in scenery – clouds, light and meteorological elements contribute to produce the scenes that we admire, remember or perhaps want to improve. The atmospheric setting often distinguishes the ways in which artists and painters capture the character of the land surface of the Earth, as shown in Constable's skies (Thornes, 1999).

The atmosphere–ocean circulation systems are driven by solar energy. Through the atmosphere the surface of the Earth receives its greatest driving force from solar radiation, some 7000 times greater than the heat flow from the interior of

Table 3.6 RATES OF EROSION IN LIMESTONE AREAS COMPARED
WITH OTHER VALUES

A Bubnoff unit, named after Sergei von Bubnoff, is a rate of erosion leading to a loss of
1 millimetre in 1000 years (equivalent to metres per million years), providing a standard way
of comparing estimates of different kinds of erosion rates (see, for example, Figure 4.14).
Overall rates can be calculated for drainage basins and range from 4 mm. ka^{-1} in the Kolyma
basin to 688 mm. ka^{-1} in the Brahmaputra basin (Summerfield and Hulton, 1994). One of the
highest rates cited is for the Tamur basin, Himalayas where steep slopes, unconsolidated
sediment, glaciated terrain and human modification gives a rate of 4700 Bubnoff units.
Drainage basin measurements may underestimate rates because some of the sediment does
not reach the basin outlet and calculations assume an average rate over the basin. Human
activity can increase the rate of denudation by up to ten times, with the highest values under
intensive land use, and globally human activity has increased the global land–ocean
suspended sediment flux by about 16% (Walling, 2006 gives a valuable world review).

Surface lowering can be measured directly by micro erosion meters or calculated from
the mass loss in rock tablets exposed to dissolution in different environments. Indirect
approaches are based on the measurements of solute load leaving the catchment area,
geomorphological investigations and methods of cosmogenic nuclides, particularly ^{36}Cl
(Gabrovšek, 2009) (see Table 5.3).

Method of estimation	Bubnoff units
Water chemistry data and discharge records to calculate denudation rates for British limestone areas (Goudie, 1995).	25–67 Chalk (range of values in several areas 35–66 Jurassic limestone 16–102 Carboniferous limestone
Limestone surface lowering by solution beneath glacial erratics in Northern England since their deposition.	40 to 50 cm figures cited in the literature (Gunn, 1986; Ford and Williams, 1989) although some sites indicate only about 5 to 10 cm and Goldie (2003) suggests maximum of 15 cm.
Dating of speleothem deposits in caves gives indication of when caves ceased to function and so gives lowering estimate.	Maximum drainage incision (Goudie, 1995): 20–50 Craven area, Yorkshire, UK 55 Manifold valley 1000 Mendips
Comparative values	
Erosion rates for selected glaciers based on suspended sediment and bedload transport.	73–30,000 with a mean value c.4000 (Goudie, 1995).
Maximum reported rate possibly for Taiwan	>10,000 (Li, 1976)
For world's 35 largest drainage basins calculated from sediment and solute transport rates (Goudie, 1995; Summerfield, 1991).	677 Brahmaputra 529 Huang He 271 Ganges 133 Chiang Yang 124 Indus 104 Shatt-el-Arab 95 Mekong 91 Orinoco 84 Colorado 70 Amazon

the Earth. Some 10^{25} joules per year of solar energy are received by the Earth's surface. The solar constant is the radiation received at the top of the atmosphere by a 1 m² surface normal to the solar beam and on average this is 1400 Watts m² (where a Watt is a measure of power or the rate of doing work, in Joules per second). Of the total amount of radiation received in the upper atmosphere approximately 24% is radiated back to space, 6% is reflected by dust particles, air and water vapour, and about 6% by the earth's surface according to albedo. A further 14% is absorbed by the air and 3% by clouds leaving just 47% to be absorbed by the earth of which only a small part reaches more than a few metres below the surface.

Energy received by solar radiation is continually redistributed by motion in the atmosphere and oceans, and the energy balance can be regarded as an open system in a steady state. During the redistribution of energy, land surface systems are coupled dynamically with the atmosphere through physical processes associated with fluxes of energy, water, biogeochemicals and sediments. Such fluxes, often involving a series of cascading system responses, illustrated by river hydrology or by glaciers, interact with atmospheric elements of temperature, pressure, humidity, precipitation and wind. These elements which combine to make up the weather that we experience are the basis for climate, often thought of as the average weather conditions at a place over a period of time. Climates over the land surface of the Earth determine the processes shaping the land surface and if that surface was flat there would be a fairly simple pattern of processes from the equator to the poles. Relief and mountain ranges complicate the surface climatic pattern, but in general terms it is the temperature and its range, and the nature and availability of precipitation that exert major influences on the processes operating in particular zones of the land surface (p. 155, Chapter 6). Certain values of climate parameters have particular significance for the land surface (Barry, 1997), including the freezing threshold which allows glaciers and permafrost to develop, the occurrence of periglacial processes, and the boundary of arid and semi-arid processes delimiting aeolian processes and normal fluvial processes (Chapter 9).

Solar radiation effectively produces an energy driving force at a rate 4000–5000 times faster than all other energy sources combined (Slaymaker and Spencer, 1998) so that the relation between climate and the land surface has prompted two major questions. First, whether average climatic conditions or events that occur relatively frequently have the greatest sustained impact on the surface features, and second to what extent the surface is a result of climates that currently exist in particular areas.

The first question relates partly to the debate between uniformitarianism and catastrophism suggested (Leopold et al., 1964; see Box 4.1, p. 120) to be analogous to the effects of a giant, a human being and a dwarf cutting down a forest. The dwarf works most of the time but produces comparatively little effect; the giant operates very occasionally and can cause dramatic consequences which are very severe; the human being is effective at regular times and is the most efficacious. This analogy was proposed to indicate how rare catastrophic events at the one extreme and regular daily events at the other, do not achieve the most work shaping the land surface. The most work is done, by analogy with the human being, by events that may occur several times each year. This notion prevailed for much of the twentieth century but in recent decades there has been greater awareness of the role that catastrophic events may have – a role which varies across the land surface in different regions. Thus, impact craters can still be identified. Extreme events may constitute hazards, which can result from windstorms, severe precipitation, temperature extremes and lightning, all potentially having an effect on the land surface (see Table 3.7) but particularly renowned for their impacts on people (McGuire et al., 2002).

Table 3.7 EXTREME EVENTS

Extreme events – phenomena differing significantly from their mean values and often associated with natural hazards which are naturally occurring geophysical conditions that threaten human activity – are reviewed by McGuire et al. (2002) and many are illustrated at http://earthobservatory.nasa.gov/NaturalHazards.

Extreme events	Possible implications for land surface
Hurricanes (typhoons, tropical cyclones) Tornadoes Lightning and severe storms Hail storms Snow storms	May be associated with strong winds and high precipitation which can induce flooding, after thaw in case of snow storms.
Tsunamis, storm surges	Damage from flooding, including coastal erosion.
Glacier surges	Glacier flows at order of magnitude higher than normal.
Frost hazards	See Chapter 7 (p. 179)
Landslides	Giant landslides transform slopes.
Avalanches	See Chapter 7 (p. 180)
Floods and flash floods	Affect river channel morphology and flood plain characteristics.
Drought	Influence upon vegetation cover affects runoff and erosion.
Wildfires	Occurrence of fire has profound effect on erosion, sediment movement, and numerous processes resulting from soil loss. In some cases increases in sediment transport following fire may be similar to sediment waves or slugs that progress downstream sometimes following clearcut.
Desertification	Increases propensity to aeolian erosion (see Chapter 9, p. 220)
Soil heave and collapse	Especially in loess and silts may have hydrocompaction giving subsidence up to 5 m.
Subsidence	Surface collapse features.
Earthquakes	Surface effects (see Table 4.15).
Volcanic eruptions	Super volcanic eruptions.
Near Earth Objects (Neos)	Comets and asteroids that have been nudged by the gravitational attraction of nearby planets into orbits that allow them to enter the Earth's neighborhood. Composed mostly of water ice with embedded dust particles, comets originally formed in the cold outer planetary system while most of the rocky asteroids formed in the warmer inner solar system between the orbits of Mars and Jupiter. See http://neo.jpl.nasa.gov/neo/.

The second question, whether the topography of a particular landscape is in balance with current climate-driven processes, or contains relict signatures of past climates, is a long-standing question in geomorphology (Rinaldo et al., 2002). Although the glaciated landscapes of the northern hemisphere contain much evidence of past climates (Chapter 8, p. 187), other parts of the world show both scenarios; some areas in balance with current climate-driven processes and others with extensive relic features.

Variability in climate means that it is not actually easy to distinguish current from past climates. Throughout the Quaternary and late Tertiary major changes of different frequencies include those shown in Table 3.8. Over the last 2.5 million years some large scale global and regional changes of climate occurred rapidly rather than incrementally (see Adams et al., 1999). During the past 150,000 years changes which appear to have occurred over periods ranging from centuries to as little as a few years have been described as a flickering switch (Taylor et al., 1993). Therefore the perception of a slowly changing environment is not always an appropriate vision – instead change is the norm rather than the exception (McGuire et al., 2002), many changes made by long-term astronomical, climatic or geological forcing or as a result of specific catastrophic geological or astronomical events can take place over relatively short periods of time (section 5.2.5). Over the last 1000 years climatic trends may have been affected by variations in solar activity, North Atlantic Oscillation (NAO), El Niño Southern Oscillation (ENSO), and by volcanic eruptions, with specific periods identified (see Table 3.8). In addition changes instigated by human activity may currently be responsible for a period of **global warming** especially through increases in carbon dioxide concentrations in the atmosphere which accentuate the greenhouse effect, leading to rising world temperatures. Such changes have implications for the land surface of the Earth because they are often implemented through short term and often dramatic events. Analysis of change at a world scale has been aided by general circulation models (GCM) designed to model climate behaviour by integrating fluid-dynamical, chemical, or biological equations either derived from physical laws or empirically constructed. Atmospheric GCMs and ocean GCMs are coupled to form an atmosphere–ocean coupled general circulation model to indicate how changes might occur in the future (see Table 5.9).

Table 3.8   SOME MAJOR CLIMATE EVENTS SIGNIFICANT FOR THE LAND SURFACE SINCE THE LATE TERTIARY (SEE ADAMS ET AL., 1999)

Type of event	Characteristics and timing
Onset of northern hemisphere glaciation	Between 4 and 2.5 Ma ice sheets started to develop in the Northern hemisphere instigating strong glacial–interglacial cycles characteristic of the Quaternary.
Mid-Pleistocene revolution	Prior to mid-Pleistocene revolution climate cycled between glacial–interglacial every 41 ka; afterwards it cycled every 100 ka.

(Continued)

Table 3.8   *(Continued)*

Type of event	Characteristics and timing
Interglacials	Closest analogues to present climate are interglacials at 420–390 ka (oxygen isotope stage 11) and 130–115 (oxygen isotope stage 5e, Eemian).
Heinrich events and Dansgaard-Oeschger cycles	Heinrich events are extreme, short duration events representing climate effects of massive surges of icebergs into the North Atlantic. Dansgaard-Oeschger cycles are rapid temperature changes at beginning of interglacials, including evidence of very rapid warming of up to 7°C in c. 50 years at the end of the Younger Dryas.
Deglaciation and Younger Dryas	About half of the warming in the Younger Dryas-Holocene was concentrated in a period of less than 15 years, and took place in a series of steps each lasting less than 5 years.
Little Ice Age	The most recent climate cooling event in the northern hemisphere, c.1450–1850.
El Niño and the North Atlantic climate oscillation	Have occurred for at least the last 1000 years; El Niño (~3–5 years), NAO (~10 years).

3.2.3 BIOSPHERE

The biosphere, including soil and vegetation, provides dynamic biogeochemical links to the Earth's surface. In the pedosphere, soil covering the land surface is an organic link because of the incorporated organic matter. Soil, defined as the thin layer of unconsolidated material on the surface of the earth, is influenced by soil-forming factors so that any soil characteristic(s) can be expressed (Jenny, 1941) in relation to climate (cl), organisms (o), relief (r), parent material (p) and time (t) in the relation:

$$S = f\,(cl \ o \ r \ p \ t...)$$

The soil medium enables plant growth contributing organic matter so that the interaction of nutrient cycles together with water from precipitation moving vertically downwards produces a soil profile. Organic residues change to humus which has been described as something like a chemical 'junk yard' because it is a repository for resistant plant material and microbial waste products, combining in humification to produce very complex molecules (Ashman and Puri, 2002). When humus is produced the mineral and organic soil constituents are changed into new compounds which have different physical and chemical properties, with the change from primary into secondary minerals including production of clay minerals. Such changes occurring in the soil profile are the basis for the horizonation observable in many soil profiles (see Figure 3.5A). Salient processes in soil formation are the accumulation of organic matter, especially in

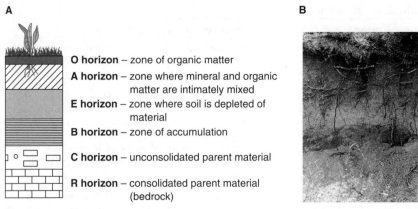

A

O horizon – zone of organic matter
A horizon – zone where mineral and organic
                matter are intimately mixed
E horizon – zone where soil is depleted of
                material
B horizon – zone of accumulation

C horizon – unconsolidated parent material

R horizon – consolidated parent material
                (bedrock)

(NB: not all soils have all these horizons)

B

Figure 3.5   (A): Soil horizons (based on Ashman and Puri, 2002)[*]; (B): an example of a soil profile in the hills of the Manisa Province, Turkey, showing a distinct horizon of accumulated red-purple metallic (manganese and/or iron) oxides (from http://earthquake.usgs.gov/learning/glossary.php?term=soil%20 profile)

[*]Reprinted with permission from M.R. Ashman and G. Puri, *Essential Soil Science*, Wiley-Blackwell.

the O horizon, the leaching of soluble ions, especially Ca++, Mg++, Na+, K+, NO_3^- and NO_2^-, and eluviation of insoluble particles including clay minerals down the profile from the E horizon (see Figure 3.5B). The soluble ions and the insoluble particles may accumulate by illuviation in the B horizon or may be washed right through the profile. The local characteristics, together with the history of soil organic matter accumulation, leaching, eluviation and illuviation, produce soil horizons which are distinctive in colour, texture and composition. Each particular mix of soil horizons provides a soil type, and a major problem confronting soil classification is to produce a scheme appropriate for the field survey level but also capable of linking to the world scale in the global classification of soil types (see Table 6.1 p. 156–8).

How is soil important for understanding the land surface? Soil properties, including soil mechanics in engineering terms, can affect the way in which the land surface is fashioned in a particular area, because regolith, which is the layer of broken and unconsolidated rock together with soil material that covers the surface of the land, affects groundwater supply and soil conservation (Ollier and Pain, 1996). Water flow through the soil is important to weathering and the hydrological cycle because flow through the soil can occur over a range of void sizes from matrix flow up to flow in subsurface pipes which can be as much as 1m in diameter. In fact it was understanding of water flow through the soil profile that revolutionized thinking about the generation of runoff in temperate environments. There are now known to be many routes that water can follow through soils; some laterally through the horizons, others through pipes, and sometimes, as the local water table rises temporarily, saturated overland flow. Soil mantling the Earth's surface has to be visualized dynamically in that over short periods of time the water content and status of the soil will change affecting weathering, water movement and other processes such as mass movement.

A repeated sequence of soil profiles associated with relief, classically from an interfluve down to a valley floor, is a soil catena. This topographic sequence of soils of the same age and often on the same parent material, usually reflecting differences in slope and drainage, has also been described as a toposequence (Bates and Jackson, 1980). Most catenas or hillslope models are thought of in two dimensions but combined with the drainage basin concept, embrace the 3D manner in which water moves through the surface layers and over the surface (see, for example, Huggett, 1975). A useful soil–geomorphological concept is the 'soil landscape' which is now the basis for soil mapping in the USA and elsewhere. Survey agencies define soil landscapes as 'areas of land that have recognisable and specifiable topographies and soils, that are capable of presentation on maps, and can be described by concise statements' (Northcote, 1978). The soil landscape system concept can be construed as a dynamic 3-dimensional pedogeomorphic model in keeping with an ecological interpretation of soil science (Thwaites, 2006).

Successive periods of soil formation produce sequences of soils which have experienced weathering and soil development for different periods of time, recognized as chronosequences which may be found on coastal or river terraces of different ages, on successive lava flows, on sand dunes, floodplains, or on till sheets of varying age. Where located on similar parent materials and developed under standard soil-forming conditions including climate, chronosequences can give data on mechanisms of soil formation and on rates of development of soil characteristics such as accumulation of organic carbon or of hardpans. Fossil soils, or palaeosols, are indicators of landform age or the chronology of landscape evolution providing important keys to (palaeo) sedimentologic environments. In south-east Australia, sequences of erosion and deposition over time periods were related to soil profile development in a series of K cycles (Butler, 1959), periods of time each composed of an unstable phase, when erosion or deposition may occur, and a stable phase when soil profile development occurs. Evidence for up to 8 K cycles has been preserved in some Australian soil landscapes.

Soil erosion is a dramatic mode of soil transformation in relation to the land surface. It includes the physical processes of moving soil particles from one site to another and involves the detachment of particles followed by their transport by wind or water. Soil erosion occurs naturally in all landscapes to some degree, so that we are familiar with headlines when rates exceed natural levels, usually as a result of poor management of the Earth's surface. Such headlines include: 20% of the world's cultivated topsoil lost between 1950 and 1990; soil erosion second only to population growth in world problems. Soils on average develop at rates of 1 t.h.a^{-1} per year whereas estimated losses from agricultural lands can be 30 t.h.a^{-1} in Africa, Asia and South America and 17 t.h.a^{-1} in Europe and the USA (Ashman and Puri, 2002) with global estimates of annual soil loss ranging from 7 to 9 billion tonnes per year with Asia and Africa contributing some 60% of this (Garland, 1999). The United States may be losing soil 10 times faster than the natural replenishment rate, while China and India are losing soil 30–40 times faster, and as a result of erosion over the past 40 years, 30% of the world's arable land has become unproductive (Pimentel, 2006). Reasons for the acceleration of normal rates of erosion include deforestation and agriculture, aggravated by ploughing of very steep slopes, greater use of machinery, removal of hedgerows and increase of field size, and reduced levels of organic matter and cultivation throughout more of the year (Goudie and Viles, 1997). The consequences of soil erosion by water on the surface can be to produce bare areas that have been denuded by sheetwash erosion or deep gullies (see, for example, Figure 10.6) produced by concentrated water flow; one can grade into the other but gullies tend to be defined as those channels that cannot be obliterated by normal ploughing operations.

Plants and animals, as a major part of the biosphere, have a controlling influence on the surface of the Earth, reflected in fields such as phytogeomorphology (Howard and Mitchell, 1985), biogeomorphology (Viles, 1988; Viles et al., 2008) and geoecology, linking physical geography and ecology to provide a new way of thinking for the whole environmental complex relating to the ways in which animals, plants and soils interact with one another (Huggett, 1995). Ecological impacts on geomorphological processes can be either stabilizing, when vegetation growth reduces erosion, or destabilizing, for example as animal burrowing enhances erosion (Viles et al., 2008). Vegetation provides a cover for the land surface more complete than the mantle provided by the soil. Dietrich and Perron (2006) asked whether landscape bears an unmistakable stamp of life apart from the obvious influence of humans, suggesting that erosion laws that explicitly include biotic effects are needed to explore how intrinsically small-scale biotic processes can influence the form of entire landscapes, and thus determine whether these processes create a distinctive topographic signature.

The biosphere is influential through biogeochemical cycles (see Table 3.2) which are global cycles of the life elements C, O, N, S, P, with H, C and O accounting for almost all living matter. Macronutrients H, C and O are accompanied by 12 elements, of which six are also macronutrients (N, Ca, K, Mg, S and P) – all required in substantial quantities for organic life to thrive (Slaymaker and Spencer, 1998). Plants, animals and microorganisms can act as controls upon earth surface processes, on interactions in environments and the character of landforms, as well as on evolution of environments and of the landforms included. The controls are effected at a range of scales extending from individuals or small communities to ecosystems and **biomes**, and some are effective passively as structures, in the way that coastal grasses can encourage coastal dune growth and sedimentation, and also actively as biological processes exercise significant effects. Vegetation affects the land surface in a myriad of ways (see Table 3.9).

Animals and organisms can produce landforms or distinctive features. Termite mounds commonly are several metres high and can exceptionally attain a height of 9 m. Their density varies according to species, can be up to >1000.ha^{-1} of the

Table 3.9 SOME WAYS IN WHICH BIOLOGICAL INFLUENCES AFFECT LANDFORMS

Biogeomorphology, introduced by Viles (1988), and referring to studies uniting biota with geomorphic form and process (Osterkamp and Friedman, 1997), is considered in relation to landscape evolution by Phillips (1995); complex systems in biogeomorphology are seen in terms of (Stallins, 2006) multiple causality and the concept of recursivity, the influence of organisms that function as ecosystem engineers, the expression of an ecological topology, and ecological memory. Disturbance regimes are reviewed in Viles et al., 2008.

Process (see Chapter 4)	Organic processes	Consequences
Weathering	Physical stresses from plant roots	Physical weathering
	Lichens, algae, fungi, decaying organic matter emit substances which trigger oxidation, reduction, complexation, etc.	Chemical weathering
Mass movement	Increase resistance to failure, organisms and animals displace material.	Animal burrowing enhances erosion; animal grazing reduces vegetation cover and thus enhances erosion.
Fluvial	Root biomass is an important determinant of erosivity, disturbances of vegetation can increase sediment yields by one to three orders of magnitude, exotic shrubs can alter riparian habitat, sediment transport, and bottomland morphology.	Beaver dams, debris dams, channel bank erosion, vegetation growth reduces erosion, enhances sediment storage, large woody debris (LWD) decrease or increase flooding and enhance sedimentation.
Coastal	Erosion of reefs by parrot fish.	Mangrove forests.
Aeolian	Stabilization of coastal dunes by algae.	Nebkhas.
Glacial	Lichens useful in suggesting ages of moraines and events, employing lichenometry.	
Periglacial	Peat formation, effects on active layer and permafrost.	Patterned ground, earth hummocks, patterned bogs or string bogs.
Subsidence and solution	Biokarst (landforms produced largely by direct biological erosion and/or deposition of calcium carbonate) and phytokarst.	Pitting of limestones, tufas, travertines and speleothems.

tropics, although some species do not produce mounds (Goudie, 1988). Nebkhas are small arrow-shaped hummocks of sand collected leeward of a clump of vegetation in arid and semi-arid areas; beaver dams are constructed across streams and rivers (Gurnell, 1998); and phytokarst is a recognizable landform produced where algal boring affects limestone weathering (Folk et al., 1973). Plants can be responsible for distinctive landscapes, with peatlands or organic terrain occupying 3% of the total land surface of the Earth. A range of terms is used to describe these landscapes associated with poor drainage, especially in areas of relatively recent deglaciation, which are dominated by species of *Sphagnum*. Peatlands include bogs which are acid, poor in minerals and raised above the ground-water level by wet anaerobic peat, being either oligotrophic (poor in minerals and species) or ombrotrophic (deriving water and minerals from the atmosphere); and fens which are developed under mineral-rich, aerated groundwater dominated by grass-like plants mainly sedges, and are either eutrophic (rich in minerals and species) or soligenous (deriving minerals largely from groundwater). Other terms include mire, a peatland where peat is currently forming and accumulating, and muskeg, a term used in Canada where it covers 13% of the country, and in Alaska where it covers some 10% and has been described as like a soggy blanket draped over the landscape. It consists of dead plants in various stages of decomposition, may include pieces of wood, many open ponds and has accumulated since the last glaciation. Mangrove forests on tropical coasts (p. 243) are further instances of the biosphere affecting the Earth's surface, and changes such as deforestation can have dramatic consequences (see Table 3.9).

3.2.4 HYDROSPHERE

In the hydrosphere water occurs in several stores and although the land surface stores contain relatively small percentages of the total world water in lakes and inland seas (0.01%), in glaciers and ice sheets (1.65%), rivers (0.0001%), groundwater (0.68%) and soil (0.0055%), substantial transfers within, and between, these various stores occur in the hydrological cycle. Estimates of rates of exchange indicate that residence time in glaciers and ice sheets can be of the order of 8000 years, but 7 years for surface water on land, 1 year for soil moisture and 0. 031 years for rivers. The water balance equation (section 4.2.3, p. 86) can be derived for any area or drainage basin, but substantial differences exist between continents and between basins; on average the percentage of precipitation which becomes runoff is as little as 11% in Australia but 44% in North America. Water volumes in the different stores of the hydrological cycle have changed in the past and some 10,000 years ago during the last ice advances there was more than twice as much ice over the earth's surface. Such changes and controls exerted by transfers of water in the different phases of the hydrological cycle affect Earth surface processes, especially by weathering, mass movement, glaciers and rivers.

3.3 THE NOOSPHERE

Human activity recognized as the **noosphere** or anthroposphere (see Table 3.1, **Anthropogeomorphology** in Glossary, p. 303), in the Anthropocene (p. 7), is the reason for major impacts and controls on the land surface. Many are obvious or direct including cloud seeding, soil conservation, deforestation, the

Table 3.10 SOME MEASURES OF HUMAN IMPACT ON THE LAND
SURFACE

In the original theory of Vernadsky (1863–1945) the noosphere is the third in a succession of phases of development of the Earth, after the geosphere (inanimate matter) and the biosphere (biological life).

Impact on	Effects
Atmosphere	Temperatures expected to rise by between 1.4°C and 5.8°C by 2100. Rise of ocean temperatures by an average of 0.5°C in 40 years.
Biosphere	In early to mid 1980s 0.3% of Earth's land surface deforested each year.
Hydrosphere	Surface river runoff km³/yr: 303,103 (1680); 253,103 (1980AD) (Wolman, 2002). 39,000 dams higher than 15 m on rivers so that few of world's rivers remain unregulated.
Pedosphere	250 million ha of the world affected by strong or extreme soil erosion.
Lithosphere	Surface subsidence due to removal of fluids or solids.
Geosphere	About 75% of Earth's subaerial surface bears the imprint of human agency.

creation of new lakes and reservoirs by the damming of rivers, or urbanization (see Chapter 11). Less obvious are the many less well known, or indirect, effects, which include changes to precipitation character and amount as a result of the existence of urban areas; the incidence of soil erosion as a result of grazing pressure; the introduction of new plant species; or the ways in which geomorphological processes have been changed, for example by altering the active layer in permafrost areas or modifying river processes downstream of dams and reservoirs (see Table 3.10).

Influences upon the land surface can include extreme events with devastating effects upon people as well as landscape. On 4 May 2008 a tropical cyclone with winds of 190 km/h affected the Irrawaddy delta in Burma and on 12 May 2008 a major 7.2 magnitude earthquake struck western China, about 225 km southeast of the city of Hotan in Xinjiang province. Both events had dramatic effects on the surface of the Earth but we naturally measure their devastating consequences in terms of human impact. Such human impacts emphasize the land surface of the Earth as a sensitive home environment for human beings, so that human impacts are stimulating new approaches such as the identification of 18 anthropogenic biomes as a basis for integrating human and ecological systems (Ellis and Ramankutty, 2008) and landscape fluidity (Manning et al., 2009) as the ebb and flow of different organisms within a landscape through time.

Despite the tendency to think of human impacts as deleterious and negative, such as increased flooding, many are positive, including ways of mitigating earth hazards. However, the greenhouse effect and global warming is now recognized to become the most serious human impact, but even when established the indirect effects may still take time to identify and assess, including those on land surface processes reviewed in the next chapter.

BOX 3.1

CHARLES DARWIN

In 2009, the 200th anniversary of Charles Darwin's (1809–1882) birth, there was much reflection on the contribution of this biologist, who was enormously influential through his ideas on natural selection and evolution. Alfred Russel Wallace (1823–1913) developed similar ideas, the two made a joint announcement of their discovery of evolution in 1858, and in 1859 Darwin published *On the Origin of Species by Means of Natural Selection*. Interpretation of many aspects of the land surface may have been influenced by evolution, shown in climatic climax vegetation communities, zonal soils and the cycle of erosion.

After studying medicine at Edinburgh University and theology at Cambridge, Darwin developed his interest in natural history and participated in the journey of the *Beagle* (1831–1836) which encouraged his knowledge of geology. On the voyage Darwin read Sir Charles Lyell's (1797–1875) *Principles of Geology* which suggested that fossils in rocks were evidence of animals living millions of years ago. He was elected a Fellow of the Royal Society in 1839, he was secretary of the Geological Society (1838–41), and by 1846 he had published several works on the geological and zoological discoveries of the voyage, works putting him amongst leading mid-nineteenth century scientists. After 1842 when he resided in Kent as a country gentleman, the ideas for his 1859 book were gradually formulated and on the basis of those he is remembered as the leader of evolutionary biology. He continued to be a prolific author, publishing at least 10 further books after 1859. He died suddenly on 19 April 1882, and was buried in Westminster Abbey. A marble statue of him was sculpted shortly after his death and placed near the entrance in the Natural History Museum in London. For a special issue of *Nature* (457, 12 February 2009) it was suggested that 'No single researcher has since matched his collective impact on the natural and social sciences; on politics, religions, and philosophy; on art and cultural relations'.

FURTHER READING

Approaches to the context of the land surface referring to global environmental change are:

Slaymaker, H.O. and Spencer, T. (1998) *Physical Geography and Global Environmental Change*. Longman, Harlow.

Slaymaker, H.O., Spencer, T. and Embleton Hamann, C. (eds) (2009) *Geomorphology and Global Environmental Change*. CUP, Cambridge.

Several books place the Earth in a systems framework including:

Huggett, R.J. (1985) *Earth Surface Systems.* Springer Verlag, New York.

and a stimulating read is:

Phillips, J.D. (1999) *Earth Surface Systems: Complexity, Order and Scale.* Blackwell, Oxford.

Particular fields are introduced by:

Goudie, A.S. and Viles, H. (1997) *The Earth Transformed.* Blackwell, Oxford.
Jones, J.A.A. (1997) *Global Hydrology: Processes, Resources and Environmental Management.* Longman, Harlow.

TOPICS

1 How many of the spheres in Table 3.1 interact to affect the land surface?
2 Landforms of limestone landscapes have intriguing names – dolines, poljes, uvulas, rinnenkarren are just a few. Compile a list of such landform names, ascertain where they were derived from, and investigate whether their origin is absolutely clear in each case.
3 Taking values of limestone erosion rates (for example, see Table 3.6) how much chalk could have been removed since the Cretaceous period – for example in southern Britain?

4

PROCESSES AND DYNAMICS OF THE LAND SURFACE

4.1 CONTROLS ON PROCESSES

In his book *A Brief History of Time* Stephen Hawking (1988) wrote 'Someone told me that each equation included in the book would halve the sales'. However, an equation can provide a context for the processes which fashion the surface of the Earth (see Table 4.1). Consideration of processes (P) requires comment on three things: the general relationships that exist for any particular process; what forms of balance may occur; and what types of processes are responsible for fashioning the land surface.

Processes involve transfers, that is flux, of energy or mass at or near the land surface. The work done in such transfers is the result of application of internally and externally applied stresses or forces which are resisted in various ways. Every earth surface process can be visualized in relation to forces which lead to stress and also characteristics which provide resistance, so that the efficacy of processes at any location depends upon the extent to which, and how often, force exceeds resistance. Hillslopes provide a good illustration because forces can be induced by gravity, by climatic events such as storm rainfall, or by earthquakes; resistance is provided by the characteristics of the soil and rock materials composing the slopes together with the binding effects of the vegetation or surface materials. A **factor of safety**, the ratio of resistance to force, can therefore be used in studying slope stability. If the factor is less than 1.0 then the slope will fail, by landsliding for example, to achieve a more stable slope angle.

Table 4.1   THE CONTEXT FOR EARTH SURFACE PROCESSES (SEE
GREGORY 1985, 2000 AND 2005b)

Morphological elements or landforms (F) of the Earth's surface result from
processes (P) operating in the physical environment on materials (M) over periods
of time t. A geomorphological equation, indicating how operating on *materials* over
time *t* produces *results* expressed as a landform, can be shown as:

$$F = f(P, M) \, dt$$

Investigations can be made at five levels:

- **Level 1: the elements or components of the equation** – study of the components
  in their own right. Some studies focus on the description, which may be quantitative, of
  landforms, of soil character, or of plant communities, including many of the controlling
  factors in Chapter 3.
- **Level 2: balancing the equation** – study of the way in which the equation is balanced
  at different scales. At the continental level may involve the energy balance relating
  available energy for environmental processes to radiation, and moisture received in
  relation to locally available materials. Studies of this kind focus upon contemporary
  environments and upon interaction between processes, materials and the resulting
  landforms or environmental conditions (see Chapter 6).
- **Level 3: differentiating the equation** – includes studies analysing how relationships
  change over time. This requires reconciliation of data obtained from different time
  scales together with a conceptual approach. Includes impact of climate change and
  human activity which may be the regulator that has created a control system (see
  Chapter 5).
- **Level 4: applying the equation** – when research results are applied to management
  problems, very often extrapolating past trends, encountering the difficulty of extrapolating
  from particular spatial or temporal scales to other scales (see Chapter 12).
- **Level 5: appreciating the equation** – involves acknowledging that human reaction
  to physical environment and physical landscape can vary between cultures, affecting
  how the Earth's surface is managed and designed (see Chapter 12).

In any location in order to manage the slope it can be necessary to increase the
resistance or to decrease the force. Such an approach can be applied to all
processes although it is not easy to compute all the forces acting, to calculate
total resistance, or to derive values in comparable units. Examples of force and
resistance as applied to specific geomorphological processes are included in
Table 4.2.

The Earth's surface must be in some form of balance or equilibrium otherwise
it would be unstable and would change or fail. **Equilibrium** is a state of balance
in any system, so that the input of energy and mass to a subsystem is the same as
the outputs from the subsystem, remaining unchanged over time unless the
controlling forces change. This steady state is one in which the average annual
input is constant, although inputs to an open system often vary through time,
for example with seasons. If the annual average input is changing through time
sufficiently slowly for the system to adjust then the condition is a dynamic
equilibrium; other types of equilibrium identified (Chorley and Kennedy, 1971;
Graf, 1988) include quasi-equilibrium which is the apparent balance between
force and resistance, and metastable equilibrium which occurs after adjusting to
a threshold value (see Figure 5.2). An influence was Le Chatelier's principle in

Table 4.2 MAJOR CATEGORIES OF EARTH SURFACE PROCESSES

Category of process	Examples of force and resistance (see also Table 4.4)	
Exogenetic	**Force**	**Resistance**
Weathering	Crystal growth, heating and cooling, tree roots	Physical and chemical bonding
Mass movement, hillslope processes, including processes of mass wasting	Gravity, increased water content, earthquake movements	Shear strength of materials, binding effects of vegetation, structures
Fluvial, drainage basin processes, hydrologic processes	Gravity, discharge reflecting precipitation stream power	Friction in fluid and between water and channel margins, obstructions in channel
Coastal processes and landforms that occur on coastal margins	Wave action, tides	Friction on coasts, strength of materials
Aeolian wind-dominated processes in hot and cold deserts and other areas such as some coastal zones	Wind action giving lift force and drag forces	Gravity, particle cohesion, friction between particles and with surface
Glacial, glaciers and ice caps, and landscapes occupied by glaciers, and those glaciated in the past	Gravity, pressure of snow and ice	Friction with bedrock
Periglacial/Nival, Cryonival, typify the processes in the periglacial zone, in some cases associated with permafrost, but also found in high altitude areas	Expansion of water on freezing	Strength of materials
Subsidence	Gravity following removal of fluids or material	Strength and cohesion of materials
Soil pedogenic processes, soil erosion	Water and wind on surface	Vegetation cover
Ecosystem dynamics	Animal burrowing	Indurated soils resist plant growth
Endogenetic		
Earthquakes and tectonic	Rock uplift	Gravity, rock strength
Volcanic processes	Magma extruded	Friction with surface

chemistry that 'if to a system in equilibrium a constraint be applied then the system will readjust itself to minimize the effect of the constraint'. This, with linear thinking, encouraged the belief that small forces produce small responses. Alternative thinking was based on non-linear systems in which relatively small events can trigger large and rapid changes. Chaos theory and catastrophe theory can account for sudden shifts of a system from one state to another as a result of the system being moved across a threshold condition (see Table 5.4). Non-linearities are characterized by periodic behaviour, sensitivity to initial conditions and switching between behaviour patterns which may result from external forcing, crossing internal **thresholds** or small random perturbations close to critical points (Thornes, 2003). Thresholds are stages, or tipping points, at which the essential characteristics of the natural system's state change dramatically, (p. 133). Well-known thresholds include the accumulation of snow on mountain sides to a critical depth above which an avalanche occurs (p. 179). The build up of snow and ice on a glacier may occur up to a threshold level at which a glacier surge occurs, such surges occurring periodically even though the precipitation as snow remains consistent over decades.

Processes responsible for fashioning the surface of the earth can be thought of as exogenetic, which occur at or near the surface, and endogenetic, which originate within the Earth and beneath the surface. One would expect that a single accepted classification of the processes affecting the land surface of the Earth would be available but this is not so! However the categories shown in Table 4.2 are chosen to be as comprehensive as possible.

4.2 EXOGENETIC PROCESSES

4.2.1. WEATHERING

Weathering is the alteration of rock or minerals in situ on or near the surface of the Earth by chemical, mechanical and biological mechanisms leading to weathering products of waste or regolith which may form the parent material for soil development. Yatsu (1988) in his extensive survey recognizes mechanical and chemical weathering, and weathering by organisms, demonstrating how a knowledge of chemistry and exchange reactions is necessary to understand weathering processes. Some landforms result almost exclusively from weathering, such as tors, weathering pits including tafoni, which are pitted surfaces, and hoodoos which are rock pinnacles formed by differential weathering and erosion classically found in Bryce Canyon National Park, Utah (see http://www.nps.gov/brca/geology_hoodoos.html). Combined with other processes, frost weathering creates distinctive landforms in high latitudes.

The main consequence of weathering is the breakdown of material so that it can be transported and changed by other processes, which fashion the land surface. Mechanical or physical weathering is the breakdown or disintegration of rock without any significant degree of chemical change of the minerals in the rock. Stresses on rock can be caused by crystal growth (ice or salt); heating and cooling by insolation and fire; prizing effect of tree roots; and unloading or pressure release which creates fractures parallel to the ground surface, admirably demonstrated by the granite domes in Yosemite National Park (see Figure 4.1).

Exfoliation may give onion-like layers of rock, so that walking on the surface produces audible cracks. It was originally thought that great daily fluctuations of temperature could cause rocks to exfoliate, literally peel off, but it now seems that this process is effective only when moisture is present, possibly involving some degree of chemical weathering. However, rocks subject to fire,

Yosemite Valley looking west at the cathedrals

Higher Cathedral
Middle Cathedral

Figure 4.1 Yosemite slopes in exposed granite of Sierra Nevada batholith demonstrating features of glaciations and exfoliation of granite

bush fires for example, may be affected by exfoliation. Frost weathering and salt weathering occur as a result of phase changes – the former by the change from water to solid water as ice, the latter by crystallization. Water on freezing expands by about 9% in volume creating maximum pressures c.2100 kg.cm^{-2} that are higher than the tensile strength of rock, which is generally less than 250 kg.cm^2 (Goudie, 1999). When freezing, water attracts other water from pores or capillaries so that growth of ice nuclei can cause further pressure. The efficacy of frost weathering depends upon the availability of moisture, frequent fluctuations around 0°C, and suitably fractured and jointed rocks. Salt weathering occurs because solutions containing salts (usually including NaCl, Na$_2$CO$_3$, Ca$_2$SO$_4$) on cooling or evaporation will crystallize producing pressures greater than the rock tensile strength. In addition some salts expand when water is added to their crystal structure, such as Na$_2$SO$_4$ and Na$_2$CO$_3$ which expand by up to 300% when they are hydrated. Salt weathering occurs in cracks and fissures in rocks but can cause problems in building materials, so that building foundations in arid and semi-arid areas need to be protected by using salt resistant cement, by avoiding any salt in the aggregates and ensuring that groundwater cannot affect the foundations (Douglas, 2005a, 2005b). Other types of mechanical weathering include hydration and swelling of clay minerals, and the alternate wetting and drying of rocks which may cause splitting or disintegration. Processes of mechanical weathering although individually small scale can be extremely effective.

Chemical weathering involves the decay or decomposition of minerals in rock and may be assisted by mechanical weathering which increases the surface area available for chemical attack. Water is necessary for virtually all chemical weathering processes: acting as a solvent in the process of solution which is particularly effective on limestones (p. 52); in hydrolysis whereby the chemical reaction with water involves H and OH ions in the way that feldspars (e.g. orthoclase feldspar $K_2O,Al_2O_3,6SiO_2$) can be transformed into clays (e.g. kaolinite $H_4Al_2Si_2O_9$); and in hydration which involves the incorporation/absorption of water into the molecular structure of minerals. In hydration some minerals such as gypsum and ferric oxide are readily changed to form new minerals, some clay minerals allow swelling during wetting and drying, and, as salts in pore spaces are hydrated, the resulting pressures can fracture rock. Carbonation arises because carbon dioxide is present in most water on the surface as a weak solution of carbonic acid. Chelation occurs where plant roots are surrounded by a concentration of hydrogen ions which can exchange with the cations in adjacent minerals, showing that weathering can be very effective in organic-rich environments: on lichen-covered surfaces lichens are known to excrete chelating agents. Oxidation takes place when elements lose electrons to an oxygen ion in solution and reduction occurs when electrons are gained. The redox potential (Eh) of the environment (measured in millivolts) favours oxidation if positive, as where iron rusts as a result of the production of iron oxide, and reduction if negative, as in waterlogged soils or below the water table for example.

Biological activity involved in many weathering processes is assisted by organisms including burrowing animals, ants, termites and earthworms; by plants including the stresses induced by the wedging effects of plant roots in cracks; and by microorganisms involved in decomposition of organic matter, in the synthesis of organic acids, and directly in some cases as decomposers of minerals and rocks (Yatsu, 1988). Bioturbation of soil by earthworms could be responsible for the mixing of soil about 5 mm in depth over the humid parts of continents (Summerfield, 1991). Biochemical processes are facilitated by organic acids; humic acids produced during decay of organic material; by fulvic acids derived from peats; by bacterial acids and microfloral acids which can decompose silicates and are particularly effective in warm humid environments.

As weathering prepares material for action by other processes, a weathering-limited situation is where the removal of weathered material takes place more rapidly than it is produced, whereas a transport-limited situation is where weathering rates exceed the rate of removal (p. 240, Chapter 10). Rates of weathering are notoriously difficult to measure because of the time involved and because of their great variability, associated with rock characteristics, climatic conditions and the biotic environment. Values measured or deduced tend to be the highest rates because many are so low that they cannot easily be measured (see Table 4.3). The term erosion has been used for more than 200 years to signify the physical breakdown, chemical solution/decomposition and transport of material, involving ways in which processes obtain and remove rock debris. Denudation, laying bare a surface by the stripping of overlying material, is a

Table 4.3   WEATHERING PROCESSES

Type of weathering	Processes involved
Physical or mechanical, disintegration processes	Crystallization – salt weathering, frost weathering
	Temperature changes – insolation weathering, fire, expansion of ice in cracks
	Wetting and drying
	Pressure release – removal of overburden
	Organic processes – root wedging
Chemical, decomposition processes	Hydration and hydrolysis
	Oxidation and reduction
	Solution and carbonation
	Chelation
	Biological–chemical changes (organic weathering)
Biological	Mechanical – growth, wedging, burrowing
	Chemical – biochemical

Rates up to 0.2 mm.yr^{-1}. Some of the most intriguing measurements have been made below perched erratic blocks (c. 0.025–0.042 mm.yr^{-1}), or on the Great Pyramid (can be as much as 0.2 mm.yr$^-$). Weathering of gravestones and public buildings provides information on weathering rates. At the global scale weathering processes produce a weathering mantle which is very small in desert and semi-desert and in tundra areas of up to 3 m, but can be greater than 100 m in the humid tropics (Strakhov, 1967).

broader term than erosion because it includes weathering as well as processes wearing down the land surface of the Earth.

4.2.2 MASS MOVEMENT PROCESSES

Mass movement processes, the detachment and downslope movement of surface material including solid rock, are often associated with the development of hill-slopes. A hypothetical nine-unit land surface model (Dalrymple et al., 1969) characterizes the components of a slope anywhere in the world (see Figure 4.2). The nine elements in this model can be combined in a variety of ways all potentially affected by two groups of processes – mass movement and water flow, although there is a transition from one to the other.

Mass movement processes are a classic example of the force–resistance situation. Resistance is provided by the shear strength of the slope materials depending on both their lithology and structure (p. 53), often reinforced by the binding effect of vegetation or by human artefacts. Shear strength (S), the maximum available resistance to movement, depends upon the friction between grains which in turn reflects the amount of grain interlocking, influenced by the angularity of the grains and their density of packing. These control the angle of shearing resistance (φ'). In addition, shear strength relates to the effective pressure transmitted between

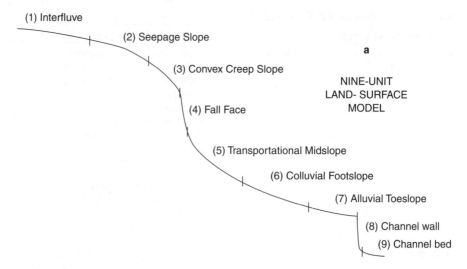

Figure 4.2   The hypothetical nine-unit land surface model (Dalrymple et al., 1969)

This model is a means of characterizing the components of a slope anywhere in the world. Predominant geomorphic processes associated with each of the units are:

1 Pedogenetic processes associated with vertical subsurface water movement
2 Mechanical and chemical eluviation by lateral subsurface water movement
3 Soil creep, terracette formation
4 Fall, slide, chemical and physical weathering
5 Transportation of material by mass movement (flow, slide, slump, creep) terracette formation, surface and subsurface water action
6 Redeposition of material by mass movement and some surface wash. Fan formation, transportation of material, creep, subsurface water action
7 Alluvial deposition, processes resulting from subsurface water movement
8 Corrosion, slumping, fall
9 Transportation of material down valley by surface water action, periodic aggradation and corrosion

particles (σ) and their effective cohesion (c), influenced by chemical bonding of rock and soil particles and adhesion of clay-sized material. These are related in the Coulomb-Terzaghi shear strength equation:

$$S = c + \sigma \tan \varphi'$$

Because strength can be expressed in this way, and measured, it is possible to derive stability analyses for hillslopes. Strength can be affected by pore water pressure, exerted by water in the voids or interstices of material which increases as the water table rises. As moisture content increases material changes from elastic to plastic condition at the plastic limit, then with further water content the liquid limit is the moisture content at which the plastic material becomes liquid. These plastic and liquid limits together with the shrinkage limit between solids and semi-solids which can crack, are collectively known as Atterberg limits.

In addition to increased water content reducing shear strength of materials in hillslopes, strength can also be reduced by weathering processes including biological mechanisms and by changes of material structure. If shear strength is reduced the slope may become unstable but instability often results from stresses on slope materials. Such stresses can be internal, due to water or ice or to swelling, or for external reasons (Table 4.4). The stress on a particular slope depends upon the existing slope angle (α) because gravity forces (g) acting on material of mass (m) will be mg sin α acting down the slope and mg cos α at right-angles into the slope. For specific materials there are threshold angles of slope or maximum slope angles which can exist in a particular environment, and these may be as much as 30° or more for sandstones but much lower for clays.

If stress exceeds shear strength in the factor of safety (p. 71), then the slope may fail by any one of a series of types of mass movements. The limit at which a failure is likely, the threshold condition, can be triggered by increases in stress (see Table 4.4) or by reduction in strength of materials composing the slope. Storm events can trigger mass movements because the introduction of a significant amount of water in a relatively short period of time is one of the ways of changing the slope conditions, leading to a different factor of safety and so to failure. Three forms of movement are slide, flow, creep/heave and the relationship between them (Figure 4.3) indicates how some are transitional in nature. Subsidence is noted later (p. 106). Falls are a special case of slides and a spread is the lateral movement of blocks of material or of material that has lost its shear strength to behave like a fluid. Some mass movements are extremely slow such as soil creep, whereas others such as landslides or rockslides are fast; some are very dry such as rockfalls whereas a mudslide is a wet case. The types of mass movement identified (see Table 4.5) vary according to the material involved,

Table 4.4   EXAMPLES OF REASONS FOR INCREASES IN SHEAR STRESS (SEE VARNES, 1978)

Cause	Specific examples
Loading increased	Natural accumulation of material from upslope
	Weight of precipitation as snow or rain
	Seepage pressures of percolating water
	Human activity such as buildings, piles of material
Removal of underlying support	Undercutting by rivers and waves
	Solution
	Mining
	Removal/squeezing out of underlying sediments
Removal of lateral support, steepening of slope	Erosion by rivers, glaciers, waves
	Earlier mass movements by falls or slides
	Weathering or progressive failure
	Human impact by quarrying, road building
Lateral pressure	Water or ice in cracks
	Swelling, especially of clays
Transitory stresses	Earthquakes
	Tree movement in the wind
	Vibrations due to human impact, explosions

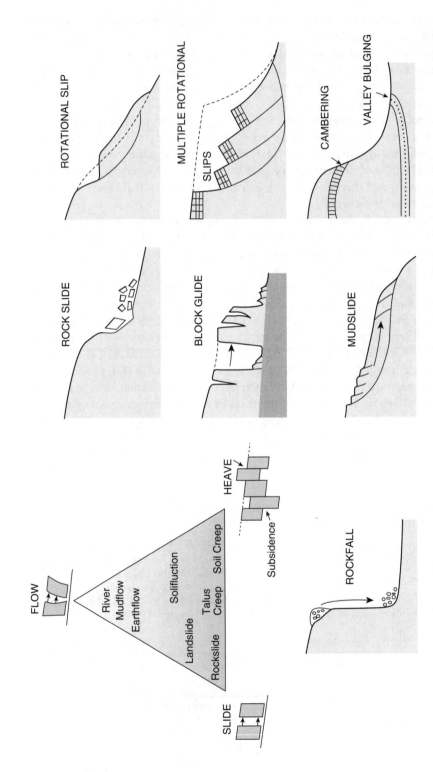

Figure 4.3 Triangular diagram indicates classification of mass movement processes according to flow, slide and heave/subsidence; sketch sections indicate several types of mass movement (after Gregory, 1979 in Derbyshire et al., 1979)

Table 4.5 TYPES OF MASS MOVEMENT AND THEIR CHARACTERISTICS (COMPILED FROM SEVERAL SOURCES INCLUDING SUMMERFIELD, 1991)

Mode of movement	Type	Strain and nature of movement	Typical rate
Fall: Free movement of material away from the slope.	Rock fall	Blocks detached from a free face fall freely to the slope foot, facilitated in well jointed rock.	Rapid, >100m.s^{-1}
	Topple	Rotational fall of slab of rock or soils, such as segment of river bank, and may be induced by wedging behind the mass.	Very rapid
Spread: lateral movement.	Flowslide	Or spreading failure, when material loses shear strength and behaves as a fluid.	
	Cambering	Lateral movement of blocks of material floating on underlying plastic or liquefied zone.	
	Sackung	Graben-like feature develops in middle of spreading mass.	
Slide – Translational: Shear displacement movement along largely planar surface.	Rock slide	Or blockglide, segments of bedrock sliding on bedding plane, joint, or fault surfaces.	80–350 km.h^{-1}
	Debris/earth slide, and debris/earth block slide	Detachment of shallow section of earth material which slides along bedding plane or shear surface.	Usually slow but debris slides can be rapid
Slide – Rotational: Often below a cap rock and involves a turning moment about a point above the centre of gravity.	Rock slump	Single slips followed by creep. Multiple rotational slips are arcuate blocks related to single failure line.	Extremely slow to moderate
	Debris/earth slump	Rotational movement of rock debris or soil, along concave failure plane.	Slow
Heave: Cycles of expansion and contraction involve expansion normal to the slope but contraction vertically under	Soil creep	Involves relatively fine material and expansion and contraction may result from wetting and drying, from frost action, or action of organisms.	1–2 mm.yr^{-1}

(Continued)

Table 4.5 *(Continued)*

Mode of movement	Type	Strain and nature of movement	Typical rate
gravity gives net downslope movement.	Talus creep	Involves coarser material and usually slow downslope movement is the result of frost action, and freezing and thawing when may be termed frost creep.	1–2 mm.yr^{-1} but under some conditions may be 300 mm.yr^{-1}
Creep: May be combined with heave.	Rock creep Continuous creep	Slow plastic deformation of rock or soil in response to stress created by weight of overburden.	Very slow
Flow: Shear occurs throughout the material but is greatest near the surface and almost zero at the base of the mass of material. Types distinguished according to water content and snow and ice in avalanches.	Dry flow	Usually contains some water and develops from falls or slides.	Can be extremely rapid
	Solifluction	Slow downslope movement of masses of material saturated with water.	Very slow
	Gelifluction	Solifluction associated with periglacial conditions transporting saturated material above permafrost.	10–300 mm.yr^{-1} on angles of 15–25°
	Mud flow	Usually poorly sorted coarse material in a clay-silt matrix, with high water content. If composed mainly of volcanic ash is a lahar and may be triggered by a crater lake bursting, or from prolonged torrential precipitation.	Slow
	Slow earthflow	Rapid movement of soil and loose material >80% sand sized, often below a landslide, after heavy precipitation event, and may occur above impermeable rocks.	Slow
	Rapid earthflow	Often after an initial slide material containing sensitive clays collapses and spreads laterally. See flowslide (under Spread above).	Very rapid
	Debris flow	Rapid movement occurs if water added to debris slide/avalanche. Viscous	Very rapid

Table 4.5 *(Continued)*

Mode of movement	Type	Strain and nature of movement	Typical rate
		debris-laden mud usually flowing along existing drainage courses.	
	Debris slide	Sudden downslope movement of debris due to increased water content resulting from heavy precipitation. May be termed debris avalanche.	Very rapid
	Snow avalanche	Rapid gravitational movement of snow and ice en masse down steep slopes generally in mountains (see Table 7.6).	Extremely rapid
	Slush avalanche	Or wet snow avalanche is the most powerful, usually occurs in spring with large quantities of water-saturated snow, often along existing drainage lines (see Table 7.6).	Very rapid
Complex	Many mass movement events involve combinations of the above types. They may occur sequentially down the slope so that a rockfall leads to a debris slide which in turn produces an earthflow.		

moisture content and rate at which movement takes place. Mass movement is ubiquitous but rates vary enormously; much mass movement is intermittent, as times between movements vary considerably. The most memorable mass movements are those that occur most rapidly in steep terrain or mountain areas, especially those subject to seismic activity, high precipitation amounts, or human impacts such as deforestation.

Catastrophic mass movement events can have very significant impacts (see Table 4.6 and Figure 4.4, Colour plate 2) and on pasture land in New Zealand rates of soil displacement of $1000–4000 m^3.km^2.yr^{-1}$ have been measured (Crozier, 1986). As mass movement provides material on the land surface sediment conveyor belt, as inputs for fluvial, coastal, or aeolian processes, characteristic landforms are produced on slopes and there are many cases where hillslopes are punctuated by fossil mass movement features so that it is necessary to be able to 'read' the scenery to identify where former mass movements have occurred (see Figure 4.4, Colour plate 2).

In addition to mass movement, water is a significant way whereby transfers are achieved and slopes are fashioned, leading to rivers which are possibly the key pathway for material transfer on the land surface and responsible for suspended load transport of $15–20$ Gt.yr^{-1}. As the arteries of the land surface, rivers are the conveyor belts for the sediment system, contributing 95% of the sediment

Table 4.6   EXAMPLES OF CATASTROPHIC MASS MOVEMENT EVENTS

Location	Characteristics of event	Reference
Turtle Mountain, Frank, Alberta, Canada 29 April 1903	Huge rock mass, more than 1 km² in area and up to 150 m thick, broke away from east face of Turtle mountain. About 30 million m³ of rock avalanched down the face of Turtle Mountain at a speed up to 100 km.h⁻¹, struck the valley floor with such force that it spread across the 3 km wide valley floor and 120 m up the other side. Event killed 70 people.	Rhodes et al., 2008: 11
Mount Toc, Italy, 80 km north of Venice, 9 Octorber 1963	Trial flooding of Vajont valley north of Mount Toc began in February 1960, as first stage of filling 5 km long reservoir behind dam 265.5 m high which was to be highest double-arched dam in the world. Massive crack, nearly 2 km long, opened on side of Mount Toc above reservoir, some 200 million m³ of rock moved downslope into reservoir over 3 years, until 22.39 on 9 October 1963 when the mountainside fell into the reservoir causing massive wave to overtop the dam and eliminating village of Longarone 2 km downstream with loss of 2000 lives. Flooding of reservoir raised water table and this augmented by heavy rain in September possibly accompanied by new rock cracking.	McGuire et al., 2002; Kilburn and Petley, 2003
Huascaran, Peru, 31 May 1970	Earthquake triggered combined rock avalanche and debris flow, up to 133 m.s⁻¹ buried two cities, causing more than 20,000 deaths.	See Chapter 7, p. 181, and Colour plate 7
Mount St Helens, Washington, 18 May 1980	Earthquake of magnitude 5.1 triggered collapse of about 3 km³ from northern	McGuire et al., 2002 See also Figure 4.13

Table 4.6   *(Continued)*

Type of mass movement	Characteristics of event	Reference
	flank of Mt St Helens, exposing gas-rich magma accumulating inside the volcano for at least 2 months. Magma gave cloud of gas and volcanic debris moving downslope at 300–500 km.h^{-1}. Devastated some 600 km^2 of land within 2 minutes.	
China, Sichuan province, 12 May 2008	Aftermath of the Sichuan earthquake caused many landslides, some of which dammed the rivers; subsequent landslides triggered by rainfall, failure of landslide dams and sediment production, and killed at least 68,000 people	See Table 4.14

entering the oceans (Walling, 2006). The four longest are the Nile (6696 km) together with the Amazon, the Mississippi-Missouri and the Yangtze. The largest drainage basin is the Amazon (6,915,000 km^2), with an average discharge of 219,000 m^3.s^{-1}, responsible for 6923 km^2.year^{-1} water reaching the sea, accounting for nearly 20% of the world's river flow.

4.2.3 Fluvial Processes

Fluvial processes, things that are of, or in, a river, are one of the most researched fields (see Box 4.1) because of the significance of rivers for political, water resources, navigation and industrial purposes. Research has become increasingly multidisciplinary, through links with ecology, engineering and hydraulics, and sedimentary geology. Fluvial processes relate to drainage basins, to hillslopes, and to river channels. As an accounting unit the drainage basin (see Table 1.6), which collects all the surface runoff flowing to a network of channels to exit at a particular point along a river, can be employed for water balance equations, for river discharge calculations and for hydrological modelling. The water balance equation relates runoff (Q), precipitation (P), evapotranspiration (ET) and changes in storage in soil or groundwater (ΔS) in the form:

$$Q = P - ET \pm \Delta S$$

At the scale of the drainage basin, input of precipitation is translated into streamflow as river discharge, with losses to evapotranspiration and infiltration to the groundwater table, all depending upon the basin or catchment characteristics which are the geological, topographical, soil and vegetation/land use features of the basin. The drainage basin unit is ideal for interpreting river hydrographs (the plot of volume of flow per unit time, or discharge, against time) derived for a period such as a year or for the time immediately following a storm, for any gauging point along a river, and also for constructing mathematical models which enable flow estimation. Progress by developments in topographic specification, including that by GIS, has enabled refinement of catchment-scale models, with a cellular automaton approach dividing a catchment into uniform grid cells to simulate the dynamic response of a basin to flood events (Coulthard et al., 1999).

Drainage basin characteristics affect how much precipitation flows over the ground surface of hillslopes, how much infiltrates into soil and down to the water table, how much runs through the soil, or rock above the water table, and how rapidly the concentrated water flow runs through the river channels in the drainage basin (see Figure 4.5A). Knowledge of the way in which water is transmitted over and within hillslopes and through drainage basins is rather like a detective story. At first, following the ideas of R.E. Horton the American hydrologist, it was thought that water could flow over the surface or infiltrate down to the water table, giving hydrographs composed of just two elements. Subsequent field research showed that water draining through drainage basins can follow a variety of routes over the surface (overland flow), through the soil (throughflow or pipeflow), through the rock above the water table (interflow) or into groundwater from which it may eventually emerge as a spring providing base flow. These routes are affected by stores of water on the surface (surface detention, lakes), in the soil, in the aeration zone, or in groundwater (see Figure 4.5B). Characteristics of hydrographs depend upon the size of the drainage area, the amount and location of precipitation, and the routes followed by the water (see Figure 4.5 C and D). Hydrographs are composed of quickflow and delayed flow components, all potentially derived from flow through different routes at contrasting rates (see Table 4.7).

Water flowing through a drainage basin and over hillslopes transports sediment and solutes, the amount transported reflecting water available, discharge and sources of sediment. Sediment may be transported in suspension or as bedload, and as solutes dissolved in the water or attached to particles, including pollutants. Sediment and solute transport can be expressed as sediment or solute hydrographs (chemographs), which may not always mirror the discharge hydrograph because sources of sediment or solutes may not be the same as the origins of the water in the drainage basin. Hillslope hydrology is dynamic because the water and sediment-producing areas expand and contract depending upon the catchment characteristics, the antecedent conditions prior to any storm event, and the character of the storm input. Precipitation on hillslopes may produce rain splash before water flows as sheet flow or in rills, subsequently concentrating in ephemeral channels which contain water after storm events, leading to intermittent channels which flow during part of the year, whereas perennial channels always contain flowing water. At any one time the drainage network is one stage of a range of values of drainage density (total length of stream channel per unit basin area), all related to stages of the hydrograph.

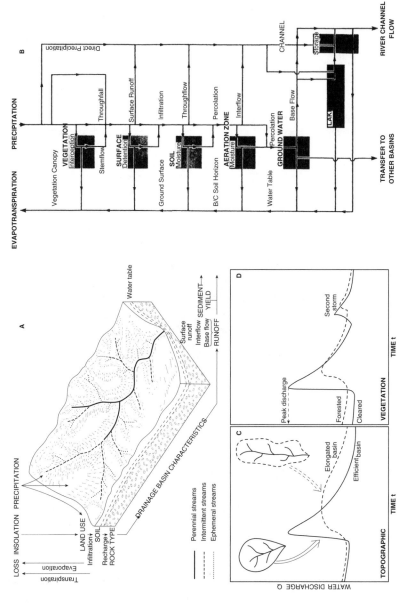

Figure 4.5 A geomorphological view of the drainage basin (A) after Gregory and Walling (1973); the components of the hydrological cycle (B); and hydrographs indicating effects of topographic (C) and vegetation (D) controls

Table 4.7   EXAMPLES OF VELOCITIES FOR WATER FLOW IN TYPES OF ROUTES THROUGH DRAINAGE BASINS (DEVELOPED FROM GREGORY, 1979: 58)

Type of flow	Velocities measured or calculated (in each case there can be a great range of values so that those given are indicative)
Overland flow	<0.1 cm.s^{-1} on low slopes with thick vegetation cover 3–15 cm.s^{-1} on slope of 0.40
Vertical percolation	<7.5 cm.day^{-1}
Saturated throughflow	0.2–37.2 cm.h^{-1}
Throughflow	50–80 cm.day^{-1}
Interflow	0.5 m.day^{-1}
Pipeflow	10–20 cm.s^{-1}
Stream channel flow	45 cm.s^{-1}

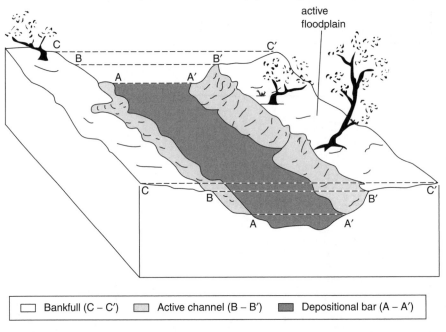

Bankfull (C – C')   Active channel (B – B')   Depositional bar (A – A')

Figure 4.6   Definition of river channel capacity (after Osterkamp and Hedman, 1982) – the channel can also be characterized by depth, width or shape (width/depth ratio) values

Flow in river channels is affected by force which is determined by gravitational acceleration, channel gradient and momentum, and by the resistance provided by friction within the fluid and between the flowing water and the channel margins and the air above. Channel cross-sectional area or channel capacity (see Figure 4.6) is an important value because at any point along a river it reflects the discharge history of the river at that location, although it is modified by the local slope, as well as by the material in the bed and banks. Relationships between channel capacity and discharge based on data from gauging stations can be used to estimate flows for ungauged sites. Open channel hydraulics provides relationships between variables governing fluid flow in channels, and regime equations have been used for discharge estimation for relatively simple river systems. Theoretical relations are also available for estimating sediment transport although not always at capacity load situations because many rivers do not have sufficient sediment supply to achieve capacity load values. Such relationships are very important for understanding entrainment thresholds or critical flow conditions at which particles are just able to move.

The rate of energy expenditure in flowing water is expressed by stream power (ω). Power is the rate of doing work (force × distance) and is expressed in Watts (Joules per second, $J.s^{-1}$). The potential energy that water possesses at a particular location is proportional to its height above some datum which can be sea or lake level; this potential energy is converted into kinetic energy as the water flows downhill under the influence of gravity. Stream power was defined in relation to the rate of sediment transport (Bagnold, 1960) as the product of fluid density (ρ), slope (s), acceleration due to gravity (g) and discharge (Q) in:

$$\omega = \rho Qgs$$

To allow comparison of different situations, unit stream power is expressed per unit length of channel or per unit channel area. Values of unit stream power (ω) range from less than 1 W. m^{-2} in flow between rills to >12,000 W. m^{-2} in riverine flood flows in India, up to 18,582 W. m^{-2} for large flash floods, and up to 3×10^5 W. m^{-2} for the Missoula flood (Baker and Costa, 1987) which occurred during the Quaternary and was the largest known discharge of water on Earth.

Although it is convenient to separate water at the drainage basin, hillslope and channel scales, continuity is emphasized by concepts, events, and landforms. The river continuum concept recognizes ecological progression with continuous transport of nutrients from headwaters to the valley floor (Stanley and Boulton, 2000) although downstream sediment transport may be far more 'jerky' with storage alternating with periodic transport; and drainage networks often possess significant 'steps' in their hydraulic geometry and habitats according to incoming tributaries. Human impacts, such as dam construction, can reduce the river continuum to a sort of giant ladder with the serial discontinuity concept (Ward and Stanford, 1983) recognizing the potential for the continuum to break down. Other concepts emphasize coupling between components such as hillslope to channel, or channel to floodplain; and elevational movement along flow pathways including the flux of water and sediment and solutes from one component to another. Formative events may occur in parts of the basin but also emphasize linkage of components. A river flood occurs when water exceeds the capacity of the river channel and flows out on to the floodplain as a result of intense or prolonged precipitation over all or part of a drainage basin, from snow melt floods which are an annual occurrence in some parts of the world, or sudden releases of large amounts of water either by natural causes or collapse of man-made structures including dams.

Table 4.8   FLUVIAL LANDFORMS AT DIFFERENT SCALES

Scale of fluvial system (Downs and Gregory, 2004)	Contemporary	Landform response to change including secular climate change or human activity
Channel cross-section environment at a point, within channel unit.	Bedforms and sedimentary structures including ripples, dunes (see Bridge, 2003).	Lithofacies, sedimentary structures
River channel, the linear feature along which surface water can flow, usually clearly differentiated from the adjacent floodplain or valley floor (see Figure 4.6).	Pool and riffle sequence, alluvial channels and bars, step-pool sequences in rock channels, levees.	Palaeochannels Abandoned channels Enlargement or reduction of channel cross sections
River reaches and valley segments, a homogeneous section of the river channel along which the controlling factors do not change significantly.	Channel planform: the plan of the river channel from the air, may be either single thread (straight or meandering) or multithread (braided), varies according to the level of discharge; anabranching, anastomosed.	Ox bows, meander cutoffs, clay plugs. Palaeomeanders Terraces Channel metamorphosis Valley fills Confined channels Incised channels Channelized rivers
River corridor: Linear features of the landscape bordering the river channel Floodplain: The valley floor area adjacent to the river channel.	Alluvial fans, deltas. A distinction may be made between the hydraulic floodplain, inundated at least once during a given return period, and the genetic floodplain which is the largely horizontally bedded landform composed of alluvial deposits adjacent to a river channel.	Terraces Valley trains
Drainage basin delimited by a topographic divide or watershed as the area which collects all surface runoff flowing in a network of channels to exit at a particular point along a river.	Drainage patterns including dendritic, parallel, trellis, rectangular, radial, annular. Drainage network, the network of stream and river channels that exist within a specific basin. The channels may be	Gullies Dry valleys

Table 4.8 *(Continued)*

Scale of fluvial system (Downs and Gregory, 2004)	Contemporary	Landform response to change including secular climate change or human activity
	perennial, intermittent and ephemeral, the network may not be continuous and connected, and the extent of the network will be affected by storm events, season, basin characteristics and human activity. Drainage density.	

Natural causes include *jokulhlaups* in Iceland, which are flood waves produced when water drains from an ice-dammed lake. Immediately after a flood-inducing storm event the drainage network is at its maximum extent including ephemeral as well as intermittent and perennial channels; as the flood wave is generated and moves downstream it may exceed the capacity of the reach, the planform and the channel cross-section so that flood water extends over the hydraulic floodplain. Analogous to the flood pulse are sediment waves or sediment pulses termed sediment slugs which are bodies of clastic material associated with disequilibrium conditions in fluvial systems over time periods above the event scale (Nicholas et al., 1995). These and other fluvial processes relate to landforms some of which, such as meanders and oxbows, are universally well-known, but landforms and characteristics associated with present environments (see Table 4.8) are complemented by features produced as a result of secular climate change or human activity.

Landforms evolve over time and recent progress in fluvial geomorphology has been fostered by increasing multidisciplinary involvement and the availability of new analytical methods, instrumentation and techniques, enabling development of new applications for river management, landscape **restoration,** hazard studies, river history and geoarchaeology (Thorndycraft et al., 2008). This resurgence has included attention accorded to fundamental core scientific issues such as hydraulics of open channel flow, investigated with more insight in relation to sediment transport and to the character of flow in meander bends, as well as to the solution of environmental management problems, some of which have arisen directly from process-based investigations, as in the way that studies of slugs or sediment waves have implications for river management. Although downstream alluviation and headwater incision can be contemporaneous, time emphasizes discontinuities that can occur within catchment processes, so that sediment released upstream from gully erosion can accumulate in floodplains as a sink so that much later the sediment in storage is released to continue through the continuum to the sea.

4.2.4 PROCESSES ON COASTAL MARGINS

Processes on coastal margins operate in some of the most dynamic environments on Earth, admired and appreciated at least as much as rivers, including the beaches and scenic cliff coasts that are associated with holidays and leisure. The land surface coastline is probably close to 1 million km long (Bird, 2000), of which about 75% is cliffed and rocky, and about 20% is sandy and backed by beach ridges, dunes or other sandy depositional terrain. Some countries have large proportions of artificial coastline, ranging from 85% in Belgium, 51% in Japan, to 38% in England. Canada has the world's longest coastline (202,080 km) although, according to length per unit land area, countries like Norway have 257 km for each square km of land, contrasting with c.22 km.km^2 for Canada.

Half the population of the industrialized world lives within 1 km of a coast so that areas where fluvial and coastal systems interface are of great significance for population and economic activity; two types of environment are deltas and estuaries. The 10 largest world deltas occupy 0.64% of the land surface (see Table 4.9). A delta (see Figure 4.7, Colour plate 2) is formed by the deposition of sediments in shallow waters, usually on the inner continental shelf but also in inland seas or lakes. Delta characteristics are affected by water densities, the hydrology of the river and calibre of its load, particularly the proportion of bed-load to suspended load; coastal processes including wave action; currents and tidal scour; as well as the geometry of the coast and tectonic activity. Three sedimentary units make up a delta:

- topset beds – floodplain sands and silts, marsh organic deposits, platform sands
- foreset beds – cross-bedded coarser sands and silts of delta slope, with channel fingers grading laterally into clays, marsh deposits, sand splays
- bottomset beds – offshore clays, toeslope turbidity silt deposits.

Classifying deltas tells us much about their character and origin (see Table 4.9). Supporting enormous populations, deltas are fragile environments, because firstly they can change by avulsion, when diversion of the river channel to a new course at a lower level is usually prompted by aggradation (the mouth of the Yellow River shifted 650 km to the north in 1851, and the Mississippi delta has switched 7 times in the last 7000 years), but secondly because of high rates of subsidence (up to 1 cm.yr^{-1}), resulting in relative sea level rise (RSLR) to be greater than eustatic sea level rise, together with increase in level or extent due to river input, and future sea level rise occurring as a result of global climate change. In the tropics many of these fragile environments can be subject to damage from typhoons or cyclones, exemplified on 2 May 2008 by Cyclone Nargis which had a path directly over the densely populated Irrawaddy delta, led to more than 90,000 deaths, caused damage estimated at $10 billion, and affected 1.5 million people.

Estuaries, the widening reaches of a river mouth, usually tidal, are also the location for dynamic interaction of fluvial freshwater and marine salt water processes.

Table 4.9 MAJOR WORLD DELTAS AND THEIR CLASSIFICATION

Delta	Receiving sea	Area (km² × 10³)
Amazon	Atlantic Ocean	467
Ganges	Bay of Bengal	106
Mekong	South China Sea	94
Yangtze	East China Sea	67
Lena	Laptev Sea	44
Hwang He	Yellow Sea	36
Indus	Arabian Sea	30
Mississippi	Gulf of Mexico	29
Volga	Caspian Sea	27
Usumacinta Grijalva	Gulf of Mexico	22
Orinoco	Atlantic Ocean	21
Irrawaddy	Bay of Bengal	21`

CLASSIFICATION according to:

- **shape or pattern** – they can be arcuate (outer margin is arc-like convex towards the sea, e.g. Nile), birds-foot (finger-like distributary pattern with river levees building out into body of water), or cuspate (symmetrical where river reaches straight coastline);
- **location** – they can be estuarine bay head or continental shelf deltas where their sedimentary structure is constructed out on to shelf;
- **origin** – whether their formative processes are fluvial (e.g. Po, Danube, Mississippi), wave (Nile, Niger), or tide dominated (Fly (PNG), Mekong).

(See also Table 4.11)

Following the Holocene rise of sea level, drowned river valleys include (Grabau et al., 2005):

- unmodified drowned river valleys, given the name of ria in SW England, and NW Spain
- barred estuaries that have been partly separated from the sea by bars or spits across their mouths (for example, much of the east coast of North America)
- fjords (drowned formerly glaciated valleys can be very deep so that Scoresby Sound in East Greenland has depths greater than 1300 m, and Sogne fjord, Norway is up to 1244 m deep) and fjards (low-lying glaciated rocky terrain drowned valleys, in south west Sweden, also forden in Germany)
- tectonic structures flooded by the sea (e.g. rias of New Zealand).

Variations occur due to tidal range and biota, but it is usually possible to recognize the fluvial upper zone, mainly freshwater dominated, the middle zone of mixing of fresh and saline water, and lower zone which is saltwater dominated. The mixing of salt and freshwater is influenced by the water densities, tidal action, river flow, wave climate and the size and shape of the estuary; a salt wedge may occur whereby the freshwater overrides the denser salt water, and the mixing process affects the sedimentation in the estuary. Estuaries act as a buffer between the ocean and the land in that they can decrease the effects of flooding and storm surges.

There are many other types of shoreline, the zone where the ocean meets the land, affected by tidal fluctuations and wave action; this dynamic interactive area

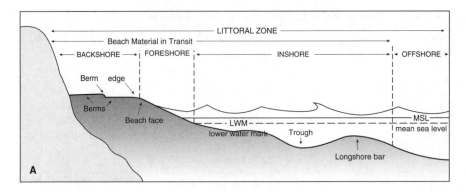

Figure 4.8 Components of the beach profile

has been subdivided in various ways (see Figure 4.8). The coastal zone includes the area between tidal limits, the continental shelf and coastal plain (Viles and Spencer, 1995) whereas the shore zone includes beaches, cliffs, tidal and brackish water wetlands and individual reef communities. A variety of characteristic features and landforms (see Table 4.11) each has specific scales and duration timespans. In this extremely energetic zone of the land surface, wave energy has much greater power than stream power. The total instantaneous wave energy of the earth is estimated at 5×10^7 Joules with 10^4 Watts dissipated along each metre of coast (Beer, 1983; Pethick, 1984), meaning that 100 km of shore could provide as much energy as a large conventional thermal power station and that the total annual dissipation around the world's shorelines is about 1.5×10^{20}J. Some coasts are of course more energetic than others – the NW coasts of the Atlantic and Pacific oceans have annual power delivery of 10^{11}–10^{13} J (Carter, 1988).

There are many bases for classification of coasts (Bird, 2000; Davies, 1980) based on either origin and processes, upon descriptive character, or evolution of the coast. Whichever classification is employed, what are the controls upon coasts? Exogenous sources of energy are the forcing functions associated with waves, tides, currents and large events (see Table 4.10) and their role is complemented by sediment factors, coast configuration and temporal change. Wave action provides the dominant driving force and is a more efficient energy transporter than the river channel where considerable loss of energy occurs through frictional drag. The rate of energy transfer or power of a wave is an important variable with rate of energy transfer as P = ECn (E = wave energy, C = velocity of the wave form, n = group velocity). In theory wave steepness is H/L (H = wave height, L = wavelength) and when >0.14 waves become unstable and collapse, and few waves are less steep than 0.056. Circular orbits of water particles in waves at sea are transformed as the waves approach the coastline and at a critical point the wave breaks, and produces one of four forms (Pethick, 1984): surging breakers associated with flat, low waves, and steep beaches; spilling breakers associated with high, short waves, and on flat beaches breaking a considerable distance from the shore and advancing as a line of foam at the wave crest, giving way to a line of surf moving onshore. In between types are when the wave breaks by curling its crest over by

Table 4.10 INFLUENCES ON COASTAL MORPHOLOGY AND COASTAL PROCESSES

Processes can be amplified from detailed texts and internet sources.

Influence	Character	Comments
Forcing factors:	Wave activity	Rate at which wave energy is transmitted in the direction of wave propagation is energy flux, or wave power (P). When waves approach coast may be refracted, or reflected (from cliff coasts). Breaking wave height (H_b) is the driving force of morphological variations in coastal morphology as wave energy is proportional to wave height squared. Longshore gradients of H_b can be related to variation in beach ridge height and beach sediment size. The temporal and spatial distribution of wave steepness (wave height/wave length) has been used as an indicator of the likelihood of a beach eroding. Steep waves tend to be destructive, eroding sediment from the beach and transporting it into temporary sinks in the subtidal zone, often in longshore bar form. Waves of low steepness tend to be constructive and rebuild the beach creating step-like berms that may be deflated into back-beach dunes.
	Tide-related: tidal range;	If the tidal range is less than 2 m then wind waves form the dominant coastal processes and beaches, spits and barrier islands will be the dominant features of the coast. Large tidal range associated with broad intertidal zone which can be 20 km wide. Microtidal <2 m, mesotidal 2–4 m, megatidal >6 m. Tidal bores, when tides drive water into funnel-shaped inlets, can be 2 m high, moving up Severn estuary at 4 m.s^{-1}; bore moves up Amazon as 2 km wide waterfall 5 m high, moving at 10m.s^{-1}. Coastal areas that experience tidal ranges in excess of 4 m are dominated by tidal landforms such as tidal flats and saltmarshes.
	Currents;	Rip currents return from breaking waves, vary with wave patterns. Tidal currents produced as tides rise and fall.
	Large events and storm surges.	Tsunami – when strong onshore winds build up coastal water to high level, very large waves can be >30 m. Hurricanes may also have similar effects.
Sediment	Coastal sediment cells source and supply	Coastal cells defined by sediment inputs to the shoreline, sediment transport by tidal currents.

(Continued)

Table 4.10   *(Continued)*

Influence	Character	Comments
	Sediment movement, corridor or pathway	Sediment transport by wave action; most coastal cells rely on erosion of terrestrial sediment by waves as the contributing source.
	Sediment sinks	Where sand is the dominant sediment type then cell sinks may be either onshore sand dunes or offshore shoals.
Organic	Coral and algal reefs Mangroves and salt marsh	
Coastal setting	Coastal irregularity	Configuration distribution of erosion and deposition, topography of the adjacent continental shelf, structure and lithology may be significant.
	Morphology of nearshore zone	Affects wave refraction (process whereby deepwater waves differentially adjust to varying bathymetric topography as the wave moves onshore) and wave energy.
	Exposure of coastline to ocean swell, storm and locally generated waves. Subaerial processes	Affects fetch (extent of water over which wind blowing). High energy coasts (wave height >2 m), low energy coasts (<1m). On cliffs
Temporal change	Oscillations in mean sea level	Eustatic changes world wide, including recent sea level rise due to global warming Isostatic, due to loading or unloading. Gulf of Bothnia sea level falling 1 cm per year
	Tectonic movements	Orogenic related to tectonic plates and to earthquakes
	Human impacts	Groundwater or oil extraction can cause land lowering and sea level rise, coastal management affects erosion and deposition cells.

Table 4.11   COASTAL LANDFORMS AND ENVIRONMENTS (SEE BIRD, 2000)

Landform environment	Landforms
Beaches – unconsolidated deposits of sand and gravel deposited by waves and currents on the shore, fringing about 40% of the world's coastline.	*Berms*, or beach terrace built up by swash deposition. Microtidal coasts typically have a single swash-built berm, on mesotidal coasts there are often two berms, and on macrotidal coasts there are multiple berms.

Table 4.11 *(Continued)*

Landform environment	Landforms
	Ridges.
	Cheniers, long, narrow, vegetated marine beach ridge or sandy hummock, 1 to 6 m high, forming roughly parallel to a prograding shoreline seaward of marsh and mudflat deposits, enclosed on the seaward side by fine-grained sediments, and resting on foreshore or mudflat deposits. Well drained, often supporting trees on higher areas. Widths range from 45–450 m and lengths may exceed several km.
	Beach lobes.
	Cusps – scoop-shaped hollows in the beach front maintained by variations in water depth in which waves diverge on either side of spurs and backwash is concentrated in the hollow.
Spits – a narrow embankment of land, above high tide level, oriented in the direction of longshore drift, commonly consisting of sand or gravel deposited by longshore transport and having one end attached to the mainland and the other terminating in open water, usually the sea, usually ending in one or more hooks or recurves.	Paired spits – often border river mouths and lagoon entrances, may result from convergent longshore drift or the breaching of a former barrier, and have grown in different directions at different times.
	Trailing spits – formed in the lee of islands, and where island destroyed by erosion flying spit may persist at right angles to the predominant waves.
	Tombolos – a sand or gravel bar or barrier that connects an island with the mainland or with another island or stack.
	Cuspate spits – form where beach sediment is deposited as protruding more or less symmetrical structures, with convergence of longshore drift from two directions.
	Cuspate foreland – cuspate spits enlarged by the accretion of beach ridges parallel to their shores.
Coastal barriers and Barrier islands – major longshore bars (barrier beaches) with lengths up to 100 km, and constituting some 13% of world's coastlines.	Formed by the deposition of beach material offshore or across the mouths of inlets or embayments, extending above the normal level of high tides and partly or wholly enclosing lagoons or swamps. Barriers are thus distinct from bars which are submerged for part of the tidal cycle and from reefs of biogenic origin.
Coastal wetlands.	Salt marsh.
	Mudflats.
	Mangrove coast/swamps.
	Sebkhas.
Tidal flat and intertidal landforms.	
Deltas – low nearly flat area where sediment accumulates instead of being redistributed by sea or lake water (see Table 4.9 and Figure 4.7, Colour plate 2)	Elongated or digitate – alluvium abundant, river can build into the microtidal sea or lake and wave action limited, e.g. Mississippi.
	Birds-foot – multiple branching elongate deltas.

Table 4.11   *(Continued)*

Landform environment	Landforms
	Cuspate – wave action dominates the distribution of sediment away from the river mouth, e.g. Tiber. Lobate – river builds into the sea but wave action redistributes sediment along coastal barriers, e.g. Niger. Crenulate – tidal currents produce numerous sandy islands separated by tidal channels along the delta front.
Estuaries – submerged sections of valleys, tidal reaches of river mouth and lagoons subject to tidal fluctuations and the meeting of fresh and salt water and receiving sediment from both river and marine sources.	Rias – inlets formed by partial submergence of unglaciated river valleys. Fjords – inlets at the mouths of formerly glaciated valleys. Fiards – inlets formed by Late Quaternary submergence of formerly glaciated valleys in low-lying rocky terrain. Calanques – mouths of steep-sided valleys that were deeply incised during low sea level stages into the limestone plateau east of Marseilles submerged by late Quaternary transgression to form cliff-edged inlets. Sharms and sebkhas – on the Red Sea coast long narrow inlets termed sharms (sherms) have formed where wadis or valleys cut by streams in wetter Pleistocene low sea level episodes flooded during late Quaternary transgression. Broader, often branched embayments are called sebkhas on the arid coasts of the Red Sea and Arabian Gulf. Lagoons.
Organic coasts – Corals and calcareous algae are important builders of reefs although molluscs and sponges construct reefs in particular situations.	Fringing reefs – attached to and extending seaward from the land. Barrier reefs – separated from the coast by a lagoon. Atolls – an annulus of reefs enclosing a central lagoon.
Cliffs – 0.75% of world's coastline is cliffed and rocky.	Cliff profiles, caves, blowholes, gorges, arches, stacks, hanging valleys and coastal waterfalls.
Shore platforms – a platform with low slope angle at the foot of cliffs.	Strandflat – on Norwegian coast up to 64 km wide on coasts rising isostatically.
Coastal dunes – formed when sand on the shore dries out and is blown to the back of the beach.	Foredunes are ridges of sand built up at the back of a beach or on the crest of a sand or shingle berm where dune grasses have colonized and are trapping sand. Parallel dunes – multiple dune ridges usually parallel to the coastline formed successively as foredunes behind a prograding beach. Blowouts and parabolic dunes – unstable dunes with little or no vegetation cover.

Table 4.11 *(Continued)*

Landform environment	Landforms
	Transgressive dunes – formed either when sand blown inland from a beach and retained by vegetation or where previously vegetated coastal dunes disrupted by blowouts until they merge into an elongated dune spilling inland. Are extensive on desert coasts. Cliff-top dunes – on some cliffed coasts where winds blow sand against cliffs and then climbing cliffs. Dunes on shingle – sand arrived as transgressive dune spilling from a nearby beach.
Machair	On coasts of Scotland and Ireland areas of almost featureless low-lying calcareous sandy plain, typically lie behind vegetated coastal dune fringe. Generally of Holocene age.

plunging (plunging breakers) or collapsing (collapsing breakers) forwards on the low water left by the previous wave. If wave crests do not approach parallel to the shore (wave approach angles greater than 10° are unusual), then return of water down the beach is normal to the beach and progressive drift occurs along the beach. Many beaches of the near shore zone are characterized by a circulatory current system in which both shore-normal and longshore currents are present and can be distinguished as cells (see Table 4.10). The development of cells can result in the characteristic landforms of cliffs, beaches, barriers and dunes, reflecting wave power, currents and tidal variations (see Table 4.11) with detailed explanations available (e.g. Bird, 2000).

4.2.5 AEOLIAN PROCESSES

Aeolian processes resulting from wind action are conventionally thought of as characteristic of hot and cold deserts (Chapter 9), but wind action also prevails along the Earth's coastlines, can be influential in many areas of the world (e.g. Seppala, 2004) and may be the agent of soil erosion over ploughed land. Aeolian processes are amenable to fluid analysis in terms of energetics – following the lead first given by Brigadier R.A. Bagnold (see Box 4.2; Bagnold, 1941).

   Gravity, friction between particles and the surface, and cohesion of particles are the forces which resist movement. These can be overcome by forces acting on a grain which can be resolved into the drag force acting horizontally and a lift force acting vertically; together these must momentarily exceed the gravitation, friction and cohesion forces holding the grain in place for aeolian movement to occur. Lift forces, analogous to the lift generated by the wings of aircraft, cause particles to jump into the airflow as saltation, whereby sand rises up to 1 m above the surface and falls again so that it follows a leaping or jumping path, representing c.80% of material movement (see Figure 4.9A). Drag forces, arising from differences in fluid pressure on the windward and leeward sides of grains in an airflow, cause particles to move downwind by rolling or sliding in surface creep, usually responsible for some 20% of the material moved. Once lift or drag have entrained the first few particles, most subsequent entrainment is by ballistic impact whereby grains are bombarded by those already in motion. Such impacts give additional forward momentum meaning that grains of a particular size can be set in motion at a velocity less than that required to initiate movement. When the wind is light and the surface is smooth, flow is laminar but, as wind speed increases, it is turbulent and more effective. At the surface, velocity is zero but increases above the ground, so that most saltating grains travel within 10 mm of surface but with leaps of 0.5–1.5 m (see Figure 4.9). In addition to wind speed, other factors affecting the rate of aeolian transport are air density, which varies slightly with altitude and temperature, particle size and land surface characteristics. Particle size is influential because particles less than 0.1 mm are either clay which can be cohesive and difficult to move, or quartz silt which is usually conveyed as loess at higher levels; whereas to move material sized above 1 mm requires winds of high velocity so that the range of 0.1 to 1.0 mm (very fine to coarse sand) is the size range that moves most frequently. Grain characteristics that influence the threshold velocity for motion include mean grain size, grain shape, grain sorting and grain mineralogy. In addition, particle cohesion can affect the shear stress required to initiate movement. Roughness of the ground surface is the most significant land surface characteristic and is affected by percentage cover of vegetation, and size and frequency of obstacles.

Investigation of aeolian transport has been aided by laboratory wind tunnel (e.g. Box 4.2) and also by field experiments so that relationships have been established between major variables in the system. Critical or threshold wind velocities are necessary to set particles of different sizes in motion so that, as wind drag velocity increases, the critical velocity at which particles are set in motion is the fluid threshold (the velocity at which particles begin to move under the influence of wind velocity alone) but, once particles are moving, there is an impact threshold because grains are already saltating (see Figure. 4.9C). The critical velocity (V^*) at which particle movement begins is given by:

$$V^* = k_1 \sqrt{\tau_o/\rho} = k_1 \frac{\sqrt{\sigma - \rho}}{\rho} gD$$

where τ_o is shear stress, σ and D are the density and diameter of the grains respectively, k_1 is a coefficient which is 0.1 for particles above 0.1 mm in diameter for the fluid threshold and 0.084 for the impact threshold (Chepil, 1945) and   is the fluid density. Experimental investigations have shown that the rate of sand transport is proportional to the third power of wind speed, so that most sand is moved during periods of high wind velocity.

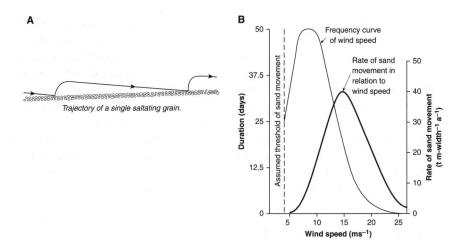

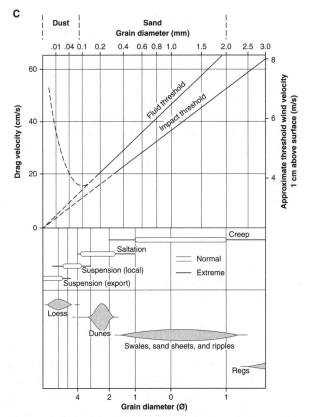

Figure 4.9   Sand grain movement (A) shows the trajectory of a single saltating grain (Summerfield, 1991); (B) shows the relation between wind velocity and sand movement (Warren, 1979); (C) shows the relationship between grain size, fluid and impact threshold velocities together with modes of transport in relation to grain size (Cooke and Doornkamp, 1990 after Mabbutt, 1997)

Reprinted from R.U. Cooke and J.C. Doornkamp, *Geomorphology in Environmental Management, A New Introduction*, by permission of Oxford University Press.

During wind transport, corrasion of a rock surface can be caused when wind-borne sand is transported across it, leading to abrasion wearing rocks down; deflation is the process by which wind removes fine dry, unconsolidated material which can be sand, silt or clay, from the surface of a beach or desert. Erosion, transport and deposition are the fundamental aeolian processes, and a range of bedforms and landforms (for example, Table 9.3) are affected by the interplay of wind velocity and turbulence, inherent roughness of the surface often with sparse if any vegetation, together with cohesion and grain size of the bed. Aeolian processes also include duststorms that are windstorms accompanied by suspended particles of dust (dry soil or sand), wide-ranging in their possible effects, and can be one type of soil erosion (see pp. 106–7).

4.2.6 GLACIAL PROCESSES

Glaciers and ice caps involving glacial processes currently affect some 10% of the land surface, most influentially in high latitude regions (see Chapter 7, p. 165). However there have been times in the past 3 million years when more than 30% of the land surface was covered by glacial ice, although before the Tertiary period there were no extensive areas of ice on the surface of the Earth. These massive spatial changes are one reason for study of glacial geomorphology, concerned to provide physically based explanations of the past, present and future impacts of glaciers and ice sheets on landform and landscape development. Glaciers on the surface of the Earth are extremely sensitive to climate change so that substantial glacier shrinkage is now associated with the effects of global warming (Global Land Ice Measurement from Space, GLIMS http://www.glims.org/, see Table 1.2).

A glacier is a mass of snow and ice which deforms under its own weight, exhibits evidence of downslope movement, and flows so that, like wind and water, it is a form of fluid flow. Glaciology is the interdisciplinary scientific study of the distribution and behaviour of snow and ice on the Earth's surface, with contributions from physics, geology, physical geography, meteorology, hydrology and biology, and glacial geomorphology.

Three major types of glacier are:

- Ice sheets or ice caps (generally less than 50,000 km² in area) in the form of broad domes which submerge the underlying topography with ice radiating outwards as a sheet. Antarctica and Greenland are examples of ice sheets and there are ice caps in Canada, Iceland and Norway. Outlet glaciers may occur at the ice sheet margin, and if fast-flowing may be referred to as ice streams.
- Ice shelves – floating ice sheets or ice caps which have no friction with the bed so that the ice can spread freely; the largest examples are the Ross and the Filchner-Ronne ice shelves in Antarctica.
- Valley or Alpine glaciers – occur in a mountain valley in polar regions and on mountains elsewhere, may be fed by cirque glaciers, have glacier flow strongly influenced by topography, with variations according to the amounts of debris cover.

Snow in the accumulation zone is transformed to firn (snow which has survived one summer melt season) and thence to glacier ice involving regrowth of ice crystals and elimination of air passages, achieving densities of 0.83 to 0.91 kg.m^3, which may occur in a year in valley glacier environments but can take several thousand years in ice sheet environments. In the accumulation zone, accumulation each year exceeds ablation which includes output processes of surface melting and runoff, as well as calving of icebergs, sublimation, evaporation and removal of snow by wind. The equilibrium line is an imaginary line separating the accumulation from the ablation zone (see Figure 4.10B). Where ice is below the pressure melting point it is known as cold or polar ice, whereas warm or temperate ice contains water and is close to the pressure melting point. Mass balance studies quantify the relationship between accumulation processes, principally from snow, and ablation loss, giving the net balance as the annual difference over a glacier (see Figure 4.10A). Such mass balance studies are analogous to the water balance equation (p. 85) for a drainage basin. Influenced by gravity, glacier flow is a response to shear stresses which vary according to surface slope and the thickness of the glacier, and is achieved by creep involving plastic deformation within and between ice crystals, basal sliding and bed deformation. Movement is characteristically up to 300 m per year but velocities can occasionally attain 1–2 km per year and surging glaciers may periodically have velocities up to 100 times greater than normal with a wave of ice moving down-glacier at velocities of 4–7 km per year.

In glacier systems, water and sediment are major components so that their sources, the routes followed and their subsequent fate are the determinants of many landscape features and landforms associated with the glacier environment (Chapter 7, p. 170, Table 7.2) and prominent in many formerly glaciated landscapes (Chapter 8, p. 188). Sediment is entrained at all levels in the ice, on, in and under it, and water can flow on, in, below and around glaciers. Glacial erosion occurs by abrasion, gradual downwearing by debris-laden ice, glacial plucking, removal of blocks and sediment from the glacier bed, and glacial meltwater erosion; characteristics of superglacial, englacial, subglacial and proglacial environment processes are summarized in Table 4.12.

4.2.7 PERIGLACIAL/NIVAL AND CRYONIVAL PROCESSES

Periglacial/Nival and Cryonival processes are also evident in high latitude regions (Chapter 7, p. 171). Periglacial refers to the type of climate and the climatically controlled surface features and processes adjacent to glaciated areas, often associated with permafrost (Chapter 7, p. 174), but also found in high altitude, alpine, areas of temperate regions. Whereas frost or cryogenic weathering, responsible for frost wedging erosion of exposed bedrock, can occur in other parts of the earth's land surface, permafrost is found only in high latitudes, significantly affecting Earth surface processes in these areas. Permafrost is ground in which a temperature lower than 0° has existed continuously for two or more years whether water is present or not; its depth can be greater than 1400 m in

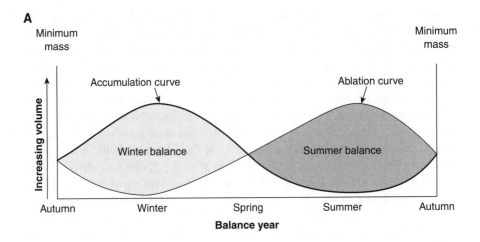

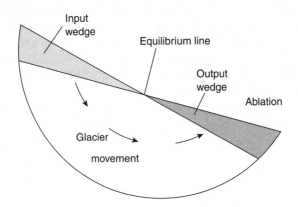

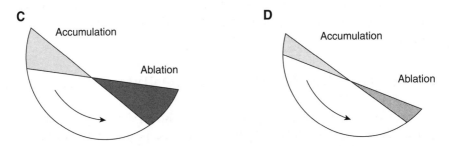

Figure 4.10 Accumulation and ablation defining glacier balance for a year
(A); idealized glacier with input wedge which is net accumulation and output
wedge which is net ablation (B); illustrated for a maritime climate (C); and a
continental climate (D). Greater ice flow occurs in the maritime situation in
response to larger accumulation and ablation (developed from Sugden and
John, 1976 and Bennett and Glasser, 1996)

Table 4.12 SOME CHARACTERISTICS OF SUPERGLACIAL, ENGLACIAL, SUBGLACIAL AND PROGLACIAL ENVIRONMENT PROCESSES (SEE BENNETT AND GLASSER, 1996)

Glacial environment	Ice	Water	Sediment erosion and entrainment, transport, deposition
Superglacial	Ablation	Development of meltwater channels depends on rate of melting, rate of ice deformation, extent of crevasses and pattern of other structures such as foliations, and ice temperature. Channels form tiny rills to canyons several metres deep and develop best on stagnant ice and cold glaciers.	Very little for ice caps and ice sheets but mountain valley glaciers can be almost entirely covered by material derived from frost action on hillsides which may also provide material by mass movement. Dark debris on surface can absorb radiation so that moraines occupy depressions in the ice, although a thick sediment cover can reduce ablation so that debris stands on a distinct ridge. Debris may be as lateral or medial moraines and proportion of debris increases towards the snout.
Englacial	Plastic flow	Small pools of water may develop on the surface and melt down into the ice.	Material buried by snow and incorporated, some material down crevasses, some carried by wind.
Subglacial	Main processes enabling basal sliding are regelation, enhanced creep around bedrock asperities, and cavitation.	Glaciers are buoyantly supported by the pressure of subglacial water. Channels can be very responsive to precipitation up glacier.	Types of till (see Table 8.2). Processes whereby debris in transport can be deposited by lodgement, melt-out (direct release of debris by melting), sublimation (vaporization of ice giving direct release of debris), subglacial deformation (assimilation of sediment into deforming layer below the glacier). Subglacial sediments tend to be poorly sorted, spanning a wide range of grain sizes.
Proglacial	Surges, margin recession.	Outwash, proglacial lakes.	Deltas, ice contact processes.

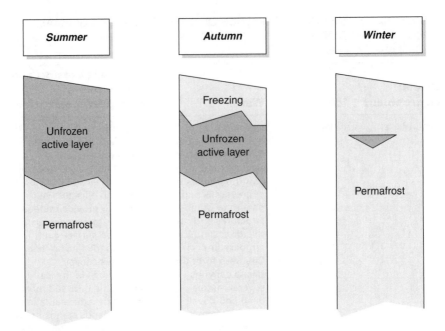

Figure 4.11   Annual cycle of permafrost thawing and freezing

Siberia (Lena and Yara river basins), more than 700 m in the Canadian Arctic islands, and more than 600 m in Alaska (Prudhoe Bay); its lower limit is determined by the increase of temperature with depth beneath the earth's surface (the geothermal gradient). Permafrost has a distinctive and influential annual cycle of permafrost melting and freezing (see Figure 4.11). In winter the ground is frozen from the surface downwards but in the spring, as thawing takes place from the surface downwards, water released cannot infiltrate because the permafrost table (the level below which permafrost does not thaw) provides an impermeable layer. An active layer develops during the spring and summer months, with a thickness typically between 0.6 and 4 m according to air temperature, slope, drainage, soil and vegetation characteristics. Subsequently in autumn and winter the active layer freezes from the surface downwards. As water expands by 9% volume on freezing, stresses are produced together with high pore water pressures which can lead to extrusion of water to the surface which will immediately freeze as an icing. If the winter is not severe then freezing may not occur down to the permafrost table so that an unfrozen layer (talik) persists until the following spring. Repetition of this pattern each year can produce complex vertical patterns of frozen and unfrozen ground. Above the permafrost table the result of spring thaw and summer melting of snow and of ground ice provides large amounts of water in the soil so that mass movement called congelifluction occurs on most slopes, even those with angles as low as 1° or 2°. Frost action and cryonival processes produce distinctive landforms (Chapter 7, p. 175).

4.2.8 SUBSIDENCE

Subsidence, the downward movement of the land surface, can occur in permafrost areas as an integral part of thermokarst development (p. 176), in karst areas (p. 54) as a result of solution, over long periods of time as a result of tectonic downwarping, but it may also occur as a result of changes to sediments and deposits, withdrawal of fluids or extraction of mining materials, both involving human activity. Solution is not confined to karst areas because solution of salt deposits, halite or rock salt, can lead to subsidence of the surface as illustrated in Cheshire, UK, where subsidence hollows can be 1000 m across and are locally known as meres. Tectonic subsidence can occur where the marginal zones of plates are subducted close to convergent boundaries so that the Nagoya region of eastern Japan is subsiding at 10 mm/year; or where stretching of a plate causes thinning and surface subsidence as in the case of the North Sea (Waltham, 2005). Progressive changes to deposits include those on rocks and deposits that are porous and deformable and hence compressible, such as clay. Venice (see Figure 4.12), Shanghai, Mexico City and Bangkok are examples of cities that have subsided as a result of the reduction of pore water pressure in clays following water abstraction by pumping (Waltham, 2005). Subsidence can also occur after extraction of other fluids such as petroleum or natural gas, or after compaction following loading: the rock sequence including clays was insufficient as foundations and produced the leaning tower of Pisa. Peat deposits also display subsidence as a result of either loading or drainage. Conversely, in some sediments, especially wind-deposited loess and some alluvial silts, there can be hydrocompaction which results in collapsing soils and subsidence of up to 5 m when water is added, illustrated by the effects of irrigation in the San Joaquin valley of California. Human activity, as in the case of mining, is a further cause of subsidence (see Chapter 3, p. 66).

4.2.9 SOIL PROCESSES AND SOIL EROSION

Soil processes (see p. 60) and soil erosion are significant land surface processes with soil usually removed at approximately the same rate as soil is formed but, when significantly affected by human activity, accelerated erosion gives loss of soil at much faster rates. Soil erosion has been suggested as one of the most important (yet probably the least well-known) of today's environmental problems (Favis-Mortlock, 2007) second only to population growth as the biggest environmental problem the world faces. More than 10% of the world's soil lost a large amount of its natural fertility during the latter half of the twentieth century so that 10.5% of the planet's most productive soils – an area the size of China and India combined – have been seriously damaged by human activities since 1945; soil globally is now being swept and washed away 10 to 40 times faster than it is being replenished (Pimentel, 2006).

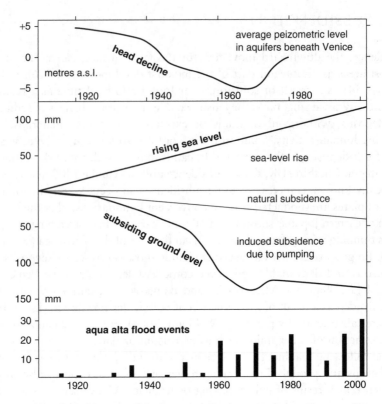

Figure 4.12   Subsidence of Venice over the last century – the graph indicates, for 5-year periods, the number of flood events when high tides reach more than 600 mm above the level that initiates flooding in the lowest part of Piazza San Marco (after Waltham, 2005)

Reprinted with permission from *Geomorphology for Engineers*, P. G. Fookes, E. M. Lee and G. Miligan (eds), Whittles Publishing.

Soil erosion, including the physical processes of removal of particles of soil horizons by water and wind, occurs if the forces or erosivity acting on the soil particles, provided by rain, water flow, or wind, exceed the resistance provided by the coherence of particles and by vegetation cover. The soil erosion process includes the detachment of soil, its transport, and its subsequent deposition. The principal way in which accelerated erosion is instigated is following removal of vegetation that reduces the resistance or erodibility of the soil; during building construction sediment yields can be up to 2000 times greater than those in undisturbed areas (US Environmental Protection Agency, 1973).

Soil erosion by water is the result of rain detaching and transporting vulnerable soil, either directly by means of rainsplash which detaches particles or indirectly by rill and gully erosion. Overland flow can be either laminar or turbulent, the latter giving up to 90% of erosion during a single rainfall event. Attempts to control soil erosion required quantification of the processes involved so that equations to estimate soil erosion and to predict soil loss were developed in the 1930s and 1940s following the creation of the dust bowl and

badlands in the USA. These culminated in the Universal Soil Loss Equation
(USLE) to predict mass of soil removed annually from a unit area (Wischmeier
and Smith, 1965). More erosive than sheet erosion or overland flow is concentration of water into rills, which are channels as small as a few centimetres in
width, but which can grow or lead downstream to gullies. Defined as features
which cannot be crossed by traffic or ploughed out, gullies are usually several
metres wide and deep, and steep-sided, have headcuts or knickpoints, can produce considerable quantities of sediment during erosion, and are of two main
types: valley floor gullies and valley side gullies. Valley floor gullies occur if
runoff is concentrated in alluvial valley floors whereas valley side gullies can
form if runoff is concentrated on hillslopes, if surface pipes collapse, or if
landslides create elongated scars (see Figure 10.6). Some weak, impermeable
rock types in high relief areas are especially prone to gullying; in certain soils,
especially those subject to cracking, soil pipes that develop can initiate gullying if the roof collapses. Soil erosion can also occur due to wind if the wind
velocity exceeds the critical value; although globally less significant than soil
erosion by water, it is dominant in arid or semi-arid areas.

4.2.10 ECOSYSTEM PROCESSES

Within the field of biogeomorphology the influence of ecosystem processes
involves plants, animals and microorganisms affecting earth surface processes
and landforms (see Table 3.9). Biological influences upon geomorphological
processes (Viles, 1988) shape the land surface in three major ways: ecosystem
processes, organic processes, changes of ecosystems. Land surface systems are
dynamically coupled with climate through the physical processes of energy,
water and biogeochemical and sediment fluxes (Chapter 3, p. 46), fluxes
which involve the principal components of carbohydrate (H,C,O), required in
substantial quantities for organic life to thrive (Chapter 3, p. 64), involving
photosynthesis, which includes fixation of solar energy in plant tissues.

Vegetation plays a major role in regulating water transport to the atmosphere
and there are a series of organic processes in which ecosystem dynamics act as
buffers between land surface processes and landforms (see Table 3.9) and may
lead to the creation of organic landforms. Whereas ecological stability is the
ability of an ecosystem to resist changes in the presence of perturbations,
resilience is the ability of an ecosystem to return to 'normal' after perturbations.
Other stability concepts include resistance which is the ability of the ecosystem
to resist changes when external factors are altered.

4.3 ENDOGENETIC PROCESSES

Originating beneath the land surface of the Earth are some of the most dramatic
earth surface processes and include earthquakes and tectonic processes (see
Table 4.2). Rock uplift, the upward movement of the rock column with respect

to some datum, arising from the vertical transfer of crust (tectonics) equates to surface uplift only if there is no denudation during the rock uplift. If rock uplift is matched by denudation there is no evident surface uplift (steady state) and if rock uplift is less than denudation the surface elevation is lowered (Bishop, 2007: 334).

4.3.1. EARTHQUAKES

Earthquakes are the result of a sudden release of energy in the Earth's crust which creates seismic waves. The resultant natural shaking or vibrating on the land surface can be associated with volcanic activity or with the breaking and movement of rocks along faults, which are fractures in rocks where differential displacement, which may be vertical, horizontal or some combination of both, occurs. Tectonics applies to the building of earth structures at the global scale where three types of plate interactions (see Figure 3.2, Colour plate 1) are *divergent* at spreading ridges, *convergent* where one plate subducts beneath the leading edge of another, and *transform* where one plate slides past another horizontally. A few large earthquakes occur at faults in the interior of tectonic plates (intraplate earthquakes) but generally the mapping of the plate boundaries has been done largely through the location of earthquakes (interpolate earthquakes). Four zones of concentration of earthquakes (Summerfield, 1991: 44) include a diffuse belt from the Mediterranean region through the Himalayas into Burma; a zone confined to relatively small areas within continents characterized by shallow earthquakes not associated with marked volcanic activity; together with two marine zones along mid-oceanic ridges, and closely related to oceanic trenches and their associated island arcs or mountain belts. When the stress exceeds the rock strength the resulting strain releases energy and is expressed as an earthquake. The focus or *hypocentre* is the point within the earth where the rupture begins; the point on the surface directly above the focus is the epicentre. When earth movement occurs the two types of seismic waves that move outwards are Primary (P, compressional) waves that travel faster than Secondary (S) waves that have particle displacement 'up and down' perpendicular to the direction of motion. The National Earthquake Information Center (NEIC) estimates that there are 500,000 detectable earthquakes in the world each year of which 100,000 can be felt, and 100 cause damage. Seismographs record seismic waves, and scales have been developed (see Table 4.13) to classify earthquakes according to their severity. Whereas the *magnitude* of an earthquake is a measurement of the earthquake size, the *intensity* is a measure of the shaking created by the earthquake, varying according to location.

Some of the largest recorded earthquakes (see Table 4.14), with enormous consequences in terms of loss of life, have effects on the land surface. Such neotectonics or morphotectonics, the study of landforms affected by recent earth movements, are not confined to the immediate areas but through the

Table 4.13 EARTHQUAKE DETECTION AND RECOGNITION

The mission of the National Earthquake Information Center (NEIC) is to determine rapidly the location and size of all destructive earthquakes worldwide and disseminate this information (http://earthquake.usgs.gov/regional/neic/); available from the website is an extensive list of world earthquakes since one in 856 AD 1222 at Damghan, Iran which caused 200,000 fatalities. The NEIC/WDC for Seismology compiles and maintains an extensive, global seismic database on earthquake parameters and their effects.

The *magnitude* of an earthquake is a measured value of earthquake size, recorded by the *Richter magnitude scale* (developed in 1935 by Charles F. Richter of the California Institute of Technology as a mathematical device to compare the size of earthquakes), which is a measure of the amount of energy released. The magnitude of an earthquake is determined from the logarithm of the amplitude of waves recorded by seismographs, so that each whole number increase in magnitude represents a tenfold increase in measured amplitude. This means that each whole number step in the magnitude scale corresponds to the release of about 31 times more energy than the amount associated with the preceding whole number value, and an earthquake that measures 5.0 on the Richter scale has a shaking amplitude 10 times larger than one that measures 4.0.

Richter magnitude	Effects and frequency
<4.0	About 8000 per day < 2.0 are not felt; about 1000 per day between 2.0 and 2.9 are recorded but not felt; whereas up to 50,000 per year between 3.0 and 4.0 are felt but rarely cause damage.
4.0–5.9	c. 6200 per year are 4.0 to 4.9 giving shaking and rattling but significant damage unlikely; c. 800 per year are 5.0 to 5.9 can cause major damage to poorly constructed buildings.
6.0–7.9	c. 120 per year are 6.0 to 6.9 which can be destructive, and c. 18 per year are 7.0 to 7.9 which can cause serious damage over larger areas.
>8.0	c. 1 per year between 8.0 and 8.9 can cause serious damage in areas several hundred km across; whereas c. 1 in 20 years for 9.0 to 9.9 can be devastating across areas several thousand km; >10.0 not known but for comparative purposes it is estimated that the impact that supposedly wiped out the Dinosaurs had an equivalent magnitude of around 16–17.

Intensity, a measure of the shaking created by the earthquake, varies according to location, and the intensity number describes the severity of an earthquake in terms of its effects on the earth's surface and on humans and their structures. Several scales exist, but the Modified Mercalli scale is frequently used.

Modified Mercalli scale

Intensity	Effects
I	Instrumental. Usually not felt.
II	Feeble. Delicately suspended objects may swing.
III	Slight. Vibration similar to the passing of a truck.
IV	Moderate. Sensation like heavy truck striking building.
V	Rather strong.
VI	Strong. Felt by all. Windows broken. Damage slight.
VII	Very strong. Difficult to stand; furniture broken; damage negligible in buildings of good design and construction; slight to moderate in well-built ordinary structures; considerable damage in poorly built or badly designed structures.

(Continued)

Table 4.13 *(Continued)*

Intensity	Effects
VIII	Destructive. Damage slight in specially designed structures; considerable in ordinary substantial buildings with partial collapse. Damage great in poorly built structures. Fall of chimneys, factory stacks, columns, monuments, walls.
IX	Ruinous. General panic; damage considerable in specially designed structures, well designed frame structures thrown out of plumb. Damage great in substantial buildings, with partial collapse. Buildings shifted off foundations. Ground cracked conspicuously.
X	Disastrous. Some well built wooden structures destroyed; most masonry and frame structures destroyed with foundations destroyed. Rails bent. Landslides considerable on river banks and steep slopes. Shifted sand and mud. Water splashed over banks.
XI	Very disastrous. Few, if any masonry structures remain standing. Bridges destroyed. Rails bent greatly. Earth slumps and landslips in soft ground.
XII	Catastrophic. Total damage. Almost everything is destroyed. Lines of sight and level distorted. Objects thrown into the air. The ground moves in waves or ripples. Large amounts of rock may move position.

generation of tsunamis can devastate wider areas, and the effects can be detected as seiches in very distant countries. Although earthquakes are largely confined to the environs of the plate margins, their effects on the land surface can be more extensive (see Table 4.14). The distinction between active and passive tectonic controls (Summerfield, 1991: 405) distinguishes those responses to ongoing tectonic activity (active) from those structural controls

Table 4.14 EXAMPLES OF LARGE EARTHQUAKES

Date and location	Magnitude	Comments
12 May 2008, Eastern Sichuan, China	7.9	At least 69,185 people killed, 374,171 injured and 18,467 missing and presumed dead in the Chengdu-Lixian-Guangyuan area. An estimated 5.36 million buildings collapsed and more than 21 million buildings were damaged. The total economic loss was estimated at US $86 billion. Landslides and rockfalls damaged or destroyed several mountain roads and railways and buried buildings. At least 700 people were buried by a landslide at Qingchuan. Landslides also dammed several rivers, creating 34 barrier lakes which threatened about 700,000 people downstream. At least 2473 dams sustained some damage and more than 53,000 km of roads and 47,000 km of tap water pipelines were damaged. About 1.5 km of surface faulting was observed near Qingchuan, surface cracks and fractures occurred on three mountains in the area, and subsidence and street cracks were observed in the city itself.

Table 4.14 *(Continued)*

Date and location	Magnitude	Comments
		Maximum intensity XI was assigned in the Wenchuan area. Seiches were observed at Kotalipara, Bangladesh.
8 October 2005, Pakistan	7.6	At least 86,000 people killed, more than 69,000 injured and extensive damage in northern Pakistan. Maximum intensity VIII. Landslides and rockfalls damaged or destroyed several mountain roads and highways cutting off access to the region for several days. Liquefaction and sandblows occurred in the western part of the Vale of Kashmir and near Jammu. Landslides and rockfalls also occurred in parts of Himachal Pradesh, India. Seiches were observed in Haryana, Uttar Pradesh and West Bengal, India and in many places in Bangladesh.
26 December 2004, Sumatra	9.1	The largest since 1964, 227,898 people were killed or were missing and presumed dead and about 1.7 million people were displaced by the earthquake and subsequent tsunami in 14 countries in South Asia and East Africa. The tsunami caused more casualties than any other in recorded history and was recorded nearly world-wide on tide gauges. Subsidence and landslides were observed in Sumatra. A mud volcano near Baratang, Andaman Islands became active on 28 December and gas emissions were reported in Arakan, Myanmar.
16 December 1920, Haiyuan, Ningxia, China	7.8	200,000 deaths. Total destruction (XII – the maximum intensity on the Mercalli scale) in the Lijunbu-Haiyuan-Ganyanchi area. A landslide buried the village of Sujiahe in Xiji County. Damage (VI–X) occurred in 7 provinces and regions. About 200 km of surface faulting was seen, there were large numbers of landslides and ground cracks throughout the epicentral area. Some rivers were dammed, others changed course. Seiches from this earthquake were observed in 2 lakes and 3 fjords in western Norway.

Source: NEIC.

which operate through the influence exerted by previous tectonic activity (passive). This distinction is classically demonstrated by fault scarps which are steep slopes in the landscape produced by faulting; and fault-line scarps which are morphologically similar but are the result of lithological differences

Table 4.15   EFFECTS OF EARTHQUAKES AND TECTONIC MOVEMENTS ON THE LAND SURFACE OF THE EARTH (SEE KELLER AND PINTER, 1996)

Process	Effect on land surface
Direct effect of earthquakes	
Change of land elevation	Alaskan earthquake of 1964 (Mercalli X–XI) caused vertical deformation over an area of more than 250,000 km^2 including uplift as much as 10 m and subsidence as much as 2.4 m. Orogenic uplift.
Violent ground motion	Surface disturbance.
Active faulting	Strike slip, normal faulting, reverse faulting, Horst grabens.
Liquefaction	Transformation of water-saturated granular material from solid to liquid.
Tectonic creep	Displacement along a fault zone not accompanied by perceptible earthquakes, may be discontinuous and variable in rate.
Effects on other processes	
Mass movement	Trigger incidence of mass movement including landslides and avalanches.
Rivers	River capture, formation of lakes.
Waves	Instigate tsunamis when ocean water displaced vertically.
Effects on existing landforms and features	
River patterns	Offset streams, drainage basin asymmetry.
Terraces	Displaced, faulted, warped, tilted.
Coral reefs	Emerged from sea.
Shorelines	Deformed, tilted, offset.
Alluvial fans	Segmented, deformed.

arising from the existence of a fault or other structural feature which is the consequence of past faulting. Some ways in which earthquakes affect the land surface are shown in Table 4.15.

In view of the substantial effects of neotectonic processes, prediction of earthquakes is of great interest. After observations of the 1906 San Francisco earthquake, an earthquake cycle was suggested (Keller and Pinter, 1996) which models the way in which elastic strain gradually builds up until rupture occurs releasing the strain by displacement as faults. Subsequently strain will build up again and in any particular location the pertinent question is how long will elapse before the next earthquake.

4.3.2 VOLCANIC ACTIVITY

Volcanic activity is the other endogenic process responsible for dramatic and often spectacular events. Whereas igneous activity connotes movement of molten rock or magma, extrusive igneous activity or volcanism occurs when magma is extruded onto the earth's surface as flowing lava or by material erupted by explosive volcanic activity. Volcano refers to the vents through which hot

Figure 4.13   Eruption of Mount St Helens, 1980

After 18 May 1980 five further explosive eruptions occurred in 1980 including
this event, which sent pumice and ash 10–18 km into the air and was visible in
Seattle, Washington 160 km to the north. Before the eruption, Mount St Helens
was about 1600 m above its base but on 18 May 1980 its top slid away in an
avalanche of rock and other debris so that on 1 July 1980 the mountain's height
had been reduced from 2950 m to 2549 m (Campbell, 2001). Aggradation in the
Toutle river valleys was up to 170 m and in Coldwater Creek and Castle Creek
blocked the rivers, forming Coldwater Lake and Castle Lake. The time since the
eruption enabled a fascinating study of the reaction of processes including
sediment supply (Major, 2004), showing that post-eruption suspended sediment
transport has been greater and more persistent from zones of channel
disturbance than from zones of hillslope disturbance. Although small-magnitude
and large-magnitude discharges locally and episodically transported
considerable amounts of suspended sediment, there was no notable change in
the overall nature of the effective discharges; moderate-magnitude flows have
been the predominant discharges responsible for transporting the majority of
suspended sediment during 20 years of post-eruption landscape adjustment.

Source: USGS photograph taken on 22 July 1980, by Mike Doukas (MSH80_st_
helens_eruption_plume_07-22-80.jpg).

igneous lava is expelled and to the landform constructed from the erupted material. Volcanoes have been described (Francis, 1993: 1) as mountains with a Jeykell and Hyde personality: passive for most of its life until the darker side of the volcano's character is manifested in a violent eruption (see Figure 4.13). At least 20 volcanoes will probably be erupting as you read this, about 550 have had historically documented eruptions, and c.1500 have erupted during the Holocene of the past 10,000 years (USGS, 2009). Most volcanoes are located at convergent and divergent plate boundaries but those located in the interior of plates are located at hot spots (unusually thick crust often associated with significant crustal uplift). Volcanic eruptions release enormous amounts of energy, typically from an individual volcano between 10^{12} and 10^{15} J, compared with the 10^{16} J released by a one megatonne hydrogen bomb (Selby, 1985); although major eruptions can be larger, with the 1980 eruption of Mount St Helens (see Figure 4.13) being equivalent to about 30 one megatonne bombs, and the eruption of Laki in Iceland in 1783 produced energy of 10^{20} J, equivalent to 10,000 one megatonne hydrogen bombs (Summerfield, 1991).

Eruptions differ according to amount, composition and physical nature of the material expelled, and the rate and style of the eruptive process. Volcanic materials include lava and tephra (pyroclastic material) but, in addition, pyroclastic flows of debris and hot gas may flow down the sides of a volcano as well as deposits from mudflows and lahars rich in volcanic debris. Lava flows usually move at a few km per hour but can achieve 100 km per hour, their speed and distance travelled being determined by viscosity and slope. Acid lavas are the most viscous, flow the shortest distances, and tend to have a jagged surface called aa, whereas the more fluid basic lavas, predominantly basaltic, are erupted at higher temperatures and tend to have a ropy surface called Pahoehoe, a Hawaian term reflecting the basic volcanoes Kilauea and Haulalai of Mauna Loa. Rate of cooling depends on depth – a 1m thick flow of basalt could take 12 days to cool from 1100°C to 750°C, the same cooling will take c. 3 years for a 10 m flow and 30 years for a 100 m flow (Selby, 1985: 136). Tephra includes all forms of hot volcanic debris which falls through the air including, according to size, airfall ash and blocks and volcanic bombs (>64 mm), lapilli (2–64 mm) and ash (<2 mm). During an eruption the gas produced may create glowing clouds of ash, pumice and pyroclasts which can move down slope faster than 100 km. per hour; glowing avalanches are known as nuees ardentes. Extremely large eruptions release large quantities of CO_2 and aerosols – indeed many of the 16,000 who died at Pompeii from the eruption of Vesuvius in 79AD were killed when engulfed in clouds of hot searing dust and gas before the town was completely buried by the fall of ash over 2 days.

Volcanic eruptions are of two basic kinds: fissure eruptions from a crack or fissure and central eruptions from a central pipe or vent, although an eruption may begin as a fissure and develop as one or more central eruptions. Eruptions differ according to the composition of the magma and the gas contained, so that mafic magmas like basalt erupt relatively passively, because they contain relatively little gas, and their low viscosity permits the relatively free escape of gas. Fissure eruptions occur when magma-filled dykes intersect the surface – they occur naturally where the crust is undergoing extension, as in Iceland which is widening 1 or 2 cm every year. Between June 1783 and February 1784, 14 km³ of lava was extruded from a fissure 25 km long in Iceland (Francis, 1993). A spectrum of eruptions ranges from effusive, dominated by passive emission of lava material, to explosive, dominated by pyroclastic material. Hydrovolcanic eruptions, involving groundwater or sea-water, tend to be the most violent. Eruption size can be measured according to magnitude (the total volume of erupted material), dispersive power (the area covered by tephra fallout), intensity (the rate at which magma and ash disgorge), and volcanoes and eruptions have been classified in a scale of increasing severity (see Table 4.16). Associated features, particularly associated with the late stages of volcanic activity, include vents called solfataras or fumaroles through which gas and steam are emitted; sinter mounds created by precipitation of silica and calcium carbonate around a vent; hot springs; geysers which are jets of hot

Table 4.16 VOLCANIC ERUPTIONS (DEVELOPED FROM MACDONALD, 1972 AND FRANCIS, 1993)

VEI, the volcanic explosivity index, is a relative scale based on volume of material ejected into air, scale of explosions, and height of eruptive cloud.

Eruption type	Magma	Effusive activity	Explosive activity (VEI)	Structures and ejecta
Icelandic	Basic, fluid	Flows from fissures	Very weak (0)	Very broad lava cones and plains
Hawaian	Basic, fluid	Extensive lava flows from central vents usually thin	Weak (0)	Very broad, flat lava domes and shields, very little ash
Strombolian	Part acid/ basic, moderately fluid	Thicker flows less extensive or absent	Weak to violent (1–2)	Cinder cones, lava flows
Surtseyan	Part acid/ basic, moderately fluid	Tephra and glass	Larger explosion over subaqueous vent (2)	Flat cone built up to water level
Vulcanian	Acid, viscous	Flows often absent, thick when present	Moderate (2–3)	Ash, block cones, explosion craters, pumice
Vesuvian or sub-Plinian	Acid, viscous	Flows often absent, thick when present	Moderate to violent (4)	Ash cones and flows, explosion craters and calderas
Plinian	Acid, viscous	Flows may be absent, ashflows small to extensive	Very violent ejection of large volumes of ash (4–5)	Generally no cone, widespread pumice, lapilli and ash beds

(Continued)

Table 4.16   *(Continued)*

Eruption type	Magma	Effusive activity	Explosive activity (VEI)	Structures and ejecta
Ultraplinian	Acid, viscous	Flows may be absent, column height may be >45 km	Violent ejection of large volumes of ash	None in historic times but Taupo NZ, in AD186
Pelean	Acid, viscous	Domes and/ or short very thick flows;	Ejection of solid or very viscous hot fragments with glowing avalanches (nuees ardentes)	Domes, ash and pumice cones
Krakatoan	Acid, viscous	Ashfalls	Highly explosive Cataclysmic (6)	Large explosion calderas
Supervolcano	Acid, viscous	Extensive	Cataclysmic (7–8)	Collapse of large calderas
Associated processes and features		Fumaroles, solfataras, sinter mounds. Mud volcanoes		

water periodically spurted into the air; and mud volcanoes which are dirty hot springs where mud arises from the heating of shallow groundwater. Many eruptions are complex, combining many of the features shown in Table 4.16; calderas are huge volcanic depressions, which can be at least 5 km across. These are often formed by collapse after catastrophic eruptions or by subsidence after a magma chamber is drained leading to collapse of an unsupported roof. It is now realized that the Yellowstone plateau has the largest caldera on earth, 70 × 45 km, which could be the location for a supervolcano sometime in the future succeeding previous events 2 million, 1.3 million and 0.65 million years ago (McGuire et al., 2002).

4.4 CONCLUSION

Processes affecting the land surface of the Earth have often been studied separately, reflecting the reductionist tendency, but although this allows consideration of the landforms produced it neither shows how processes combine nor indicates their relative effectiveness in shaping the land surface.

Combinations of processes, emphasized in world zones (Chapters 6–10) where spatially identifiable areas are characterized by distinct suites of geomorphic processes, have been described as **process domains** (Montgomery, 1999;

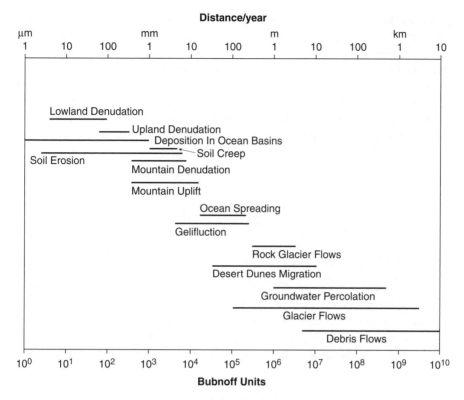

Figure 4.14 Rates of operation of some geomorphic processes from Smil, 1991; also see Tables 3.6 and 5.3, and Selby, 1985

Thornes, 1979). To compare the relative rates of different processes we can separate erosion, which is those processes whereby debris or rock material is loosened or dissolved and removed from any part of the Earth's surface by the processes considered in this chapter, from denudation, which is usually thought of more broadly to include weathering as well as all processes which wear down the surface of the Earth (see p. 73). Denudation rates therefore provide a measure of the lowering of the land surface by erosion expressed in Bubnoff units (see Table 3.6). Rates of operation of geomorphic processes (see Figure 4.14) require comparison with rates of orogenic and epeirogenic uplift because, in some areas, rates of uplift can be eight times the rates of denudation (Schumm, 1963) and one long-term consequence of denudation can be that uplift occurs because of reaction by the crust, as included in the next chapter.

Human impacts also have substantial effects upon denudation rates – indeed human activity (Chapter 3, pp. 66–7) could have been included as one of the processes considered above, because the effects of human society on the development of the Earth's relief already exceed the effects of natural geomorphological processes (Goudie, 1986; Goudie and Viles, 1997; Gregory, 2000: ch. 7) and it has been shown that, in the United States, rates of movement of material by humans are greatest in the east whereas the effect of rivers is greatest in the west (Hooke, 1999). Although rates of basin denudation derived from present-day mass flux estimates are not, overall, significantly different from estimates of long-term rates based on sediment volume and thermochronologic data (Summerfield and Hulton, 1994), the degree of present modified and often accelerated rates requires consideration in relation to changes over time addressed in the next chapter.

BOX 4.1

LUNA LEOPOLD (1915–2006)

Luna Leopold was lead author with M.G. Wolman and J.P. Miller of *Fluvial Processes in Geomorphology* published in 1964 – an extremely influential book instrumental in demonstrating how geomorphology could, and should, become more process-based. In 2006 the Benjamin Franklin Medal in Earth and Environmental Science was awarded jointly to Luna Leopold and M. Gordon Wolman for

> advancing our understanding of how natural and human activities influence landscapes, especially for the first comprehensive explanation of why rivers have different forms and how floodplains develop. Their contributions form the basis of modern water resource management and environmental assessment.

Leopold has been described as the 'intellectual father' of geomorphology, as an individual who had a profound influence on nationwide efforts to restore and protect rivers, as one who shaped the water resources programmes of the USGS, achieving a unification of the three subdisciplines in water: surface water, groundwater, and water quality. Few have written papers spanning 68 years, and fewer still have had such a great influence on a scientific field or on society.

Born in 1915, the son of Aldo Leopold, he received his BS in Civil Engineering from the University of Wisconsin in 1936, an MS in Physics Meteorology from UCLA in 1944 and a PhD in Geology from Harvard in 1950. This training provided the excellent foundation reflected in his career 1950–1972 with the US Geological Survey (USGS), rising eventually to the level of chief hydrologist of the Water Resources Division, and subsequently at UC Berkeley 1972 to 1986, serving as chair of the geology and geophysics department in his final year. Of his 200 publications, that with Thomas Maddock, *The Hydraulic Geometry of Stream*

Channels and Some Physiographic Implications, published in 1953 as a USGS professional paper, effectively initiated a new quantitative study for rivers – an approach reinforced by the 1964 book, galvanizing other branches of geomorphology. In addition to these contributions he is credited as an instigator of the Environmental Impact Statement through its design and early application, he wrote for the more general reader in *Water: A Primer* (1974) and *A View of the River* (1994), and really influenced how society approaches environmental problems and conducts environmental science in the service of people and the **natural environment**. He was a member of the National Academy of Sciences and recognition included honorary degrees from six universities, the Penrose Medal of the Geological Society of America, as well as the Warren Prize of the National Academy of Sciences, the Busk Medal of the Royal Geographical Society, and the National Medal of Science in 1991.

Luna Leopold had great vision, vitally necessary in the development of any science. I particularly valued his contribution in 1954 (Leopold and Miller, 1954) which included one of the earliest uses and definitions of the word palaeohydrology. His vision was also evident in his monitoring of urbanizing channels – the subject for his last paper in 2005. He was a great field worker, floated on a raft through the Grand Canyon to measure the depth of the Colorado River, and his many interests included hunting and fishing, horse riding, composing piano and guitar music, and building furniture. At the end of his address as retiring President of the Geological Society of America, in November 1972, he concluded with a song in calypso time: 'Better get the garbage before it gets you'. That is included in the paper (Leopold, 1973) and his stimulating publications are listed at eps.berkeley.edu/people/lunaleopold.

BOX 4.2

BRIGADIER RALPH A. BAGNOLD (1896–1990)

Ralph A. Bagnold was educated at Malvern College and the Royal Military Academy at Woolwich, served in the army from 1915 to 1935 and from 1939 to 1944, rising to the rank of Brigadier. A special military educational leave programme enabled him to study engineering at Cambridge University, gaining an honours degree in 1921. As a young boy he was inventive and curious so that when he was a British Engineering Officer stationed in Egypt in the 1930s he used his leave time to explore the desert to the west of the Nile using an old Ford car and rolls of chicken wire to help get through loose sand areas (Chorley et al., 1984). His group

(Continued)

(Continued)

began the practice of reducing tyre pressure when driving over loose sand. He developed a sun compass which, unlike a magnetic compass, is unaffected by the large iron ore deposits found in desert areas or by metal vehicles.

After his retirement from the army in 1935, he began scientific research at Imperial College, London using a home-made wind tunnel for experiments. These experiments and his careful field observations led to his classic book *The Physics of Blown Sand and Desert Dunes* published in 1941. From 1949 his research at Imperial College, London focused on the fluvial transport of solids, leading to collaborative work with L.B. Leopold (see Box 4.1). He was an internationally recognized authority on the transport of blown sand. According to J.R. Underwood, Jr he once remarked that he was not a very keen soldier and that he would rather be a Fellow of the Royal Society than a Brigadier General. In fact he became both, being elected a Fellow of the Royal Society (FRS) in 1944. He also received the Founder's Gold Medal of the Royal Geographical Society (1934), the Wollaston Medal of the Geological Society of London (1971), the G.K. Warren Prize of the US National Academy of Science (1969), the Penrose Medal of the Geological Society of America (1970), the Sorby Medal of the International Association of Sedimentology (1978), and the Linton award of the British Geomorphological Research Group (1981). His research is classic, showing how engineering training can be a foundation for research in one field (aeolian transport) leading to others (fluvial transport). He expressed great pleasure that his fundamental work on movement of particles on Earth by wind could be applied to other planetary bodies with atmospheres, such as Mars. His autobiography *Sand, Wind and War: Memoirs of a Desert Explorer* was published in 1990 and in the Foreword by Luna Leopold, Paul Komar and Vance Haynes they commented that 'It is unusual to find a professional soldier who is known primarily for his contributions to scientific thought' and that 'The position of importance he has earned in science stems from his insistence on critical experimentation combined with the use of pure physics. His goal was to eliminate the empiricism that so long dominated the problem of transportation of debris in water'.

FURTHER READING

An interesting read is:

Yatsu, E. (2002) *Fantasia in Geomorphology.* Sozosha, Tokyo.

Many books deal with specific processes including:

for weathering:

Bland, W. and Rolls, D. (1998) *Weathering.* Oxford University Press Inc, New York.

for mass movement:

Selby, M.J. (1993) *Hillslope Processes and Materials.* OUP, Oxford, 2nd edn.

for fluvial processes:

Knighton, A.D. (1998) *Fluvial Forms and Processes.* Arnold, London.

for coastal processes:

Bird, E.C.F. (2000) *Coastal Geomorphology: An Introduction.* Wiley, Chichester.
Viles, H. and Spencer, T. (1993) *Coastal Problems: Geomorphology Ecology and Society at the Coast.* Arnold, London.

for aeolian processes:

Cooke, R., Warren, A. and Goudie, A. (1993) *Desert Geomorphology.* UCL Press, London.

for glacial processes:

Evans, D.J. (2004) *Glacial Landsystems.* Arnold, London.
International Glaciological Society website. www.igsoc.org.

for periglacial processes:

French, H.M. (2007) *The Periglacial Environment.* Wiley, Chichester.

for soil processes and soil erosion:

Morgan, R. (2005) *Soil Erosion and Conservation.* Wiley Blackwell, Chichester.

for earthquakes:

Keller, E.A. and Pinter, N. (1996) *Active Tectonics. Earthquakes, Uplift and Landscape.* Prentice Hall, Upper Saddle River, NJ.

and for volcanoes:

Ollier, C.D. (1988) *Volcanoes.* Blackwell, Oxford.

TOPICS

1 Decide what weathering processes occur in a particular area/location, deduce what processes might be involved (including chemical reactions), and consider how you could assess weathering rates in that area.
2 Investigate the incidence and impact of a recently reported specific mass movement event.
3 Why have wind tunnel experiments increased understanding of aeolian processes and why are they particularly useful for studying aeolian environments?
4 Derive a mass balance for a specific glacier and investigate the extent to which variations occur from year to year and whether there are any consistent trends, for example reflecting global warming.
5 Use internet resources to compile details of a recent eruption such as the birth of a volcano (Paracutin, Mexico, 1943) or a complex eruption (e.g. Mount St Helens, from March 1980). Does material obtained give a dispassionate view of the surface of the Earth or does it emphasize the dramatic?

PART III
LANDFORM EVOLUTION

5

THE CHANGING SURFACE – EVOLUTION OF LANDSCAPES

The past is not dead. It is not even past.

William Faulkner (cited in Carey, 2001)

Environmental processes acting in a particular landscape are associated with landforms over periods of days, months or years – often related to our personal time frame or to the periods of time for which measurements are available. However, in visualizing landforms in a particular space and time position how can we relate space and time? A way of looking at geologic time is as a spiral (see Figure 5.1, Colour plate 3), with the face of the spiral representing now, and geologic time shown by the descending spiral. Scientists have long appreciated the age of the Earth; Leonardo da Vinci (1452–1519) declared in his notebooks (see Table 2.1) that the river deposits of the River Po had required 200,000 years to accumulate, presuming that this was less than the whole of geologic time (Clements, 1981), but in 1654 Archbishop Usher contended that the Earth was created at 9:00 am on 23 October 4004 BC (Tinkler, 1985). This notion of a 6000 year old Earth was gradually overcome between 1785 and 1800 as the idea of geologic time evolved, based upon the ideas of James Hutton, Charles Lyell and William Smith, embracing what came to be known as the Huttonian view and uniformitarianism (Chapter 2, p. 22). In 1896 a French physicist, Henry Becquerel, discovered the natural radioactive decay of uranium, and in 1905 the British physicist Lord Rutherford defined the structure of the atom and suggested using radioactivity to measure geologic time, so that in 1907, Professor B.B. Boltwood, a radiochemist of Yale University, published a list of geologic ages based on radioactivity, showing the duration of geologic time in hundreds-to-thousands of millions of years. It is now known that ancient rocks exceeding 3.5 billion years in age are found on all of Earth's continents, that the age of the Earth is around 4.6 billion years, with an age of about 14 billion years for the Universe. The oldest dated rocks found so far are at Mt Narryer and Jack Hills in Australia (Wilde et al., 2001).

5.1 TIME SCALES AND THE LAND SURFACE

Within the geological time scale (see Table 5.1) how much time was necessary to create the present land surface of the Earth? Although some areas of the land surface are much older than others, it is unlikely that most of the present shape has any specific features older than 60 million years and most is very much younger. Taking the tallest man-made structure, the Burj Dubai sky-scraper which reached 707 m on 26 September 2008, as equivalent to geologic time, it is just the top 1m (0.14 %) in which most of the landforms and land-scapes of the present surface of the earth have been fashioned; much has been produced in the last 2 million years (equivalent to about 3 cm, 0.004%), humans have been on the earth for time equivalent to 6 mm (0.0008%), and

Table 5.1   THE GEOLOGICAL TIME SCALE (SEE GRADSTEIN ET AL., 2005 AND OGG ET AL., 2008)

An International Stratigraphic chart can be downloaded from The International Commission on Stratigraphy at http://www.stratigraphy.org/upload/ISChart2008.pdf, accessed 15 April 2009.

Eon	Era	Period	Series/Epoch	Beginning Mya (million years ago)
Phanerozoic	Cenozoic	Quaternary	Holocene	0.011430
			Pleistocene	2.588
		Tertiary	Pliocene	5.332
			Miocene	23.03
			Oligocene	33.9
			Eocene	55.8
			Palaeocene	65.5
	Mesozoic	Cretaceous		145.5
		Jurassic		199.6
		Triassic		251.1
	Palaeozoic	Permian		299.0
		Carboniferous	Pennsylvanian	318.1
			Mississippian	359.2
		Devonian		416.0
		Silurian		443.7
		Ordovician		488.3
		Cambrian		542.0
Proterozoic	Pre Cambrian			2500
Archean				3800
Hadean				c.4570

the Holocene, during which much of the landscape detail has been fashioned, is equivalent to 0.15 mm.

This extended analogy illustrates the enormity of geological time compared with the time in which geomorphological processes have fashioned and evolved the landforms that we see today. Separate time scales have been identified (see Table 1.3) and a composite view of time scales (see Table 5.2) identifies five time spans of interest – analogous to different magnifications of a microscope. These range from now or instant time, usually applied to a range from a few seconds to several days, but could extend to several weeks; through steady time which is the time during which a steady state may obtain or a particular event may occur; to management or engineering time up to 100 years which is the scale for which engineering structures are often designed and sustainable development applies. Graded time could cover several centuries when a graded condition or dynamic equilibrium exists typically post-Pleistocene with significant impact of climate and sea level changes, whereas cyclic, geologic or phylogenetic time is millions of years required to complete an erosion cycle, or to achieve continental displacement. These five time spans demonstrate the need for alternative magnifications in investigations of the land surface. Different spatial scales are associated with temporal scales as illustrated in Figure 5.2.

Table 5.2 TIME SCALES FOR THE LAND SURFACE OF THE EARTH

(This table incorporates elements of time scales suggested by Cowell and Thom, 1997; Driver and Chapman, 1996; Harvey, 2002; Schumm and Lichty, 1965; and Udvardy, 1981). See also Table 1.3.

Instant time, now time, instantaneous time	Steady, secular, event time	Management, engineering time	Graded, millennial scale time	Cyclic, geologic, evolutionary time, phylogenetic time scale
Few seconds to many days or weeks, human years.	An individual event, through few days, seasonal variation, a year or less, to many years.	Years to centuries.	Millennial scale, covers at least post-Pleistocene time and spatial scales up to 1000 km.	Millions of years – may be up to 500 million years.
Involve the evolution of morphology during a single cycle of the forces that drive morphological change (waves, tides). Onshore	When a true steady state may exist, the seasonal closure of an estuary by a sand bar, within-hillslope	The migration of tidal inlets and the development of a foredune ridge; zonal coupling, between major zones of the system or regional coupling,	Graded condition or dynamic equilibrium exists, where climate and sea level change are major factors	Spatial extent may reach 40,000 km so that continental displacement may be important; infilling of a tidal basin or estuary, the switching

(Continued)

Table 5.2 *(Continued)*

Instant time, now time, instantaneous time	Steady, secular, event time	Management, engineering time	Graded, millennial scale time	Cyclic, geologic, evolutionary time, phylogenetic time scale
migration of an intertidal bar over a single tidal cycle.	coupling, hillslope-to-channel coupling, and within-channels, tributary junction and reach-to-reach coupling.	relating to complete drainage basins.	operating.	of delta lobes, long-term landscape evolution.

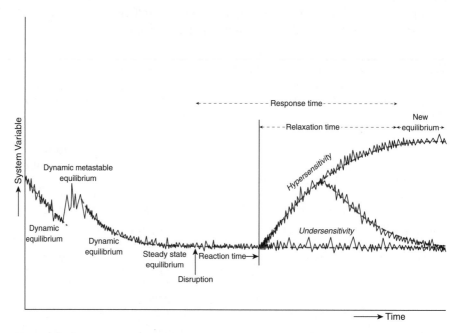

Figure 5.2 Conceptualizations of system responsiveness (see Graf, 1977; Gregory, 2005b)

Several types of equilibrium are indicated. See also Viles et al., 2008 and Table 5.9.

One of the most difficult tasks when undertaking a geomorphological research investigation is to transfer the results from one of these time scales to another. This is necessary to obtain rates of erosion (see Figure 4.14) and, somewhat tongue in cheek, Mark Twain (see Box 5.1) implied how extrapolating short-term measurements to a longer-term situation might give nonsensical results. Whereas rates of erosion were based upon measurements of present processes and extrapolated retrospectively into the past, or by using historical methods, collation of such rates is not easily accomplished because of different units used, the tendency to obtain results from the most dynamic areas and the dangers of extrapolating from small to larger scales, as well as the variations which occur over time and the impact of human activity (Goudie, 1995).

In recent years, a variety of new tools has enabled significant advances in dating landforms and sediments, and in quantifying the rates and timing of geomorphological processes in ways not previously possible (Heimsath and Ehlers, 2005). Cosmogenic isotopes are formed by high energy cosmic ray reactions in the uppermost 10–20 m of the crust. Interactions between cosmic rays from deep space and target minerals on Earth lead to the accumulation of certain isotopes (e.g. ^{10}Be and ^{26}Al) reflecting a sample's shielding history in the uppermost crust. Analysis of concentrations of these cosmogenic nuclides can provide estimates of the timing and rate of geomorphic processes (Cockburn and Summerfield, 2004). The concentration of cosmogenic isotopes produced within mineral grains varies with both the exposure age and erosion rate of the rock surface. In principle therefore, exposure age and erosion rate may be determined by analysing two cosmogenic isotopes from the same sample, provided the erosion rate is constant. Calculated erosion rates and exposure ages depend strongly on the models used to interpret isotopic data, the validity of assumptions inherent to these models, and the geologic surroundings in which the samples were collected. Although such rates have to be interpreted with caution, the results advance understanding and interpretation of landscape development (see Table 5.3). Fallout-derived, short-lived isotopes (^{7}Be, ^{210}PB and ^{137}Cs) and optically stimulated luminescence (OSL) can also provide a way of tracking tagged sediments over both short (days to weeks) and medium (several decades) time scales.

Changes in landforming processes, or in land surface environments may occur as a result of either changes in controlling conditions, especially climate (see section 5.2 below), or as a consequence of progressive sequential change, providing pointers for the future (see section 5.3). Whatever the reason for change, a conceptual model is required of how change might have operated. Early interpretations envisaged change taking place in a continuous, linear manner but experience subsequently demonstrated many complications that did not accord with this interpretation. An important general concept (Graf, 1977) used the analogy with the radioactive decay rate law to model the change from one equilibrium situation to another. This notion (see Figure 5.2), which can embrace several types of equilibrium, indicates that after a disruption, such as a change in climate, a reaction time commences when the system absorbs the impact, but subsequently during the relaxation time there is adjustment of the system towards a new equilibrium condition. Recent thinking about to disturbance regimes (Viles et al., 2008) offers promising approaches.

Table 5.3   Illustrations of the application of cosmogenic dating

'The past decade has seen a rapid growth in applications of cosmogenic isotope analysis to a wide range of geomorphological problems, so that the technique now plays a major role in dating and quantifying rates of landscape change over timescales of several thousands to several millions of years' (Cockburn and Summerfield, 2004).

Area studied	Conclusions	Reference
Rio Puerco Basin, New Mexico	Analysis of *in situ*-produced ^{10}Be and ^{26}Al in 52 fluvial sediment samples shows how millennial-scale rates of erosion vary widely (7 to 366 m Ma^{-1}). Using isotopic analysis of both headwater and downstream samples, determined that the semi-arid, Rio Puerco Basin is eroding, on average, about 100 m Ma^{-1}.	Bierman et al., 2005
Granite dome, north-east of Seoul, South Korea	Measured concentrations of ^{10}Be and ^{26}Al produced *in situ* at bare bedrock surface and calculated exfoliation rate of sheeting joints to average 5.6 cm.ka^{-1}.	Wakasa et al., 2006
North-west Tibet	Concentrations of *in situ*-produced cosmogenic nuclides ^{10}Be and ^{26}Al in quartz measured basalts and sandstones giving effective exposure ages between 23 and 134ka (^{10}Be) and erosion rates between 4.0 and 24 mm ka^{-1}.	Kong et al., 2007
Colorado River, Glen Canyon, Utah	Calculated episodic incision rates between c.500 ka and c. 250 ka to be 0.4 m. ka^{-1} and between c.250 ka to present to be c.0.7 m. ka^{-1}. These rates more than 2x rates reported in Grand Canyon.	Garvin et al., 2005

'One of the most stimulating of recent conceptual advances has followed the consideration of the relationships between tectonics, climate and surface processes and especially the recognition of the importance of denudational isostasy in driving rock uplift ... the broader geosciences communities are looking to geomorphologists to provide more detailed information on rates and processes of bedrock channel incision, as well as on catchment responses to such bedrock channel processes' (Bishop, 2007).

Sensitivity (see Figure 5.2) describes the ratio between the magnitude of adjustment and the magnitude of change in the stimulus causing the adjustment (Downs and Gregory, 1995). The concept of geomorphological sensitivity was defined as '...the likelihood that a given change in the controls of a system will produce a sensible, recognizable and persistent response' (Brunsden and Thornes, 1979: 476). As this involves both the propensity for change and the capacity of the system to absorb change, this means that the risk of response (i.e.

adjustment) is variable both in space and time. Thresholds are the boundary conditions or tipping points separating two distinct phases or equilibrium conditions. Although not easy to define and isolate, thresholds potentially indicate where and how much change will occur and may give clues about when and why. Three types suggested (Schumm, 1979) are:

- *extrinsic* – threshold condition at which the landform responds to an external influence, for example, the way in which a river channel cross-section may change abruptly from a single-thread to a braided channel following a large-storm event;
- *intrinsic* – where geomorphic response is caused by exceeding a threshold in an internal variable without the need for an external stimulus, such as long-term weathering reducing the strength of hillslope materials until slope failure occurs, potentially providing a significant change in sediment production to an upland river channel;
- *geomorphic* – in which abrupt landform change occurs as a consequence of progressive intrinsic or extrinsic adjustments, for example, where progressive river bank erosion eventually causes meander cut-off which prompts further changes related to bed-level adjustments in the channel.

Some systems show greater sensitivity than others, with the most sensitive often closest to threshold limiting conditions. An apparently disproportionately large adjustment resulting from a small change is described as *hypersensitivity*, whereas *undersensitivity* denotes a disproportionately small response (Brown and Quine, 1999). Because temporal dimensions of sensitivity and the propensity to change can be defined by the ratio of the event's recurrence interval (average length of time separating events of a particular magnitude sufficient to cause change) to its relaxation time (the elapsed time between the start and completion of the landform response), the same outcomes can be achieved from different sequences of development described as 'complex response' of geomorphological systems (Schumm, 1979). Some conceptual ideas developed over the last three decades related to changes through time (see Table 5.4), question assumptions underlying concepts included in Figure 5.2.

5.2 CHANGES IN CONTROLLING CONDITIONS

New ideas about change and the **resilience** of landform systems to absorb the impact of change lead us to ask how change has been effected, bearing in mind that it is artificial to separate present and past. Different orders of magnitude of change can be identified (e.g. Brown, 1991):

- individual years or events can have effects that may persist for many years and may cause thresholds to be crossed to new states. These are increasingly

Table 5.4   SOME RECENT DEVELOPMENTS IN CONCEPTUAL THINKING

Development	References
Whereas linear relationships involve small forces producing proportionately small responses, *geomorphic systems are typically non-linear*, owing largely to their threshold-dominated nature, so that change in geomorphic systems can progress along multiple historical pathways rather than proceeding towards some equilibrium state or along a cyclic pattern. Therefore geomorphic change may not require large external forcings and geomorphic systems may have multiple potential response trajectories or modes of adjustment to change.	Phillips, 2006a
Most environmental systems are described as chaotic, giving rise to responses which do not settle down to a fixed equilbrium condition or value. *Chaos theory* has been used to explain the fact that complex and unpredictable results can and will occur in systems sensitive to their initial conditions.	Phillips, 2006b
Because some geomorphic systems rarely achieve equilibrium, *catastrophe theory* has been used as a basis for theory building, providing an alternative to thresholds as zones of transition where two states of equilibrium are possible. The mathematical formulations of catastrophe theory can be used to account for sudden shifts of a system from one state to another, as a result of the system being moved across a threshold condition.	Graf, 1988
As geomorphic systems have multiple environmental controls and forcings, and degrees of freedom in responding to them, a *perfect landscape perspective* can be envisaged leading towards a worldview that landforms and landscapes are circumstantial, contingent results of deterministic laws operating in a specific environmental context, such that *multiple outcomes are possible*.	Phillips, 2006c
The geomorphic character of places is the result of *historical and spatial contingency*. **Historical contingency** means that the state of a system or environment is partially dependent on one or more process states or upon events in the past. It arises from inheritance, conditionality and instability. Inheritance relates to features inherited from previous conditions. Conditionality is when development might occur by two or more different pathways according to the intensity of a particular phenomenon, for example whether a threshold is exceeded to instigate different trajectories of development. Instability refers to dynamical instabilities whereby small perturbations or variations in initial conditions vary or grow over time giving divergent evolution. Spatial contingency occurs where the state of an earth surface system is dependent on local conditions which relate to local histories, landscape spatial patterns and scale contingency.	Phillips, 2001
A system subject to change has a resistance to change, described as *resilience*, which is the capacity of a system to absorb disturbance and reorganize while undergoing change but still retaining essentially the same function, structure, identity, and feedbacks.	Walker et al., 2004
The **adaptive cycle** (see Figure 5.3) as a fundamental unit of dynamic ecosystem change comprises a forward and backward loop in four phases: *rapid growth* (r), *conservation* (K), *release* (Ω) and *reorganization* (α). In the forward loop, systems self-organize through rapid growth in which processes exploit and accumulate free energy (r) towards a point of maximum conservation and connectedness (K) epitomized by complex or mature states, such as forest ecosystems.	Holling, 2001

Table 5.4   *(Continued)*

Development	References
In geomorphic terms, this process may be viewed as the evolution of landforms to a point of incipient instability where intrinsic or extrinsic thresholds are more easily exceeded. These include the concept that systems may exist in *multiple steady states*, flipping from one state to another as thresholds are transgressed. *Panarchy* is a conceptual term for a nested set of adaptive cycles that cross multiple spatial and temporal scales. It focuses the need to understand different scales of change in order to explain the causation of modern states, and can be applied in geomorphology. Resilience theory was used to reconstruct landscape system behaviour for the past 3000 years in the Erhai lake-catchment system, Yunnan, SW China, showing the possibility of alternative steady states in the landscape, as expressed by the relationship between land use and erosion in phase space. A period of agricultural expansion ~1400 cal. BP triggered rapid gully erosion that continued to accelerate for 600 years until the formation of a 'steady' eroded landscape state that has existed since ~800 cal. BP.	Schumm, 1979 May, 1977 Scheffer et al., 2001 Thornes, 2008 Dearing, 2008

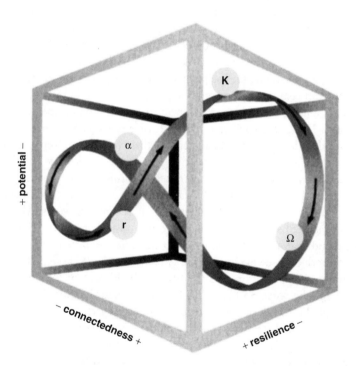

Figure 5.3   The adaptive cycle (from Dearing, 2008, based on Gunderson and Holling, 2001)

The vertical axis is the capacity or potential in the accumulated resource of biomass or nutrients, the horizontal axis is the connectedness and the third dimension is the resilience. The cycle begins at bottom left and proceeds through phases of rapid growth (r), conservation (K), release (Ω) and reorganization (α).

Original work from *Panarchy* by Lance M. Gunderson and C.S. Holling.
Copyright © 2002 Island Press. Reproduced by permission of Island Press, Washington D.C.

'readable' from the sedimentary record. The effects of a major flood can have long-term effects upon the landscape or lead to a threshold being crossed in order to move to a new state.

- over periods of decades or centuries, change can be induced by climate fluctuations, geomorphological adjustment (e.g. earth movement in Japan, Turkey or New Zealand) or human activity (e.g. deforestation);
- climatic variations such as interstadials with a periodicity of either 40,000 or 25,000 years or fewer can effect temporal change;
- major 'cyclic' time, including Milankovitch cycles (100 ka, 44 ka and 23 ka), causing variations in climate involving large fluctuations in the water balance; such changes include changes from glacial to interglacial climate conditions which occurred more than 20 times during the Quaternary.

Controls of various magnitudes can induce land surface alterations, classified as climatic, tectonic, eustatic or anthropogenic. The climatic have occasioned the largest **environmental changes** in the recent geological history of the land surface, producing successions of glacial and interglacials of the Quaternary over a period of 2 to 3 million years. Originally four major glaciations were proposed for the Alps of Europe (Penck and Bruckner, 1901–9), and for Scandinavia, Britain and North America they provided the framework for Quaternary investigations. Refined dating techniques, including the use of varves in lake deposits in Sweden since 1912; fossil pollen from peat deposits prompting palynology after 1916; and radiocarbon dating (^{14}C) applied to wood, charcoal, peat, organic mud and calcium carbonate in bones since 1949, all contributed to progressive refinement of glacial history. In the last three decades of the twentieth century these techniques were supplemented by enormous advances in the identification of past time scales (see Box 7.1). Deep sea cores, dated by radiometric and other means, furnished an uninterrupted stratigraphical record for the last 900,000 years; ice core evidence from Greenland and Antarctica established that the Quaternary began 2.8 million years (Ma) ago, that the Holocene started some 10,000 years ago but possibly 6500 or 13,500 according to area, and that a new time scale of 26 chronozones based upon isotope ages should be introduced (see Figure 5.4; see Gibbard and Kolfschoten, 2005).

The Quaternary glaciation was the fifth in the history of the Earth, the previous one being between 350 and 250 Ma ago. When the Earth's climate began to cool about 25 Ma ago, polar ice caps began to expand, triggering climate change. This included a series of sudden changes in climatic conditions over the last 2.6 million years as the climate alternated between cool conditions, when the massive polar ice caps expanded in glacials or cold stages, and warm interglacials or temperate stages when the ice caps contracted. Such fluctuations had dramatic consequences for the land surface in terms of the increased glacial extent and volume, the associated fluctuations of sea level, and the tectonic and climate effects. At the maximum extent of glacial ice some 30% of the Earth's surface was covered by ice, three times the present cover of ice caps and glaciers, a volume three times greater than at present, and maximum thickness of 3300 m over the location of the present Hudson Bay, with typical thickness of 2000–3000 m and a maximum possibly of 4000 m. The enormity of the ice

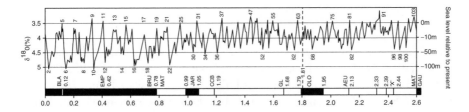

Figure 5.4 2.6 million years of climate change – peaks represent a warm earth, troughs a cold earth (Quaternary Research Association). See Quaternary chronology (Glossary p. 306).

extent and volumes is indicated in Table 5.5. In addition the depression of the land surface due to the loading by ice during the maximum glacial extent could be as much as 800 m, and resultant eustatic sea level fluctuations occurred over a range of nearly 300 m.

The significance of Quaternary environmental change for the land surface of the Earth is an ever-unfolding story, meriting many substantial publications (e.g. Williams et al., 1998; Mathews, 2001; Siegert, 2001; PAGES at http://www.pages.unibe.ch/); QRA at http://www.qra.org.uk/. Those parts of the world still largely dominated by the effects of Quaternary glacial advances include paraglacial environments (Mercier and Atienne, 2008; see Chapter 7, p. 176; Chapter 8, p. 188). Key features introduced here include glacial impacts, loading effects of ice, sea level changes, recurring climate shifts, and the potential rapidity of change.

Table 5.5 PAST AND PRESENT EXTENT OF GLACIERS AND ICE CAPS (SEE EMBLETON AND KING, 1968; GOUDIE, 1992; WILLIAMS AND FERRIGNO, 1999)

Area	Present area km^2.10^6 (% of total)	Volume km^3.10^6 (% of total)	Last glaciation km^2.10^6 (% of total)	Quaternary maximum km^2.10^6 (% of total)	Potential sea level rise m
Antarctica	13.59 (84.6%)	30.11 (91.49%)	13.6 (32.2)	13.6 (28.0)	73.44
Greenland	1.78 (11.12)	2.62 (7.96)	2.16 (5.9)	2.16 (4.6)	6.55
Laurentia			12.74 (31.6)	13.79 (29.3)	
North American Cordillera			2.20 (5.5)	2.5 (5.3)	
Scandinavia			4.09 (10.1)	6.67 (14.1)	
Siberia			1.56 (3.9)	3.73 (7.9)	
Other northern hemisphere			3.45 (8.6)	4.07 (8.6)	
Other southern hemisphere			0.9 (2.2)	1.02 (2.16)	
TOTAL	16,051,094 (100)	32,909,800 (100)	40.3 (100)	47.14 (100)	80.44

5.2.1 GLACIAL IMPACTS

During the last 1.6 million years 17 glacials/interglacials have occurred, each major cycle characterized by the build up of ice volume over a period of c. 90,000 years followed by an abrupt termination in as little as 8000 years (see Figure 5.5, Colour plate 3). The last glaciation began about 110,000 years ago, the maximum (LGM) was approximately 20,000 years BP and the glaciation ended between 15,000 and 10,000 years BP. As in earlier glaciations the timing was not synchronous from continent to continent. In each case, however, in addition to ice in Antarctica and Greenland, there were ice sheets in North America (Cordilleran and Laurentide), in Europe (Fenno-Scandinavian and Alpine) and Siberia in the northern hemisphere; whereas ice was less extensive in the southern hemisphere but included extension of glaciers in the Andes and in New Zealand with ice caps developed elsewhere including Tasmania and the Snowy Mountains, Australia. The Cordilleran ice was as much as 2300 m thick in the Sierra Nevada area. In Europe extension of Alpine glaciers with thicknesses up to 1500 m did not reach the Fenno-Scandinavian ice, thus leaving an unglaciated area between the two; the Scandinavian ice extended over the North Sea basin, over-ran most of the British Isles (but not all) and was probably more than 3000 m thick over Norway; the Siberian ice sheet, although coalescing with the Scandinavian ice, was less substantial as a consequence of the lower precipitation received further east.

Such glaciations had major effects on the landforms, creating some of the major characteristics of the present land surface. Some of these resulted during glaciations but others were created during deglaciation when the ice melted and released large amounts of meltwater that had dramatic effects on the landscape. An area known as the Channelled Scabland in Washington, USA was thought by Bretz (1923) to be the result of catastrophic flooding after the drainage of late Pleistocene proglacial lakes which filled the pre-existing valleys. However, Bretz encountered great resistance to his explanation because some of the features produced, including giant current ripples up to 15 m high, were not believed. It was much more recently, particularly as a result of work by Vic Baker (1978a, 1978b; 1981; also see Box 5.2), aided by satellite imagery, that the nature and scale of landscape development was appreciated and it was realized that flood discharges could have been as large as 21.3 million $m^3.s^{-1}$. Other examples of dramatic landforms and effects upon the land surface are included in Table 5.6.

5.2.2 LOADING EFFECTS

During glaciations the weight of ice sheets was sufficient to depress the land surface – below sea level in the case of the area around Hudson Bay in Canada and around the Baltic Sea in Europe. Glacial isostasy is the response of the land surface to loading and unloading by ice. As the density of glacier ice is approximately a third of average rock density, so depression is about 0.3 times

Table 5.6 EXAMPLES OF MAJOR EFFECTS ON THE LAND SURFACE
AS A RESULT OF, OR DURING, GLACIATION

Area	Major system	Comment
Cordilleran ice	Channelled Scabland	Flooding from Lake Missoula occurred between approximately 17,000 and 12,000 BP (see Baker and Bunker, 1985; also Box 5.2).
Laurentide ice sheet	Great Lakes	The Great Lakes formed at the end of the last glaciation about 10,000 years ago, during deglaciation. Glacial Lake Agassiz and the present distribution of the Great Lakes were created (see Chapter 8, p. 206).
Fenno-Scandinavian ice	Baltic Ice Lake	The Baltic Ice Lake c. 10,000 BP was succeeded by the Yoldia Sea, Ancylus Lake and Litorina Sea as precursors of the Baltic Sea (see Figure 5.6).
	Urstromtaler Pradoliny	Blocked north flowing rivers including the Vistula so that major meltwater systems directed along pradolinas to the North Sea via the Elbe valley.
Siberian ice	Southern Siberia, Altai Mountains	Ice-dammed lakes of south Siberia with total area at least $100.10^3 km^2$ dammed by pulsating glaciers and outburst discharges were comparable with those from Lake Missoula (see Rudoy, 1998).

ice thickness. Recovery of the land surface after deglaciation took place in stages: the first fairly rapid, termed elastic, occurred as the ice was melting, and the second involving slow viscous flow rather more slowly. Rates of uplift have been surveyed for many years but could be refined with GPS systems; they are largest in the areas where the centre of the ice cap had greatest loading effect and could be as much as 20 cm per year, totalling several hundred metres in such areas. Present rates of uplift are of the order of 1 cm per year, with a remaining 150 m of recovery to occur in the centre of the Laurentide ice sheet; it is estimated that recovery could continue to occur for the next 10,000 years. Isostatic uplift of 300 m has already occurred in North America and more than 300 m in Scandinavia (Goudie, 1992). In Scandinavia the pattern of recovery from the last glaciation involved a well known sequence of lakes (see Figure 5.6) which gradually led to the production of the present Baltic Sea.

However the classical glacial isostatic uplift paradigm of Fennoscandia may be in need of a thorough revision (Nils-Axel Mörner, 2007) and total uplift may amount to 830 m (not 300 m) with the centre of uplift beginning to rise at 13,000 BP (not 9300 BP). Isostatic recovery has also affected fjords, some of which were converted to lakes, and rivers where waterfalls introduced are often now the sites for hydro-electric power generation. In areas of recovery, tilted raised beaches record the pattern of recovery and uplift, with the tilt greatest on the oldest shorelines. The amount of tilt on shorelines in Scotland ranges from 1.26 m.km^{-1} for the oldest and 0.60 m.km^{-1} for the youngest, and there is a consistent and unchanging uplift pattern during the middle and late Holocene (Smith et al., 2000), incorporating evidence of tsunamis including one after ca. 7215 radiocarbon years BP (Tooley and Smith, 2005).

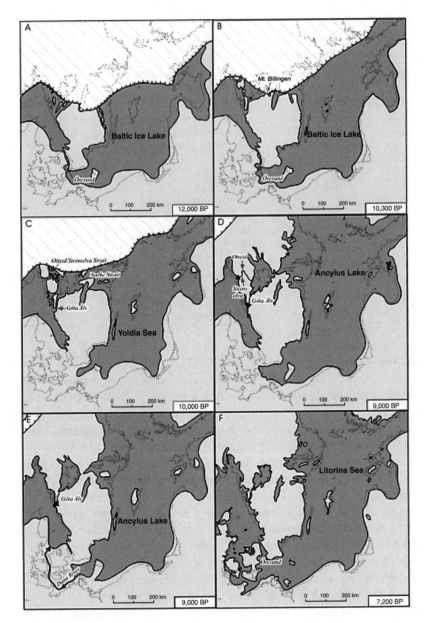

Figure 5.6   Stages in the development of the Baltic 12,000–7200 BP

Following deglaciation, the freshwater Baltic Ice Lake (12,600–10,300 BP) built up against the ice margin, followed by the mildly saline Yoldia Sea stage (10,300–9500 BP), in turn succeeded by the Ancylus Lake (9500–8000 BP) as the connecting channels had risen above sea level, followed by a greater influx of saline water creating the Litorina Sea stage (7500–4000 BP). About 62% of Finland's current surface area has been covered by the waters of the Baltic basin at some stage (from Tikkanen and Juha, 2002).

5.2.3 SEA LEVEL CHANGES

Sea level relative to the land has changed as a consequence of land uplift. However, a major cause of eustasy (global sea level changes) has been the growth of glaciers and ice caps associated with lower sea levels, and their melting giving higher sea levels. Melting of the present ice in Antarctica and Greenland could lead to world sea level rise of about 80 m; since the last glaciation sea level has risen by 100–150 m. Many factors contribute to eustatic changes of sea level, including tectono-eustatic, which can be related to plate tectonics and is of the order of 300 m, as well as glacio-eustatic, usually of the order of 50–150 m. In addition, hydro-isostatic changes arise from the effects of loading on the ocean floors by greater volumes of sea water as well as from other factors (Goudie, 1992), including infilling of ocean basins by sediment (4 mm per 1000 year rise) and decantation of water from major lakes to the sea – possibly due to isostatic uplift in the case of the Baltic and Hudson Bay (0.63 m rise). Effects of sea level fluctuations on landforms include the drowned valleys called rias in south-west England, Brittany and Spain and drowned glacial troughs termed fjords in Norway, Greenland, western Canada, the south island of New Zealand and Chile.

5.2.4 CLIMATE SHIFTS

In addition to the direct influence of glaciation and deglaciation on the land surface, the pattern of global climates also changed with each glacial/interglacial fluctuation. Areas without the direct effect of glacial ice could have very different conditions including periglaciation, which refers to the type of climate (mean annual temperatures below 0°C) and the climatically controlled surface features and processes adjacent to glaciated areas. Periglacial processes (Chapter 7, p. 172) occur in high latitude environments, in some cases associated with permafrost, but also in high altitude, alpine, areas of temperate regions. Approximately a third of the Earth's land surface has been subject to periglacial conditions at some time, and much of the permafrost of today was created during the Quaternary, although areas covered by glacial ice may not have developed permafrost. The Lodz upland (see Figure 5.7) was a classic periglaciated area located between Fenno-Scandinavian ice to the north and Alpine ice from the Tatra and Alps to the south. The University of Lodz is renowned for periglacial geomorphology through Professor Jan Dylik (1905–1973) who initiated *Biuletyn Peryglacjalny* in 1954. In periglacial areas, in addition to now relict periglacial features including patterned ground of polygons and stone stripes, blockfields or felsenmeer, there were solifluction deposits given national or local names such as *head* or *coombe rock* in southern England, together with changes to the drainage network so that dry valleys and associated landforms may have been produced.

Figure 5.7 Lodz upland in Central Poland

This area, between Fenno-Scandinavian ice from the north and ice from the
south experienced periglacial morphogenesis during the Quaternary giving
different generations of valleys reflecting formerly more extensive drainage
networks when permafrost existed.

In the extraglacial areas extensive windblown deposits were produced south
of the former ice margins in North America and Europe, especially loess, a
largely unstratified silt which has particle sizes between 0.01 and 0.05 mm in
diameter, being finer than aeolian sand, which blankets former landforms of and
covers extensive areas at least 1.6 million km² in North America and 1.8 million
km² in Europe, with thicknesses typically 10–20 m. Extensive loess deposits in
the Huang He basin of northern China cover 6.63% of the total land area of
China (640,000 km²), can be more than 200 m thick and up to 330 m thick near
Lanzhou. The loess of China may have started to accumulate 2.4 million years
BP and was thought to have been derived from the margins of Pleistocene ice
sheets in western and north-west China, but the Gobi desert (Whalley et al.,
1982) together with steppes (Goudie, 1992) may have produced the loess that
accumulated in the Chinese Loess Plateau. This Loess Plateau area has subsequently been affected by landslides, having what has been described as the most
erodible soil on Earth, leaving large gullies and giving rise to extremely high
sediment yields of the Huang He and its tributaries (see *CATENA*, 2005).

Major changes of climate elsewhere included those in lower latitudes where
periods of greater moisture (pluvials) alternated with drier periods (interplu-
vials). Some deserts, such as the Sahara, expanded during the glacial periods
and contracted during the interglacials. About 10% of the present land area

between 30°N and 30°S is covered by active sand deserts but sand dunes and associated deserts were much more extensive 18,000 years ago than at present, representing almost 50% of the land area between 30°N and 30°S (Sarnthein, 1978). These changes mean that fossil dunes exist in northern India, the Kalahari of Africa, in North America, including the Sandhills of North and South Dakota, as well as in South America including the Llanos of the Orinoco river and areas in the Bahia state of Brazil (Goudie, 1992).

In addition to those in dune systems, changes also occurred in lake extent and in associated stores of the hydrological cycle, with more extensive lakes occurring during pluvial phases, some being very large, and the 100–120 depressions in the Basin and Range Province of the American South West are probably the greatest concentration of pluvial lakes (Goudie, 1992: 109). Lake Bonneville extended over an area of 51,640 km² at its maximum extent but just 10% of that area is today occupied by saline water. Most of these lake levels were high between 24,000 and 14,000 BP, there were rapid fluctuations that may not have been synchronous between 14,000 and 10,000 BP, but most of the basins were low and dry between 10,000 and 5000 BP (Smith and Street Perrott, 1983). Elsewhere more extensive lakes are illustrated by the highest shoreline of the Caspian Sea at 76 m above the present level – recording the time when the Caspian Sea was combined with the Aral Sea and the Black Sea. In the Dead Sea rift valley a much greater lake extended south from Lake Tiberias to a location some 35 km south of the present south shore of the Dead Sea (Goudie, 1992: 109) with a highest shoreline at −180 m compared with the −400 m of the present. More extensive lakes occurred in many parts of Africa, including Mega-lake Chad in West Africa and those in the rift valley of East Africa. Palaeohydrological changes in Australia include those of the Lake Eyre basin, one of the Earth's largest internal drainage systems covering 17% of the continent, which demonstrated abrupt transitions from dry phases to wet phases (ca. 125 and 12 ka), was three times the size of the present playa in the past, including the last high-level lacustrine event at 65,000 to 60,000 BP (Magee et al., 2004; Nanson et al., 1992).

It is not easy to determine how the pattern of glacial and interglacials related to the pattern of pluvial phases with more extensive lakes, or with more arid phases when dunes were more extensive. Research continues to establish exactly when and how environmental change occurred on different parts of the land surface, although morphological and sedimentological indications of extensions of arid and semi-arid zones (see Table 5.7) assist reconstructions of landscape change.

The late Glacial is usually regarded as the years including minor stadials and interstadials between the glacial maximum and the beginning of the Holocene. Withdrawal of glacial ice resulted in the development of the present-day drainage basins of the Missouri and Ohio rivers, the development of the Great Lakes, and a global rise in sea level of up to 30 m as glacial meltwater returned to the sea. Warming climates resulted in the poleward migration of plants and animals. The Holocene, a term derived from the Greek literally meaning 'completely recent', is a time of transition from glacial conditions to the present, necessarily involving multidisciplinary research shown by contributions

Table 5.7 SOME MORPHOLOGICAL AND SEDIMENTOLOGIC INDICATORS OF CHANGES OF ARID AND SEMI-ARID ZONES (DEVELOPED FROM THOMAS, 1997, WHO CITES SPECIFIC RESEARCH INVESTIGATIONS)

Proxy is a substitute or an agency of one entity acting for another. Palaeoclimate proxies include ice cores and deep sea cores and tree rings. Landforms and sediments emplaced under past environmental conditions provide sign language to be interpreted (see p. 145 and Baker, 1998).

Landform type	Characteristic	Indicative of
Morphological evidence of arid and semi-arid zone extensions		
Slopes	Pediments in humid environments	Former arid slope systems
		Former high sediment-water ratios
	Dissected alluvial fans	
	Alluvial fan construction	Reduction of vegetation in source areas
Rivers	Clogged drainage	Former river blocked by dunes
	Incised channels	Devegetation and lower sea levels
	Aggraded rivers	Lower and more seasonal discharge
Dunes	Vegetated linear dunes	Former drier climates
	Drowned barchans	Former aridity
	Vegetated parabolic dunes	Drier conditions
	Gullied dunes	Former aridity
	Lithified dunes	Former aridity
Morphological evidence of former more humid conditions in drylands		
Rivers	Underfit modern channels	Greater discharges in the past
	Dry valley networks	Higher precipitation and channel flow or higher ground-water tables
	Valleys crossed by dunes	Wet–arid climate changes
Lakes	Strandlines	Higher lake levels in the past
Slopes	Fossil screes	Colder, possibly moister conditions
	Cave and sinter deposits	Greater local rainfall
	Alluvial fan aggradation	Increased effective precipitation
	Deep weathering profiles	Wetter climates

from fields including palaeoecology, palaeohydrology and geoarchaeology (see Benito et al., 1998; Mackay et al., 2003; Roberts, 1998) and in the journal *The Holocene* published since 1991. The Holocene which began about 11,500 BP (Roberts, 1998) included times when the present climatic conditions were established although the transition featured warmer phases such as the climatic optimum or hypsithermal from the Boreal to the Sub-Boreal (see Table 5.8) when climatic conditions were c. 1–2° warmer than at present, and cooler phases such as the Little Ice Age which peaked around 1700 and ended towards the end of the nineteenth century. Widespread decadal to multi-century scale

Table 5.8 CHRONOZONES OF THE LATE GLACIAL AND HOLOCENE (SEE ROBERTS, 1998)

Zone	Duration	Some effects
Older Dryas	14,300–13,800	Cold.
Allerod	13,800–13,000	Cool.
Younger Dryas	13,000–11,500	Cold.
Pre Boreal	11,500–9450	Sudden rise in temperature. Cool and dry.
Boreal	9450–7450	Warmer and drier conditions, similar to present, forest spread in Europe.
Atlantic	7450–4450	Warmer than present. Warm and wet. Climatic optimum, rising sea level.
Sub-Boreal	4450–2500	Cooling c. 4450, warm, dry.
Sub-Atlantic	2500–present	Cool, wet.

cooling events, accompanied by advances of mountain glaciers in Scandinavia and the European Alps occurred mainly during the Little Ice Age (LIA) between AD 1350 and 1850 (Wanner et al., 2008). When reconstructing past conditions proxies are used as a substitute or an agency of one entity acting for another, with ice cores, deep sea cores and tree rings acting as palaeoclimate proxies. In addition signs from landforms and sediments emplaced by past hydrological processes constitute a 'sign language' (Baker, 1998: 9) which has to be interpreted. Geomorphological signs related to Holocene phases are shown in Table 5.7 and 5.8.

5.2.5 RAPIDITY OF CHANGE

Why some of the major environmental changes in the Quaternary appear to have occurred quite rapidly was not easy to understand. The most detailed information available is for the rapid change over a few decades around 11,500 years ago from the Younger Dryas to the Holocene (see Table 5.8) but this may be similar to earlier, less well-studied events (Adams et al., 1999), which include Heinrich events/stadials named after a German marine geologist (Heinrich, 1988) who discovered six detritus layers in North Atlantic sediments (see Table 3.8). These layers are interpreted to have resulted from material carried by groups of icebergs that broke off from glaciers, generally suggested to have derived from the Laurentide ice sheet but some may have derived from European sources. Such events may have begun over a few years and occurred during a cold phase over about 250–750 years, and there may have been a sudden major cold event during the Eemian interglacial about 122,000 years ago. In addition changes in isotopic evidence and dust content in ice cores (e.g. GRIP from Greenland) indicate rapid changes over short periods of time, such as warming events or interstadials known as Dansgaard-Oeschger events (see Table 3.8), which may occur on a 1500 year cycle; and, in the later part of the

Holocene, transitions such as the end of the Little Ice Age probably occurred over a few decades (Adams et al., 1999). Evidence indicates that long-term climate change occurs in sudden jumps rather than incrementally which may not bode well for the future (Maslin et al., 2001).

In 2007 a team of scientists proposed that 12,900 years ago a comet broke up and exploded over the Laurentide ice sheet with the force of millions of atomic bombs causing extinction of the North American mega fauna and massive melting of the ice sheet leading to the disruption of the thermohaline circulation of the North Atlantic and to the inception of the Younger Dryas cold period (Firestone et al., 2007; Hanson, 2007). This event was described in a paper with the striking title 'Did the Mammoth Slayer Leave a Diamond Calling Card?' (Kerr, 2009), reporting nanodiamonds as evidence of a cosmic catastrophe 12,900 years ago. As research continues to establish the nature of such rapid changes and their causes, there is no doubt that dramatic short-term changes had considerable significance for the incidence of geomorphological process events and therefore for landforms, including evidence that past hydrological events occurred over short periods of time (Gregory et al., 2006). The Younger Dryas (or Loch Lomond) stadial saw the production of distinctive glacial and proglacial landforms in Britain including the 'parallel roads' of Glen Roy which are the shorelines of a former ice-dammed lake.

5.3 MODELLING CHANGE: PAST, PRESENT AND FUTURE

Such rapid short-term changes, together with detailed investigations of the Holocene, indicate how the interpretation of recent land surface history relates to present processes – when does one end and the other begin? The recognition of the complexity of the Quaternary has produced many separate investigations and subfields, recently supported by studies of contemporary processes. It has been suggested that recent high resolution data from ice and marine sediment cores do not support slow and continuous change but rather indicate that most of the temperature rise at the last termination occurred over a few decades in the northern hemisphere (Paillard, 2001); generally the Earth climate system appears much more unstable, jumping abruptly between different quasi steady states, reflecting the existence of thresholds. Such abrupt changes make modelling very difficult, and original qualitative models of land surface evolution constructed on premises of supposed long-term development, for example by Davis (p. 35), and King (p. 36), could not encompass the Quaternary so that alternative model approaches have been developed. The range of models available (see Table 5.9) is of particular interest in relation to estimates of future global warming. However general circulation models (GCMs) have to be related to the land surface and require process models which link the outputs of GCMs to their inputs to suggest

Table 5.9 MODELLING APPROACHES FOR LAND SURFACE CHANGE
(SEE ALSO TABLE 5.4)

Type of model	Purpose and method	Example of citation
Surface Process Models (SPMs)	Developed to be used both as a 'stand alone' model (a set of specific routines for the calculation of the input data is provided) and as the surface boundary subroutine of an atmospheric circulation model (all input data are taken by the atmospheric model itself). see http://personalpages.to.infn.it/~cassardo/lspm/lspm.html.	Bishop, 2007
Tectonic models (TMs)	Represent mass flux of crust in one or two dimensions. Linear systems analysis used to investigate the response of a surface process model (SPM) to tectonic forcing. The SPM calculates subcontinental scale denudational landscape evolution on geological time scales (1 to hundreds of million years) as the result of simultaneous hillslope transport, modelled by diffusion, and fluvial transport, modelled by advection and reaction. When tectonic time scales are of the same order as the landscape response time and when tectonic variations take the form of pulses (much shorter than the response time), evolving landscapes conform to the Penck type (1972) and to the Davis (1889, 1899) and King (1953, 1962) type frameworks, respectively.	Kooi and Beaumont, 1996
General Circulation Models (GCMs)	Designed to determine climate behaviour by integrating various fluid-dynamical, chemical, or biological equations that are either derived from physical laws or empirically constructed. There are atmospheric GCMs (AGCMs) as well as ocean GCMs (OGCMs). The two models can be coupled to form an atmosphere–ocean coupled general circulation model (AOGCM). A recent trend in GCMs is to extend them to become earth system models to include submodels for atmospheric chemistry or a carbon cycle model to better predict changes in carbon dioxide concentrations from changes in emissions.	http://www.ipcc-data.org/ddc_gcm_guide.html, accessed 16 June 2009
Inversion model	The preserved record in landforms and sediments is used in a strategy called	Kleman and Borgstrom, 1996

Table 5.9   *(Continued)*

Type of model	Purpose and method	Example of citation
	an inversion model to extract 'hidden' information available to reconstruct palaeo-ice sheet flow patterns and mass distribution (see Chapter 7). Glacial landscapes are classified according to the three criteria of internal age gradients, presence or absence of meltwater traces aligned to flow traces, and basal condition (frozen bed/thawed bed) inferred from morphology.	
Disturbance regimes	Conceptual models to describe biogeomorphological responses to disturbance within fluvial and aeolian environments allowing for singularity of behaviour in response to external forcing of geomorphological regimes.	Viles et al., 2008
Sediment budget studies	Techniques now available (e.g. TL, OSL, ESR) to produce accurate sediment budget estimations at spatial scales greater than that of zero order basins and over time periods greater than those covered by direct observations.	Brown et al., 2009

how processes may work. Attractive as it is, the notion of using knowledge of past change to give clues about present and future behaviour is extremely difficult, although it can be reinforced by knowledge of present processes used in retrodiction when reconstructing palaeoflood hydrology (Baker, 1998). Understanding potential effects of global warming on the land systems and landforms of the Earth's surface is urgently required, although some have suggested that any trend in global warming will be easily outweighed by much more important overlying trends and climatic feedback systems; human interference may also trigger cooling, a possibly less pleasant scenario, as a long-term consequence of initial warming – this possibility appears more likely on the evidence of the last 750,000 years. The jury is still out (Chapter 12, p. 300).

Phillips (2009) has suggested that in recent decades geomorphology has progressed from normative standards such as characteristic (steady-state) equilibrium, zonal, and mature forms to recognize that some systems may have multiple potential characteristic or equilibrium forms, and others having no particular normative state at all (see Table 5.4). This means that conceptual frameworks emphasizing single-path, single-outcome trajectories of change have been supplemented by multi-path, multi-outcome perspectives. A framework for the assessment of geomorphic changes and responses can be based on what Phillips (2009) described as the four Rs: response (reaction and relaxation times), resistance (relative to the drivers of change), resilience (recovery ability based on dynamical stability), and recursion (positive and/or negative feedbacks). This framework linking conceptual ideas (see Table 5.4) potentially links processes in the previous chapters to world landscapes in the following ones.

BOX 5.1

MARK TWAIN AND THE DANGERS OF EXTRAPOLATING EROSION RATES

Mark Twain (Samuel Langhorne Clemens, 1835–1910) had experience as a riverboat pilot, journalist, travel writer and author of *Adventures of Huckleberry Finn* and *The Adventures of Tom Sawyer*. Described by William Faulkner as 'the father of American literature', he was awarded honorary degrees by Yale University, by the University of Oxford (1907, D Litt) and by the University of Missouri. His prolific incisive writings contain many memorable quotations. Particularly cited by geomorphologists is 'There is something fascinating about science. One gets such wholesale returns of conjecture out of such a trifling investment of fact'. This quotation from *Life on the Mississippi* (page 120) published in 1883, followed:

> In the space of one hundred and seventy-six years the Lower Mississippi has shortened itself two hundred and forty-two miles. That is an average of a trifle over one mile and a third per year. Therefore, any calm person, who is not blind or idiotic, can see that in the Old Oolitic Silurian Period, just over a million years ago next November, the Lower Mississippi River was upwards of one million three hundred thousand miles long, and stuck out over the Gulf of Mexico like a fishing rod. And by the same token any person can see that seven hundred and forty years from now the Lower Mississippi will be only a mile and three quarters long, and Cairo and New Orleans will have joined their streets together and be plodding comfortably along under a single mayor and a mutual board of aldermen.

BOX 5.2

PROFESSOR V.R. BAKER

Vic Baker is Regents' Professor, Professor of Geosciences and Professor of Planetary Sciences in the Department of Hydrology and Water Resources at the University of Arizona, Tucson. He obtained his PhD from the University of Colorado

(Continued)

(Continued)

in 1971 and his career has embraced geomorphology, palaeohydrology, planetary geology and policy and philosophy issues as they relate to the environment. For a single individual to make internationally recognized contributions in all of these areas requires outstanding vision and perception together with field experience in many areas of the world.

His research on the Washington Scabland, site of one of the great controversies in the history of geology, demonstrated how massive discharges in the Missoula flood from glacial lakes were responsible for landforms on the Washington plateau. He initiated research on palaeohydrology by recognizing how slackwater deposits (SWD) in the southwestern United States could be used as palaeostage indicators (PSI) of former flood events. His research approach, subsequently used in investigations in Australia, China, India, Israel, South Africa, Spain and Thailand, can be used to modify and amend flood frequency curves based on the period of instrumented records, with considerable significance for management and planning and for risk analysis in hazard assessment. He showed (Baker, 2008a) how paleoflood hydrology has generated its share of controversy, in part because of the differing viewpoints and attitudes of the two scientific traditions from which it emerged: Quaternary geology/geomorphology versus applied hydrologic/hydraulic engineering. His research approach really does combine timebound and timeless parts of geomorphology and he has been a great advocate of the need for a global view. In addition to his palaeohydrology contributions (President of the Global Commission of INQUA, 1996–2000), his significant contributions to the understanding of science have related to the retrodictive approach, the nature of geological reasoning in relation to environmental science, and the history and development of thought in geomorphology and Quaternary science. As a member of various national and international scientific panels and committees, he has been concerned with issues at the interface between science and society, including science education for the public, the appropriate role for models and predictions in science policy, and the nature of geological reasoning in relation to environmental science. His contributions to planetary geology have included the early geological evolution of Mars, particularly in relation to the history of water and the possible presence of life.

At scientific conferences Vic's keynote lectures are great crowd pullers and he is always accessible for discussions with individuals. His exceptional output, including 15 books, more than 340 papers and over 400 abstracts and short reports, includes many innovative ideas and approaches, showing the importance of being able to 'think outside the box'. He has been President of the Geological Society of America (1998), has received many international awards including the David Linton Award of the British Geomorphological Research Group (1995), the Don J. Easterbrook Distinguished scientist award of the Geological Society of America (2001), the Grand Prix International de

Cannes de l'eau et de Science (2001); and he is Honorary Fellow of the European Union of Geosciences, Foreign Member of the Polish Academy of Sciences, and Fellow of the American Association for the Advancement of Science.

Vic Baker's constant willingness to challenge existing ideas and approaches vividly demonstrates how a positively critical approach can advance understanding of the land surface; his impact, extending beyond geomorphology to TV series, has contributed fundamental new insights for interpreting the land surface of the Earth – and he is not afraid to tackle problems of other planets as well.

FURTHER READING

Broad background is provided by:

Mathews, J.A. (ed.) (2001) *The Encyclopaedic Dictionary of Environmental Change.* Arnold, London.

A well presented approach to the Holocene is:

Roberts, N.M. (1998) *The Holocene: An Environmental History.* Blackwell Publishing, Oxford.

Comprehensive coverage is provided in:

Ehlers, J. and Gibbard, P.L. (eds) (2004) *Extent and Chronology of Glaciation. Volume 1: Europe; Volume 2: North America; Volume 3: South America, Asia, Africa, Australia, Antarctica.* Elsevier Science, Amsterdam.

TOPICS

1 Using Figure 5.2 endeavour to position particular landforms, processes or sequences of landscape development.
2 With reference to the five time spans suggested in Table 5.2 can you think of any geomorphological process–response systems that do not accord with the suggestions?

PART IV
ENVIRONMENTS OF THE LAND SURFACE

6

WORLD LAND SURFACE LANDSCAPES

Some land surface processes are concentrated in particular areas of the Earth's land surface, in the way that earthquakes and volcanoes and their associated landforms are particularly located at plate boundaries. Process domains (pp. 118–9) recognize associations of particular processes, but are there zones of the land surface that have particular features and processes in common? Early studies of the land surface were conditioned by personal experience of the home environment – in the days before the internet, media information and cheap flights! Thus Dokuchaev (Box 6.1) appreciated not only the way that landscapes were arranged across the continent but also how his field experience, especially of the black earth Chernozem soils, could create a foundation for pedology. The Earth's land surface could be classified either globally or locally: the global approach by taking the world and subdividing it, whereas the local vision described places in detail and amalgamated them to show how they fitted into broader regional, national and even world patterns (Gregory, 2009). The former included regional geomorphology (Chapter 1, p. 11) but the alternative strategy was needed so that small field mapping units could be amalgamated to make world regions. For more than 100 years a variety of methods have been employed to distinguish regions or zones of the earth's land surface (see Table 6.1). Whatever method is employed, climate has a fundamental role influencing the world distribution of landforms and land-forming processes.

A recurrent challenge has been to establish if, and how, climate controls the existence of distinctive zones or regions of the earth's land surface. This has engendered a debate about the merits of climatic geomorphology – a debate

Table 6.1 SOME METHODS RECOGNIZING ZONES OF THE EARTH'S LAND SURFACE ILLUSTRATING THE RANGE OF APPROACHES USED, MANY INTERRELATED, AND EVOLVED THROUGH SEVERAL STAGES

Basis for classification	Basis for zonation	Zones recognized
Climate: three types include *empirical* based on measurable climate characteristics with climates grouped according to annual averages and seasonal extremes; *genetic* based on the cause of the climate, synoptic types, using information about the climatic elements of air masses, pressure systems, solar radiation; *applied* are those created for a climate-associated purpose such as agriculture.		
Koppen, 1931 – method proposed 1884, several subsequent modifications	Empirical system recognizing five major climate types, each designated by a capital letter, by relating plant communities to annual and monthly averages of temperature and precipitation. Subgroups given a lower case letter distinguishing specific seasonal characteristics of temperature and precipitation.	A – Moist Tropical Climates B – Dry Climates with two subgroups: s – semiarid or steppe; w – arid or desert. C – Humid Middle Latitude Climates D – Continental Climates E – Cold Climates
Thornthwaite, 1948	An applied system water budget technique that assesses water demand under varying environmental conditions, using the physical interactions between local moisture and temperature based on local surface water balances. This rational approach followed by Budyko (1958) energy budget approach.	Specific indices include the moisture index (MI) and the potential evapotranspiration (PE) rate for a location, and a Thermal Efficiency Index (T/ET) of the ratio of temperature (T) to a calculated evapotranspiration (ET) value.
Budyko, 1958, 1974, 1986	An energy budget approach using net radiation = sensible heat flux + latent heat flux to classify vegetation zones.	1986 version recognized 16 biogeographical zones.
Soil: World soil distributions were developed by assembling knowledge of soil profiles, such as the chernozem studied by Dokuchaev, into a world distribution. The World Data Center (WDC) for Soils provides world soil information (http://www.isric.org/).		
FAO/AGL: WRB map of world soil resources.	Issued for the 14th International Congress of Soil Science Kyoto, Japan 1990, updated and adopted (1998) by the International Union of Soil Sciences as the standard for soil correlation and nomenclature. See http://www.fao.org/ag/agl/agll/wrb/soilres.stm#maj accessed 6 February 2009.	Map of World Soil Resources at 1:25.000.000 includes 30 reference groups.

Table 6.1 *(Continued)*

Basis for classification	Basis for zonation	Zones recognized
A Comprehensive System 10th edition, 2006 (US Department of Agriculture)	System for soil classification developed in the USA in the 1950s and evolved through approximations – the tenth in 2006. See: US Department of Agriculture, 2006, ftp://ftp-fc.sc.egov.usda.gov/NSSC/Soil_Taxonomy/keys/keys.pdf accessed 6 February 2009	Has 12 soil orders such as Alfisols and Vertisols, subdivided into suborders (c. 55), great groups (c. 238), subgroups (>1200), families (>7500) and series (>18,500, which is the field mapping unit.

Vegetation: World vegetation maps originally developed from climate classifications have evolved to use environmental characteristics to indicate vegetation distributions (e.g. Holdridge life zones, 39 identified), biogeographical zones (Budyko, 1986), ecosystems (Olson et al., 1985 identified 46), biomes (Prentice et al., 1992 identified 17), ecological regions or ecoregions. Databases, remote sensing and models now employed to maintain climate–vegetation classifications, which may be employed for carbon in relation to climate change.

Holdridge, 1947, 1967	System became most widely used in tropical areas, but could be applied globally and provided a first simple global biome model that could be used to assess the sensitivity of global vegetation distribution to climate change (Emanuel et al., 1985)	39 life zones recognized in 1967.
Prentice et al., 1992	Primary driving variables are mean coldest-month temperature, annual accumulated temperature over 5°C, and a drought index; biomes arise from combinations of potentially dominant types. Predictions of global vegetation patterns.	17 biomes recognized. Model predicts which plant types can occur in a given environment, agreeing well with mapped distribution of actual ecosystem complexes.

Morphogenetic regions/zones: Regions having characteristic assemblage of geomorphological processes resulting from particular climatic conditions and producing distinctive landscapes.

Peltier (1950)	Gave particular attention to the periglacial and unfortunately implied a cycle of erosion so that his scheme attracted all the criticism levelled against the Davisian cycle of erosion.	Recognizes nine morphogenetic regions according to temperature and moisture conditions, each zone characterized by a dominantly climatically controlled geomorphological process.

(Continued)

Table 6.1 *(Continued)*

Basis for classification	Basis for zonation	Zones recognized
Morphoclimatic zones follow the ideas of Dokuchaev and from research in pedology, biogeography and climatology.		
Tricart, 1957; Tricart and Cailleux, 1965	Zonal phenomena associated with climatic characteristics of latitudinal belts; azonal phenomena are non-climatic; extrazonal phenomena occur beyond their normal climatic limits; and polyzonal phenomena operate in all areas of the Earth's surface according to the same physical laws.	14 zones based primarily on major world climatic and biogeographical zones which are then further subdivided according to other climatic and biogeographical criteria as well as palaeoclimatic factors.
Climatogenetic zones each characterized by particular landscape-forming processes and by relief features related in a distinctive way to past landscape development.		
Budel, 1963, 1969, 1977; Fischer and Busche, 1982.	*Dynamic* geomorphology concerned with the study of processes; *climatic* geomorphology concerns the total complex of present processes in their climatic framework; and *climatogenetic* geomorphology involves the analysis of the entire relief embracing features adjusted to the contemporary climate as well as those produced by former climates. Recognized five climatomorphological zones in 1963, eight climatomorphogenetic zones in 1969, and ten in 1977.	1977: glacier zone; zone of pronounced valley formation; Taiga zone of pronounced valley formation on permafrost; ectropical zone of abated valley formation; subtropical zone of mixed morphogenesis: Mediterranean realm; humid subtropical; winter cold dry zone with transformation of planation surfaces by pediments and glacis; warm dry zone of planation preservation and traditional further development, especially by aeolian processes; wet–dry tropical zone of pronounced planation; equatorial zone of partial planation.

which really centres on whether we are seeking ways in which particular climates produce, over a period of time, distinctive regional assemblages of land-forms transcending regional differences in lithology, structure and relief (Stoddart, 1968), or whether we are looking for ways in which earth surface processes are broadly affected by global climates, present and past. The latter is easier than the former! Climatic impacts are not denied by Twidale and Lageat (1994: 330) although they suggest that climate has been overestimated because it is just one of several factors determining the shape of the Earth at regional and local scales. However, it is necessary to identify specifically how climatic regimes are relevant to the imprinting of those geomorphological and soil characteristics that survive in the present landscape (Barry, 1997), thus indicating how climate, in relation to other factors, is influential in determining zonation of the land surface of the Earth.

Following the initial impetus from Dokuchaev (see Box 6.1), the climatic geomorphology approach found favour in Europe and Russia because it could embrace the way that soil and vegetation types are associated with par-ticular zones, exemplified in morphoclimatic zones recognized in France (Tricart, 1957). In qualitative terms phenomena could be regarded as zonal if they were associated with climatic characteristics of latitudinal belts, whereas azonal phenomena are non-climatic such as endogenetic processes; extrazonal phenomena are those occurring beyond their normal climatic limits such as sand dunes on coasts; and polyzonal phenomena are those which can operate in all areas of the earth's surface according to the same physical laws. Such zonality provided the basis for 13 morphoclimatic zones (Tricart and Cailleux, 1965, 1972). However a disadvantage of that approach was that it lacked a secure energy balance foundation (see Gregory, 2000: 128) so that there were Russian attempts (for example, Ye Grishankov, 1973) to provide a quantitative climatic basis for geographic zonality. Morphoclimatic zones identified by the French school of geomorphology (see Table 6.1) also do not directly allow for changes which have occurred over time. This was remedied by Professor Julius Budel who recognized three generations of geomorphological study, namely *dynamic*, concerned with the study of processes; *climatic* which concerns the total complex of present processes in their climatic framework; and *climatogenetic* which involves the analysis of the entire relief embracing features adjusted to the contem-porary climate as well as those produced by former climates (Budel, 1963). This system, more widely known after a paper by Holzner and Weaver (1965), was gradually refined culminating in ten climatomorphogenetic zones (see Table 6.2). A scheme of nine morphogenetic systems (Peltier, 1950, 1975), each distinguished by a characteristic assemblage of geomorphic processes, was well-known in the US, stimulating interest in periglacial environments but also attracting some criticism because it was sometimes interpreted to imply nine different cycles of erosion – each having the disadvantages levelled against the Davisian cycle approach.

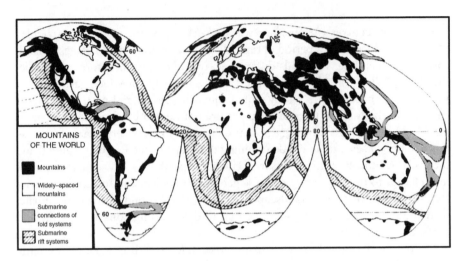

Figure 6.1   Distribution of mountains (see Gerrard, 1990)
Reprinted with permission from the MIT Press.

When dividing the land surface of the Earth into major zones as a basis for considering how different landforms occur in world landscapes, it is inevitable that climate pervades the various types of classification (see Table 6.1). However the climatic basis can be a good servant but a bad master. In the latter case climate is perceived as dominant for understanding and interpreting landforms, perhaps stifling other approaches. However in the former case it provides a way of distinguishing major environments of the Earth's land surface, which is what is required here. The subdivision adopted could be over simplified. Indeed the Greek philosopher Aristotle (384 BC–322BC) recognized frigid, torrid (between the Tropic of Cancer 23.5°N and the Tropic of Capricorn 23.5°S) and temperate zones giving a basis to which we can now add urban areas, which are such prominent environments of the land surface. A simple classification reminds us that in any subdivision there are gradations between the zones and some areas are very transitional in their characteristics.

Mountains could be recognized as a further distinct environment (see Figure 6.1). Any zonal landscape with its associated landforms may include mountain environments (Gerrard, 1990), for which a terrain model has been proposed (Charman and Lee, 2005: 506) reminiscent of the nine-unit hypothetical land surface model (see Figure 4.2, p. 78) and that suggested for drylands (see Figure 9.2, p. 222). However not only is it difficult to separate mountains from highlands and hills, which collectively cover 36% of the land surface of the earth (Fairbridge, 1968), but they are inextricably linked to processes and landforms that occur in the zonal systems where they occur.

The zonal classification used for Chapters 7–10 is summarized in Table 6.2, and the urban environment in Chapter 11 conveniently links to management of the land surface in Chapter 12.

Table 6.2 ZONES OF THE EARTH'S LAND SURFACE DESCRIBED IN
CHAPTERS 7–11

| Zone | Mean annual | | Major geomorphological processes | Threshold limits which may be critical |
	Temperature (°C)	Precipitation (mm)		
Arctic, Antarctic and High Latitudes	<–1–2	0–1000	Mechanical weathering, especially frost action, high; chemical weathering low; mass movement and fluvial processes seasonal, aeolian activity significant in some areas; permafrost influence varies with season.	Freezing threshold determines the occurrence of periglacial processes and periglacial hydrologic regimes.
Temperate and Mediterranean environments	0–20	100–1800	Chemical erosion limited, mechanical weathering includes some frost action, high angle slopes can be stable where still covered by forest, alternation of wet and dry conditions can induce landslides, seasonal streamflow regimes can give high seasonal discharges.	Where vegetation removed or degraded and other cases of human impact potentially significant.
Arid environments	10–30, can be lower in cool winter areas	0–300	Mechanical weathering rates high, especially salt weathering, generally low rates of chemical weathering, mass movement and	Vegetation extent affects aeolian processes.

(Continued)

Table 6.2 *(Continued)*

Zone	Mean annual Temperature (°C)	Mean annual Precipitation (mm)	Major geomorphological processes	Threshold limits which may be critical
			fluvial processes, although storm events have impact and in some areas may have seasonal frost action.	
Humid and seasonally humid tropics	10–30	300–>1500	Chemical weathering high in wet season, mechanical weathering low, episodic mass movement and fluvial processes vary with storm rainfall.	Incidence of mean temperatures >=20°C in all months of the year.
Urban landscapes	Occur throughout the above, heat island effect.		Processes accelerated during building activity, many reduced under impervious urban surfaces.	Removal of ground cover.

BOX 6.1

VASILY VASIL'EVICH DOKUCHAEV (1846–1903)

Often described as the father of pedology, Dokuchaev visualized soil as a product of soil-forming factors (parent material, climate, age of land, plant and animal organisms, topography) later contributing in 1900, with his student Sibirtsev, the zonal theory of soils which was especially pertinent in Russia where soils occurred in broad zonal groups. By relating the five soil-forming factors to spatial variations in soil type he showed that they could be the basis for soil classification. He undertook comprehensive surveys of Russian soils and after his research on the Nizhedorodskaya [Nizhni Novgorod] (1882–1886) region, instigated a new complex method of studying landscape and qualitatively valuing land.

He originally trained for the priesthood, obtaining a diploma from the seminary in Smolensk in 1867 but then turned to the faculty of physics and mathematics, graduating from St Petersburg University in 1871. When asked by his Professor: 'Tell me, young man, what are you occupied with primarily?', he answered: 'Playing cards and drinking', and received the reply: 'Great! Continue and do not spoil life with dry science'. Fortunately, he progressed to become Professor of Geology at St Petersburg University, 1883–1897, and head of the Department of Mineralogy and Crystallography in 1883. He produced *Various Ways of Formation of River Valleys in European Russia* in 1878, but it was his *Russian Chernozem* in 1883 that made him world famous. His contribution is universally known not only for soil science but also for the zonation of landscapes in Russia, later influencing other workers. In addition to the Dokuchaev museum established in St Petersburg in 1904 containing soil monoliths that he collected, a prestigious Dokuchaev award is presented by the IUSS (International Union of Soil Sciences).

FURTHER READING

Gregory, K.J. (2008) Place and the management of sustainable physical environments. In N.J. Clifford, S.L. Holloway, S.P. Rice and G. Valentine (eds), *Key Concepts in Geography.* Sage, London, 2nd edn, pp. 173–198.

An introduction to the work of Professor Julius Budel is:

Fischer, L. and Busche, D. (eds) (1982) *Climatic Geomorphology.* Princeton University Press, Princeton, NJ.

Mountains are reviewed in:

Owens, P.N. and Slaymaker, H.O. (eds) (2004) *Mountain Geomorphology.* Arnold, London.

TOPICS

1 Basic information on Dokuchaev is produced in Box 6.1. Explore the internet to find background information on other key scientists who have been concerned with the way in which the Earth's land surface is zoned and the way in which landforms and processes are associated (suggestions include Jean F. Tricart and Julius Budel).
2 Some methods for identifying world zones are included in Table 6.1. What other methods could be used?

7

POLAR REGIONS: ARCTIC, ANTARCTIC AND HIGH LATITUDES

The polar regions are usually delimited by the Arctic Circle (66°33'North) in the northern hemisphere and the Antarctic Circle (66°33'South) in the southern hemisphere. They include the Arctic, which can be delimited as the region north of 60°N or north of the timberline, and the Antarctic, usually defined as south of 60°S. The two differ in that the Arctic is a frozen ocean surrounded by land, whereas the Antarctic is a frozen continent surrounded by ocean. In addition to being known as the frigid zones these zones are unified by ice and snow, and are integral parts of the cryosphere, the sphere characterized by the presence of ice, snow and permafrost. The dramatic scenery and landscapes can be seen from a tour of the cryosphere offered by NASA (see http://nsidc.org/cryosphere/) and by websites giving the state of the cryosphere today (for example, http://arctic. atmos.uiuc.edu/cryosphere/ http://nsidc.org/cryosphere/).

These areas are the most recently created landscapes of the Earth's surface because, although the Antarctic sheet has existed for at least 34 million years, extensive ice and permafrost did not exist prior to the Pliocene. They are also some of the most recently explored and discovered areas: the Antarctic peninsula was first seen in 1820, the first landing and the first overwintering made in 1895, and the South Pole was first reached by Amundsen in November 1911 and by Scott in January 1912. Recently this part of the Earth's surface has been extremely important for interdisciplinary investigations which have led to significant advances in understanding Quaternary time by scientists such as Professor Sir Nicholas Shackleton (see Box 7.1).

The zone has been recognized as that most sensitive to global change, as the Earth's super-sensitive early-warning system, referred to as 'the miner's canary' by Chris Rapley when he was Director of the British Antarctic Survey. Documented effects of global warming include those in Antarctica in 2005 when a mass of ice comparable in size to the area of California briefly melted and refroze. This may have resulted from temperatures rising to 5°C. The US space agency NASA reported it as the most significant Antarctic melting in the past 30 years. The International Polar Year (IPY, 2007–09) is an internationally co-ordinated campaign of research in both polar regions (http://www.nasa.gov/vision/earth/environment/arcticice_decline.html); the first International Polar Year was held in 1882–83.

The Arctic includes the ice-covered Arctic ocean, parts of Canada, Greenland, Russia, Alaska, Iceland, Norway, Sweden and Finland, a total area of 14,056,000 km², equivalent to twice the size of Australia. Antarctica is the largely ice-covered continental land mass surrounding the South Pole, having a total area of 13,900,000 km², equivalent to an area twice the size of Europe. Periglacial environments at present cover approximately 20% of the Earth's land surface, and approximately a third of the Earth's land surface has been subject to periglacial conditions at some time.

Five types of area can be recognized:

- Glacial, including Antarctica, the Arctic, and Greenland (see section 7.1 below)
- Proglacial (section 7.2)
- Periglacial, including permafrost terrain (section 7.3)
- Paraglacial, dominated by inherited glacial features (section 7.4)
- High altitude land systems, some extra polar including many with cryosphere characteristics (section 7.5).

These land systems (Evans, 2003) provide a convenient vehicle for describing the land surface of this zone but they are not mutually exclusive, grading into one another and combining in various ways, dependent to some extent upon the definitions of periglacial, proglacial and paraglacial (Slaymaker, 2007). Slaymaker and Kelly (2007: 159) refer to the way in which glaciated landscapes are analogous to a **palimpsests** which is a manuscript page from a parchment or book which has been written on, scraped off and used again (see also Chapter 8, p. 187): glacial landforms are 'written' on the landscape beneath, so that new landforms combine with remnants of the original surface below. In addition, a more recent glacial impact, for example glacial striae, can be a palimpsest upon larger features such as drumlins which may indicate a different direction of ice movement. Three large scale geomorphological zones are recognized in North America: shield terrain which has little drift cover over bedrock; sedimentary lowlands including subglacial lodgement till plains and supraglacial moraine complexes; and glaciated valley terrain (Eyles et al., 1983), all 'overwritten' to some degree by proglacial, periglacial or paraglacial systems.

7.1 GLACIAL LANDSCAPES

Glacial landscapes principally include:

- Antarctica, some 98% of which is covered by ice with an average thickness of 1.6 km
- the Arctic, where Greenland has the principal ice sheet with an area of 1,755,637 km^2 covering 80% of Greenland's surface area, with a thickness of up to 3200 m. Its volume of some 2.85 million m^3, representing about 10% of the earth's glacierized surface area, loads the central area of Greenland so that it forms a basin lying more than 300 m below sea level
- individual valley glaciers, such as those of Vatnajökull which is the largest glacier in Iceland and is up to 1000 m thick.

   The variety of glaciers is detailed at http://nsidc.org/glaciers/quickfacts.html. At present 91.4% of the volume of glacier ice is in the Antarctic ice sheet for which development has been reconstructed (see Table 7.1) and 8.3% is in Greenland, leaving just 0.3% for the remaining glaciers (Bennett and Glasser, 1996: 17). Three major types of glacier (see Chapter 4, p. 102) are responsible for characteristic landscapes, some currently dominated by ice and many now ice-free but bearing the recent imprint of glacial processes. The East Antarctic ice sheet forms a broad dome over 4000 m in altitude, drained around its margins by a series of radial ice streams with some leading into rock-bound glacial troughs. The West Antarctic ice sheet comprises three contiguous domes rising to 2000 m, partly founded on bedrock below sea level; it is also drained by ice streams with many flowing to ice shelves, which are floating sheets of ice occupying coastal embayments and flowing seawards at speeds of several hundred metres per year. The Ross ice shelf, an area as large as France, is fed by outlet glaciers from both East and West Antarctica. Valley glaciers include surge-type glaciers which can have a dramatic increase in flow rate, can exceed 100 m/day, thus being as much as 10–100 times greater than normal flow rates. It is often characteristic for surge-type glaciers to experience a short-lived episode of accelerated flow, followed by a period of stagnation or retreat. Surging glaciers typically occur in maritime climates, such as on Iceland, Svalbard and in Alaska. Structurally controlled subglacial lakes and the interconnectedness of these lakes with subglacial rivers and with wetlands, emphasizes how meltwater streams and glaciohydrology are significant in the present process assemblage.
   There are exposed rock areas not covered by ice in Antarctica and Greenland where mountains rise above the ice surface; islands joined by ice make up the western part of Antarctica which has significant mountains, with the highest in the Ellsworth Mountains, Vinson Massif, at 4892 m. Small areas of rock emerging above ice sheets and glaciers known as nunataks are usually angular and jagged because of freeze-thaw processes. The Trans-Antarctic Mountains that stretch across the continent appear through the ice cap as

Table 7.1 THE CHRONOLOGY OF LANDSCAPE EVOLUTION IN ANTARCTICA BASED MAINLY ON TERRESTRIAL EVIDENCE (AFTER JAMIESON AND SUGDEN, 2008)

Reprinted with permission from the National Academies Press, Copyright 2008, National Academy of Sciences

Age	Characteristics
>55–34 Ma	Passive continental margin erosion of coastal surfaces, escarpments, and river valleys, removing 4–7 km of rock at the coast and 1 km inland since rifting. Cool temperate forest and smectite-rich soils, at least at coast.
34 Ma	Initial glaciation of regional uplands with widespread warm-based ice, local radial troughs, and tills. Climate cooling.
34–14 Ma	Local, regional, and continental orbital ice-sheet fluctuations associated with progressive cooling, declining meltwater, and change to tundra vegetation. Local warm-based glaciers in mountains.
~14 Ma	Expansion of maximum Antarctic ice sheet to edge of continental shelf linked to sharp temperature decline of 20–25°C. Change from warm-based to cold-based local mountain glaciers. Selective erosion of continental-scale radial and offshore glacial troughs and meltwater routes.
13.6 Ma to Present	Ice sheet maintains hyperarid polar climate. Slight thickening of ice-sheet margins during Pliocene warming in East Antarctica. Outlet glaciers respond to sea-level change, especially in West Antarctica. Extremely low rates of subaerial weathering. Glacial erosion restricted to outlet glaciers and beneath thick ice.

nunataks. Cirque landscapes include cirque glaciers partly enclosed by steep headwalls and may be separated from the main valley glaciers or ice caps. Fjord and strandflat landscapes include troughs carved by ice which become fjords if drowned. The largest troughs on Earth, the Thiel and Lambert troughs, at present occupied by outlet glaciers from the Antarctic ice sheet (Slaymaker and Kelly, 2007: 162) are 1000 km long, up to 3.4 km deep and more than 50 km wide. Long, deep fjords occur along the coasts of Greenland, Norway British Columbia, Chile, Iceland, Svalbard, and New Zealand. In many of these areas there are extensive, undulating rock platforms up to 50 km wide and cut across geological structures close to sea level (Slaymaker and Kelly, 2007: 164). These features are the culmination of processes of frost action, marine erosion, subaerial erosion and subglacial erosion acting at different times in the Quaternary. The Dry Valleys region of Antarctica (ADV) is generally classified as a hyper-arid, cold-polar desert with areas of extensive rock exposure and is the largest relatively ice-free region in Antarctica. Sometimes known as Antarctic 'oases' they are best developed in the McMurdo area of East Antarctica, being some of the coldest and driest places on the Earth's surface. The major Dry Valleys have certain characteristics in common, and some have unique features. They are generally 5–10 km wide (between ridge crests) and 15–50 km long. Only the Taylor and upper Wright Valleys have glaciers at their heads, which connect with the ice of the polar plateau; the other valleys have either barren upper reaches or small alpine glaciers. The valley floors are covered with loose gravelly material in which ice wedge polygons may occur.

The relative roles of fluvial and glacial processes in shaping the landscape of Antarctica have been debated since the expeditions of Robert Scott and Ernest Shackleton in the early years of the twentieth century but the sequence of land-form development has now been reconstructed (Jamieson and Sugden, 2008). This is in the context of the way in which Antarctica has changed since the Oligocene (see Table 7.1) as the world switched from a greenhouse to a glacial world and the Antarctic ice sheet evolved to its present state, including a deduction of continental-scale river patterns beneath the present ice sheet (see Figure 7.1). Landscape change by large ice sheets involved superposition of a continental-scale radial flow pattern on the underlying topography which embraced erosion in the centre, and wedges of deposition beneath and around the peripheries, with continental shelves deeper near the continent and shal-lower offshore as a result of erosion near the coast, often at the junction between basement and sedimentary rocks. A radial pattern of large 10 km-scale troughs breach and dissect the drainage divides near the coast and may con-tinue offshore, a radial pattern of ice streams with beds tens of kilometres

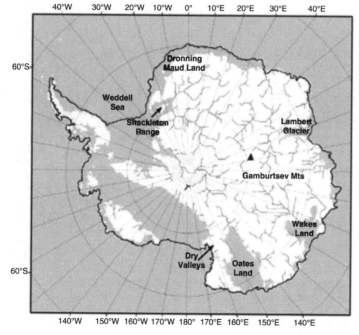

Figure 7.1   Reconstruction of the continental-scale river patterns beneath the present ice sheet of Antarctica (after Jamieson and Sugden, 2008)

The reconstruction assumes isostatic adjustment for the loss of ice load and possible drainage basins are based on a hydrological model.
Reprinted with permission from the National Academies Press, Copyright 2008, National Academy of Sciences.

wide, streamlined bedforms in bedrock and drift, and sharply defined boundaries, and a radial pattern of meltwater flow occur crossing regional interfluves, as revealed, for example, by the pattern of eskers in North America (Jamieson and Sugden, 2008).

Landscapes produced as a result of glacial erosion and deposition include classic landforms resulting from erosion, deposition and fluvioglacial processes (see Table 7.2)

7.2 PROGLACIAL LANDSCAPES

Proglacial literally means in front of the glacier and so includes all landforms adjacent to ice margins, although French and Williams (2007) contend that the

Table 7.2   CLASSIC LANDFORMS RESULTING FROM EROSION, DEPOSITION AND FLUVIOGLACIAL PROCESSES (FOR FURTHER DETAILS SEE BENNETT AND GLASSER, 1996)

Glacial erosion	Glacial deposition	Glaciofluvial landforms	Glaciolacustrine	Glaciomarine (Not directly related to land surface at present but may have affected land surface in the past)
Striations	Seasonal push moraines	Outwash fans	Ice-contact deltas	Moraine banks
Friction cracks	Composite push moraines	Outwash plain	Deltas	Submarine push moraines
Micro crag and tails	Thrust moraines	Kames	Delta moraines	Grounding line fans
p-forms	Dump moraines	Kame terraces	De Geer moraines	Till deltas
Whalebacks	Ablation moraines	Eskers	Lacustrine fans	Trough mouth fan
Roches moutonnes	Hummocky moraines	Braided eskers	Lake shorelines	Plough marks
Subglacial meltwater channels	Flutes			Boulder pavements
Ice-marginal meltwater channels	Megaflutes			
Regions of areal scour	Drumlins			
Glacial troughs	Rogens			
Cirques	Mega-scale glacial delineations			
Giant stoss and lee forms	Crevasse-squeezed ridges			
Tunnel valleys				

proglacial environment, referring specifically to ice-marginal conditions, is a periglacial environment in the original sense of Lozinski (see section 7.3 below). According to where the ice margin terminates, the environment includes combinations of glacial, fluvial, lacustrine and sometimes marine processes, often indicating the way in which the proglacial zone has expanded and contracted with recent fluctuations of the ice margin. A classification of proglacial environments is proposed in Table 7.3. Terrrestrial ice marginal environments can include proglacial lakes, created when glacial meltwater is dammed by a barrier such as an ice mass blocking a valley or terminal or lateral moraines. Outburst floods ('jökulhlaups', Icelandic for glacier-burst) from such systems can constitute high magnitude, high frequency fluxes of meltwater relative to normal ablation-controlled discharge. In addition to their significance in englacial and subglacial systems, most jökulhlaups transport sediment to proglacial sandur, with jökulhlaup deposits forming distinctive sedimentary assemblages, coupled with suites of high-energy erosional landforms (Russell et al., 2006). Meltwater inputs to proglacial systems, which are low-magnitude–high-frequency, primarily controlled by ablation inputs from the source glacier, can produce glacial drainage channels. Moraines produced at ice margins can be recessional if a sequence occurs and may be push moraines if affected by ice movement. Sometimes masses of ice detached from the glacier or ice sheet are stagnant so that kettle holes are produced where isolated blocks of ice in moraine deposits gradually melt leaving a substantial depression. Where the ice margin terminates in the sea or in a lake, depositional features produced include morainal banks and De Geer moraines or kame deltas together with fans, wedges or moraines at the grounding line which is where the ice meets land below the water level. In this way the West Antarctic ice sheet is a marine sheet partly grounded on land below sea level.

Transitional environments can be from ice marginal to fluvial (see Table 7.3), with meltwater erosion producing glacial drainage channels, and meltwater deposition creating valley trains in Alpine valleys or broad extensive outwash plains called sandur, for example in Iceland (see Figure 7.2). A sandur (pl. sandar) is a gently sloping extensive outwash plain of glaciofluvial sands and gravels deposited in front of an ice margin by a system of braided or anastomosing meltwater streams which migrate across the sandur surface, which may be pitted due to the melting of buried blocks of ice. Where the outwash is confined by steep valley sides the sandur is termed a valley train. Where large flood events discharge across the sandur the most significant characteristics include large flood bars and mid-channel jökulhlaup bars, hyperconcentrated flows, large gravel dunes, and the formation of ice-block kettle hole structures and rip-up clasts. Sandur architecture and stratigraphy show that the main controls on the sedimentary record of proglacial regions are the discharge magnitude and frequency regime, sediment supply, the pattern of glacier advance or retreat, and proglacial topography (Marren, 2005). The name sandur is derived from *Skeiðarársandur* in south Iceland, the largest sandur in the world, covering 1300 km² between the Vatnajökull ice cap and the sea (see Figure 7.2). Volcanic eruptions under the ice cap have triggered many jökulhlaup events: a peak discharge of 53,000 $m^3.s^{-1}$ in November 1996 was one of the largest historical floods, greater than the Mississippi River flood of

Table 7.3   PROGLACIAL ENVIRONMENTS AND LANDFORMS (DEVELOPED FROM BENN AND EVANS, 1998 AND EMBLETON-HAMANN, 2004)

Environment		Process	Landforms
Ice marginal	–Terrestrial	Meltwater erosion	Glacial drainage channels Urstromtaler
		Meltwater deposition	Eskers Kame and kettle topography
		Mass movement and meltwater deposition	Moraines
		Glacitectonics	Stagnant ice landforms
	– Sub-aquatic	Meltwater deposition and mass movement	Morainal banks De Geer moraines Grounding line fans
		Debris flows	Grounding line wedge
Transitional	– From ice marginal to fluvial	Meltwater erosion	Glacial drainage channels/ spillways Scabland topography
		Meltwater deposition	Sandur (Outwash plain) Outwash fan Valley train Pitted outwash Kettle holes
	– From ice marginal to lacustrine and marine	Meltwater deposition/mass movement	Deltas
		Deposition from suspension and iceberg activity	Cyclopels, cyclopsams, varves Dropstone mud and diamicton Iceberg dump mounds Iceberg scour marks

1993 (Magilligan et al., 2002) and an event which deposited some 12.8 million cubic metres of sediment, raising some parts of the level of the sandur by up to 10 m. Environments transitional from ice marginal to lacustrine or to marine include deltas and other forms of deposition including varves.

7.3 PERIGLACIAL LANDSCAPES

Periglacial, used by Lozinski (1909) to describe frost weathering conditions in the Carpathians, is generally used for the type of climate and the climatically controlled surface features and processes adjacent to glaciated areas (see p. 103). However the definition of 'periglacial' may have drifted from the original meaning as areas peripheral to Pleistocene ice sheets and glaciers to one stressing the distinctive processes of freeze–thaw and permafrost formation (Slaymaker, 2007), so that periglacial environment applies to non-glacial processes and features of cold climates, including freeze–thaw processes and frost action typical

Figure 7.2 Vatnajökull ice cap (top) with valley glaciers and Skeiðarársandur (below) in south Iceland

Sources: (top) (image taken 4 August 1999 by Landsat 7 from NASA Earth as Art); (below) (Stockphoto 4341611). Top image reprinted with permission from USGS.

of the processes in the periglacial zone, in some cases including the processes associated with permafrost. Many periglacial areas, such as large parts of northern Canada, are recently emerged from beneath glaciers and ice sheets and so have had comparatively short periods of time to develop periglacial landforms (French, 1996). Approximately 50% of the land surface of Canada currently experiences periglacial conditions, including intense frost action, the presence of permafrost, or both. Some areas experienced periglacial conditions through-out the Quaternary including central and eastern Siberia, where precipitation amounts are comparatively low and glacial ice was less extensive, interior Alaska, the northern Yukon and some Canadian Arctic islands.

Permafrost regions occupy nearly a quarter of Earth's land surface: permafrost is ground in which a temperature lower than 0°C has existed continuously for two or more years whether water is present or not (see p. 103). Spatial varia-tions in permafrost occur so that, in the northern hemisphere in the northern part of the zone, *continuous permafrost* occurs everywhere except under deep lakes; a *discontinuous permafrost* zone occurs, with mean annual soil surface temperature between −5 and 0°C, covering between 50 and 90% of the land-scape, and north of about 55°N in Canada; and further south a *sporadic permafrost* zone is found where permafrost occurs under less than 50% of the landscape and tends to be preserved at increasingly scattered sites such as on north-facing slopes or peat bogs. Permafrost has existed in Arctic areas for large parts of the Quaternary and it is estimated that it takes 100,000 years for permafrost to develop to depths greater than 500 m.

Periglacial areas can therefore be recognized according to these continuous, dis-continuous and sporadic zones with further variations according to the available relief and type of rock outcrops. The distinctive processes of terrain underlain by permafrost reflecting the annual cycle of permafrost melting and freezing (Figure 4.11) contrive to create the most distinctive periglacial landforms, includ-ing patterned ground, pingos and palsas (see Table 7.4). Patterned ground is clas-sified according to shape and sorting, with the most widespread, tundra polygons, formed by thermal-contraction cracking, dividing the ground surface up into polygonal nets 20–30 m across (see Figure 7.3, Colour plate 4). Water often pen-etrates the cracks to form ice wedges several metres deep and up to 1–2 m wide near the surface. In drier environments, mineral soil infills the cracks and sand wedges result (French, 1996). Pingos are perennial permafrost mounds formed by the growth of a subsurface body of ice, formed either when water moves to the freez-ing plane under a pressure gradient that may be hydrostatic (closed system), mean-ing that the water is derived locally and driven by hydrostatic pressure resulting from the local advancement of permafrost, or hydraulic (open system) whereby at least part of the water is derived from elsewhere, and delivered to the growing pingo by groundwater movement. Whereas hydrostatic pingos often occur singly, hydraulic pingos often occur in groups and are found in east Greenland, in Alaska and Spitsbergen. In the light of published investigations, Gurney (1998) suggests that a third category of 'polygenetic' (or 'mixed') pingos can also be recognized. Pingos do not occur in all periglacial landscapes because they result from specific geomorphic and hydrologic conditions, they can be up to 70 m high and up to 600 m in diameter; Tuktoyaktuk in the Mackenzie delta area of the Northwest Territories has some 1350 pingos (see Figure 7.4, Colour plate 5), almost all hydro-static. Pingos ultimately decay, with the best age estimate for mature pingos about 1000 years. An animated pingo development is shown by http://arctic. fws.gov/permcycl.htm (accessed 3 December 2008). Palsas are small mounds of

Table 7.4 PERIGLACIAL FEATURES AND LANDFORMS

Patterned landscape (may be associated with permafrost)	Slope features (may not require permafrost)
Thermokarst – including oriented lakes or thaw lakes – resulting from melt out of ice in unconsolidated permafrost sediments, orientation related to prevailing wind direction (Seppala, 2004). Includes decaying pingos, meltwater gullies and pipes, and many forms of ground subsidence. As active layer becomes deeper as ice wedges thaw, depressions form and progressively enlarge to become alases. Thaw lakes or oriented lakes may develop.	Steps and stripes.
Involutions and cryoturbation structures in soil horizons and regolith.	Altiplanation or cryoplanation terraces or goletz terraces.
Pingos, large perennial ice-cored mounds.	Gelifluction terraces.
Palsas, mounds of peat containing perennial ice lenses.	Cryopediments, wash or gelifluctional slopes with a thin veneer of soil and rock debris over eroded bedrock.
Earth hummocks or thufur.	Solifluction stripes.
Seasonal features – icing mounds, ice blisters, frost blisters.	Blockfields (felsenmeer).
Polygonal ground classified according to its geometric form and the absence or presence of sorting including ice wedge polygons, stone polygons, nets, sorted circles, sorted stripes and non-sorted circles.	Stone pavements or boulder pavements – stones may be part of alluvial, glacial or colluvial deposit and underlying material usually fine-grained.
String bogs or muskeg (p. 66) can be in permafrost areas.	Rock glaciers. Nivation hollows – frost action around snow patches. Large ones called thermocirques. Protalus rampart may form at foot of snow patch.

peat rising out of mires in the discontinuous permafrost zone, usually 1–7 m high, 10–30 m wide and 15–150 m long, containing a permanently frozen core of peat and/or silt, small ice crystals and thin layers of segregated ice (Seppala, 2004; Slaymaker and Kelly, 2007: 181).

Antarctic permafrost differs in several ways from that in the northern hemisphere; in some places it is much older than elsewhere, there are extensive areas of dry permafrost, and active layers are often thin. A multidisciplinary investigation of Antarctic permafrost and the periglacial environment is taking place 2004–2010 (http://www.geoscience.scar.org/expertgroups/permafrost/index.htm) (accessed 3 December 2008).

Annual changes in permafrost are complemented by trends over a number of years and, for example, in the Yukon the margin of the zone of continuous permafrost has moved poleward by 100 km since 1899, and the years since 2000 have produced record thawing of permafrost in Siberia and Alaska.

In addition to features associated with the growth of permafrost, including ice wedge polygons, ice cored mounds, pingos, palsas and rock glaciers, there are also surface features associated with thawing of ground ice in the periglacial zone. When permafrost thaws, distinctive features can develop, collectively as thermokarst because of their similarity with solution features on the surface of limestone landscapes. Although thermokarst can occur as a result of natural melting of ground ice, during a series of warmer years for example, it is often associated with human activity. The present distribution of permafrost and ground ice is therefore subject to frequent monitoring (e.g. Brown et al., 2001). Ground ice represents 47.5% of the total volume of perennially frozen ground in the upper 10 m of the surface on Richards Island in the Mackenzie delta and as much as 55% in the upper 5 m on the Yukon coast (French, 1987).

Once initiated thermokarst processes are difficult to stop: they occur as a result of the increase in the depth of the active layer, include subsidence, thermal erosion and surface ablation, and are affected by many local slope and sediment characteristics. Modes of thermokarst activity have been recognized to include (1) active-layer deepening, (2) ice-wedge melting, (3) thaw slumping, (4) groundwater flow, (5) shoreline thermokarst and (6) basin thermokarst (Murton, 2008). In Siberia a sequence of surface features were suggested (Czudek and Demek, 1970) to progress from small depressions, linear and polygonal troughs, to elongated thaw lakes and oriented lakes and thence to much larger features called alases which are thermokarst depressions with steep sides and a flat grass-covered floor, ranging in depth from 3–40 m and in diameter from 100–15,000 m in Yakutia. In the central Yakutian lowland up to 50% of the Pleistocene surface has been destroyed by alases, many originating 9000–2500 years ago but some more recently as a result of human activity (Czudek and Demek, 1970). Global warming initiated widespread thermokarst during glacial-to-interglacial transitions and, to a smaller degree, during the last 100–150 years. Projected warming during the next century will generally cause thermokarst to intensify and spread (Murton, 2008).

Periglacial phenomena independent of permafrost can result from frost wedging and the cryogenic weathering of exposed bedrock. Frost wedging is associated with the freezing and expansion of water which penetrates joints and bedding planes. Coarse, angular rock debris (block fields), normally attributed to frost wedging or cryogenic weathering, occur together with frost-heaved bedrock blocks and extensive talus (scree) slopes, frost-shattered tors, and flat, erosional surfaces known as cryoplanation terraces. Processes of frost creep and congelifluction can produce lobes, sheets and terraces which are well developed above the treeline and below sites of perennial snowbanks. The wide range of sorted and non-sorted patterned ground adds distinctive features to the landscape with non-sorted circles or nets extensive in the Mackenzie Valley, and features in saturated sediments in Keewatin in Canada termed mud boils. Periglacial features and landforms summarized in

Table 7.4 include those developed in conjunction with permafrost presence but many are affected by other processes such as wind action (Seppala, 2004).

7.4 PARAGLACIAL LANDSCAPES

The term paraglacial was introduced by Ryder (1971) and defined (Church and Ryder, 1972) as an environment in which non-glacial processes are entirely conditioned by glaciations, although Benn and Evans (1998) preferred paraglacial period to paraglacial environment because the processes that occur are not unique to such environments so that it is better to refer to a time period. Thus the paraglacial is effectively a period of recovery from glaciation. A U-shaped valley created by glaciations is very different from the V-shaped valley that existed before glaciation and the paraglacial period is when the valley, involving gravitational, fluvial and aeolian processes, gradually begins to return to the pre-glaciation condition. Without being too semantic this therefore connotes a period in which other processes, gravitational, fluvial and aeolian, are re-establishing their influence on a system dominated by inherited glacial landforms. Ballantyne (2002a) noted that paraglacial geomorphology, as the study of earth surface processes, sediments, landforms, landsystems and landscapes that are directly conditioned by former glaciation and deglaciation, had been largely ignored outside North America between 1971 and 1985. However he suggested a working definition of paraglacial as 'non-glacial earth surface processes, sediment accumulations, landforms, land systems and landscapes that are directly conditioned by glaciation and deglaciation' (Ballantyne, 2002a; Ballantyne, 2003), so that the paraglacial is the period of readjustment from glacial to non-glacial conditions (Church and Slaymaker, 1989; Slaymaker and Kelly, 2007: 167).

As the withdrawal of glacier ice exposes landscapes that are in an unstable or metastable state, and consequently liable to modification, erosion and sediment release occur at rates greatly exceeding background denudation rates. Relaxation of landscape elements to non-glacial conditions operates over time scales of $10^1->10^4$ years, with the unifying concept of glacially conditioned sediment availability (Ballantyne, 2002a). Six paraglacial landsystems have been identified (Ballantyne, 2002a): rock slopes, drift-mantled slopes, glacier forelands, and alluvial, lacustrine and coastal systems, each containing a wide range of paraglacial landforms and sediment facies. Collectively these landforms and sediments (e.g. talus accumulations, debris cones, alluvial fans, valley fills, deltas and coastal barrier structures) can be conceptualized as storage components of an interrupted sediment cascade with four primary sources (rockwalls, drift-mantled slopes, valley-floor glacigenic deposits and coastal glacigenic deposits) and four terminal sediment sinks (alluvial valley-fill deposits, lacustrine deposits, coastal/nearshore deposits and shelf/offshore deposits). The rate of sediment reworking can be described by an exhaustion model (Ballantyne, 2002b). In the case of primary reworking of glacigenic sediment, the rate of reworking declines approximately exponentially through time, though extrinsic perturbation may rejuvenate paraglacial sediment

Table 7.5 RATES OF OPERATION OF SOME PROCESSES IN PARAGLACIAL PERIODS (DEVELOPED FROM RESEARCH RESULTS COMPILED BY BALLANTYNE, 2002b)

All results given in years – the rate of change is the rate of decline of sediment availability; the half life is the time over which 50% of available sediment is exhausted; and the duration indicates the length of the period of paraglacial sediment transfer.

System	Process	Rate of change	Half life	Duration
Modification of rock slopes	Rock slope failure	1.8×10^{-4}	3850	25,000
	Rock mass creep	$0.23–1.5 \times 10^{-2}$	45–300	300–2000
Modification of drift-mantled slopes	Formation of mature gully systems	$2.3–9.2 \times 10^{-2}$	8–30	50–200
	Formation and stabilization of small debris cones	$1.5–9.2 \times 10^{-2}$	8–45	50–300
Paraglacial modification of glacier forelands	Reduction of moraine gradients	$2.3–4.6 \times 10^{-1}$	1–3	10–20
	Frost-sorting	$0.8–1.3 \times 10^{-1}$	5–9	35–60
	Infiltration of fines from near-surface till	$0.7–1.4 \times 10^{-1}$	5–10	35–70
Accumulation of large alluvial fans	Stabilization	$0.77–1.9 \times 10^{-3}$	365–900	2400–6000

flux long after termination of the initial period of paraglacial adjustment. Landscape-scale systems, particularly alluvial and coastal ones, may exhibit intrinsically complex responses due to reworking of secondary paraglacial sediment stores. The long relaxation time of such systems implies that many areas deglaciated in the late Pleistocene or early Holocene have still not fully adjusted in terms of sediment supply, to non-glacial conditions. Research data on rates of operation of some paraglacial systems have been compiled (Ballantyne, 2002a) and a condensed summary is provided in Table 7.5.

7.5 HIGH ALTITUDE CRYOSPHERE

As the cryosphere is that part of the Earth's surface where water is in a solid form, usually as snow or ice, with processes that involve freeze-thaw including frost wedging (by shattering, riving, scaling and splitting), frost heaving, frost creeping, frost sorting, nivation, and solifluction or gelifluction, high altitude mountain environments have many similarities. Mountains of course are the elements of scenery most highly acclaimed and recognized (see, for example, Figure 8.7) and they include the Alps of Europe, the Rockies of North America, the Andes in South America, the Himalayas of Asia and the Alps in New Zealand. High altitude provides some climatic conditions reminiscent of the Arctic but there are several differences. First in the energy available, they are one of the most dynamic environments in the world and, although they can have low temperatures reflecting high altitude, they can have more sunlight and greater variation in temperatures than high latitude areas. In addition mountain cryospheric zones are linked vertically to zones below, the high angle slopes and relief donate particular landform characteristics, and affect processes of flood generation and sediment production and include avalanches, which can move downslope to other zones. The highest land unit in a mountain system classification (Charman and Lee, 2005: 507) is therefore described as high altitude glacial and periglacial areas subject to glacial erosion, mechanical weathering, rock and snow instability and solifluction movements, with thin rocky soil, boulder fields, glaciers, bare rockslopes, talus development and debris fans.

Glaciers occur at the highest altitudes and, for example, in New Zealand (see Figure 7.5, Colour plate 6). In the central Canadian Rocky Mountains, cirque glaciers are the most abundant together with a few ice fields with outlet and valley glaciers (Rutter, 1987) such as the Columbia ice field at 3000 m with outlet glaciers including the Saskatchewan and Athabasca glaciers. Also present in this area are rock glaciers, features which look like glaciers in appearance but composed of rock debris, range from a few hundred metres to several kilometres in length, and have a distal area marked by a series of transverse, arcuate ridges. The rock material may contain large boulders, is usually derived from a cirque wall or steep valley walls, and movement is slow, often <1 m per year which is an order of magnitude less than most true glaciers. They are thought to be either rock debris containing ice in interstitial rock spaces (ice-cemented) or rock debris covering a decaying glacier (ice-cored). A rock glacier in Alaska is illustrated at http://earthobservatory.nasa.gov/IOTD/view.php?id=8267. Other forms of mass movement are very significant in such areas. The Hindukush and Karakoram Mountains are characterized by their high volume and variety of debris accumulations with the valley floors occupied by expansive debris-flow cones and alluvial fans with escarpment heights of over 100 m. A key role in this glacially controlled landscape system is played by the slope moraines, which cover the valley flanks up to several hundred metres above the valley floor so that landforms formerly classified as 'periglacial talus cones' can be considered as glacial or glacially controlled-landforms – which are paraglacial (Iturrizaga, 2008).

Below mountain glaciers a periglacial zone exists which is broadly similar to polar permafrost in that sporadic, discontinuous and continuous permafrost zones occur although the active layer may be thicker, the moisture content lower, and significant variations occur with aspect (Harris, 1988). Considering which landforms found in high latitudes also occur at high altitudes, Harris (1988) suggests that rock glaciers and macroscale sorted patterned ground may occur but other landforms are limited in distribution. However in the Tien Shan pingos, ice-wedge polygons and patterned ground occur and blockfields are evident.

Avalanches that are dramatic, sudden and rapid movements of ice, snow, earth or rock down a slope, occur in high altitude cryosphere areas; when force exceeds resistance (p. 77), when either the shear stress on a potential slide surface is increased (due to more snowfall especially if followed by rain, increased loading such as skiers moving across a slope) or the shear strength is reduced (increased pore water pressure due to thaw, growth of weak snow crystals) or a combination of the two. Several types of avalanche can be recognized (see Table 7.6). Many slopes in mountain areas have chutes along which avalanches may move, snow avalanches are a major agent of debris transfer, for example above the treeline in the Canadian Rockies, and avalanches constitute a major hazard in mountain areas. Below the Nevados Huascarán, the highest peak in the Peruvian Andes at 6768 m above sea level, Glacier 511 was the source of two major ice avalanches in 1962 and 1970 (see Figure 7.6, Colour plate 7).

7.6 SENSITIVITY AND FUTURE CHANGE

The sensitivity of this zone, referred to by Chris Rapley as the Earth's super-sensitive early-warning systems – 'the miner's canary' – was noted in the introduction to this chapter. Scientific reports continue to reinforce such sensitivity, one reporting that average temperatures in the Arctic have risen almost twice as rapidly as elsewhere in the world over the past few decades (Arctic Climate Impact Assessment (ACIA), 2004) meaning that melting of Arctic glaciers is a contributing factor to world sea-level rise. For example, the largest single block of ice in the Arctic, the Ward Hunt ice shelf, had existed for 3000 years, started cracking in 2000, and within two years had split all the way through, breaking into pieces (NRDefense Council). Annual satellite monitoring has shown how the amount of Arctic sea ice in the Arctic Ocean, typically lowest in September each year, has since 2002 been 20% less than the average amount 1979–2000 (http://www.nasa.gov/vision/earth/environment/arcticice_decline.html), resulting in a dramatic contrast. Each year, Greenland is losing about 80 km³ of ice (total ice sheet volume estimated at 3 million km³) and, if the annual average temperature in Greenland increases by more than about 3°C, ice is likely to be eliminated except for residual glaciers in the mountains, which could raise the global average sea-level by 7 m over a period of 1000 years or more (Gregory et al., 2004). Since 1945 the Antarctic Peninsula has warmed about 2.5°C and, although the West Antarctic ice sheet has waxed and waned in size during the past five million years (Huybrechts, 2009), imagery from 1973–2002 shows changes in the Larsen ice shelf map area including the retreat and disappearance of Wordie ice shelf, the disappearance of Jones ice shelf, and the retreat of Müller ice shelf (Ferrigno et al., 2008). These measurements showed an overall advance on the majority of the smaller ice fronts and glaciers from the 1940s

Table 7.6   TYPES OF AVALANCHE

Type		Characteristics
Snow	– Snow avalanches usually instigated by instability of snow cover	Size of a snow avalanche can range from a small shifting of loose snow (called sluffing) to the displacement of enormous slabs of snow. Loose snow avalanches, which start at or near the surface with only surface and near-surface snow involved, can be 200–300 km.hour^{-1} under favourable conditions.
	– Slab avalanches initiated at depth in the snow cover	Mass of descending snow may reach a speed of 130 km (80 miles) per hour and is capable of destroying forests and small villages in its path.
	– Powder avalanches	Aerosol of fine snow carried in the air as a snow cloud. At front can be 20–70 m.s^{-1}.
	– Dry flowing avalanches	Involving dry snow over steep irregular terrain with particles ranging from powder grains to snow blocks. Tend to follow well-defined channels such as gullies. Typical speeds from 15–60 m.s^{-1} up to 120 m.s^{-1} when travelling through the air.
	– Wet flowing avalanches	Wet snow or a mass of sludge. Relatively slow moving (5–30 m.s^{-1}), can have high density and due to large volumes of debris and boulders can cause considerable erosion along the track.
	– Slush avalanches	Essentially same as slush flows which are the mud-like flowage of water-saturated snow along stream courses. Water saturated, do not require steep slopes. In Norway associated with weak cohesionless snowpacks, hard layers or crusts of ice in snow or on ground; intense rain falling on cohesionless new snow on these layers.
Ice	– Ice avalanches	Typically occur in the vicinity of a glacier, initiated by unstable ice on steep mountain slopes through the calving and free fall of ice from a hanging glacier or the detachment and sliding of tabular masses of ice. May be caused by crevasse formation and melting or a trigger such as an earthquake.
Debris	– Debris avalanches	Contain a variety of unconsolidated materials, such as loose stones and soil.
Rock	– Rock avalanches	Large segments of shattered rock.

to about 1960, followed by retreat in the 1960s and 1970s; but, beginning in the 1990s, retreat was more pronounced and became more rapid in the late 1990s, and 82% of the measured coastlines showed average overall retreat, ranging from hundreds of metres to kilometres. There have also been changes in permafrost distribution as the extent of subarctic permafrost has been reduced significantly during the past century, leading to widespread subsidence and damage to roads and buildings (Nelson, 2003) and showing that an outstanding challenge is to separate climate-induced impacts from local anthropogenic influences.

Some related aspects are expanded in Chapter 12 but what are the emerging themes? Slaymaker and Kelly (2007) consider aspects of the cryosphere in the context of a changing global environment including future change, involving panarchy, resilience and vulnerability (see Chapter 5, p. 135). This is the zone where both dramatic and subtle changes are very apparent in the fluctuations of glaciers, in permafrost and in the associated landforms, where quite a small change can trigger crossing of a threshold and unleash a major geomorphological event with a landform consequence, and where the last two million years have shown major changes in the zone – exemplified by the adjusting changes demonstrated in the paraglacial zone.

Current sensitivities therefore include:

- Permafrost extent and character; the active layer increases and thermokarst features may increase. Possible greater release of carbon from wetlands in permafrost regions.
- Glaciers may reduce; since the Little Ice Age, in the period 1850–1990, 48 glaciers in nine different regions have shown consistent reduction in glacier length (Oerlemans, 2001). Portions of ice sheets melt and break up.
- River flows change with permafrost variations and length of winter as well as human activity. Annual duration of river and lake ice reduces by as much as 30 days.
- Precipitation may include less snow because summer warming gives longer snow-free season, although there may be increased snow at higher levels and rain-on-snow events at lower altitudes.
- Sea level change will probably continue following the 10–25 cm rise over the past century.

Media reports continue to highlight the scientific advances in monitoring that are reported and the implications of a high CO_2 world are introduced in Chapter 12 (p. 300).

BOX 7.1

PROFESSOR SIR NICHOLAS JOHN SHACKLETON FRS (1937–2006)

Sir Nicholas Shackleton made enormous contributions, demonstrating how understanding of the land surface of the Earth is dependent upon many multidisciplinary and interdisciplinary sciences, and showing how knowl-edge of past climate change can assist in simulating future change more reliably. Variously categorized as a geologist, climatologist, and palaeo-oceanographer, he was described by the International Association for Quaternary Research (INQUA) as 'a giant in the field of Quaternary science'. He was President of INQUA from 1999 to 2003; at the Congress in Reno, Nevada in 2003 he diversified his presidential address by playing the clarinet. In fact he was described as one of the best amateur clarinettists in the UK, being a collec-tor of woodwind instruments for over 40 years, particularly clarinets, with a col-lection of more than 800 instruments now housed at the University of Edinburgh.

INQUA established the Sir Nicholas Shackleton medal in recognition of his numerous contributions. These include the use of mass spectrometry to determine climate changes from the oxygen isotope composition of calcareous microfossils; high-resolution isotopic studies of deep sea sediment which showed that Quaternary ice sheet fluctuations have been much more numerous than had been supposed, and that those fluctuations showed periodicities matching those of the principal variations in the Earth's orbital geometry, validating the 'Milankovitch hypothesis' as well as extending astronomical calibration of the geological record. His research demonstrated the benefits and implications of multidisciplinary research and contributed to a critical advance in Quaternary understanding.

This outstanding scientist graduated with a BA in physics and received his PhD for a thesis titled 'The Measurement of Palaeotemperatures in the Quaternary Era' in 1967, both from Cambridge University. In 1995 he became head of the Godwin Institute of Quaternary Research at Cambridge subse-quently receiving many honours including Fellowship of the Royal Society in 1985, a knighthood in 1998 in recognition of his services to earth sciences, and many awards and medals including the Crafoord Prize (1995), the Milankovitch Medal (1999), the Wollaston Medal (1996), the Vetlesen Prize (2004) and the Blue Planet Prize (2005). His contributions facilitated the more accurate analy-sis of ice sheet fluctuations, with profound significance not only for high latitude zones but for others as well.

FURTHER READING

A useful context is provided by:

Slaymaker, O. and Kelly, R.E.J. (2007) *The Cryosphere and Global Environmental Change.* Blackwell Publishing, Malden, MA.

Two authoritative surveys are:

French, H.M. (2007) *The Periglacial Environment.* Wiley, Chichester.

Hansom, J.D. and Gordon, J. (1998) *Antarctic Environments and Resources: A Geographical Perspective*. Longman, London.

A more detailed study, integrating many aspects of cold climates is:

Seppala, M. (2004) *Wind as a Geomorphic Agent in Cold Climates*. Cambridge University Press, Cambridge.

TOPIC

1 Use the internet to explore sensitivities of this zone as a consequence of a high CO_2 world.

8

TEMPERATE AND MEDITERRANEAN ENVIRONMENTS

The temperate zone presents several paradoxes. First, although they are most familiar, temperate landscapes have often been visualized as those remaining when other, more distinctive world environments have been identified; hence the characteristics of the temperate zone have not been explicitly described, merely thought of as what is left! Second, despite being the most intensively studied, and the source of many concepts, inspiring the normal cycle of erosion and the importance of the drainage basin as the basis for hydrological understanding of runoff generation, until relatively recently we did not have an adequate understanding of the zone. Third, although the zone bears the clear imprint of many former climates and environmental conditions, it has often been taken as the norm against which other landscapes should be considered. Although it contains landscapes which owe much to development over the Cainozoic, it is, fourthly, an area where the direct and indirect influences of human activity are so profound that many of the typical inherited characteristics have been obliterated. This means that, fifthly, it is a zone about which many organizations and individuals have strong views, regarding preservation, conservation or restoration of its landscapes without necessarily possessing sufficient appreciation of what should be preserved, conserved or restored. The final paradox is that whereas internet search engines could be expected to deliver numerous links to temperate landforms or temperate landscapes, they appear to be confounded by them and produce relatively few co-ordinated responses.

Despite these paradoxes, over the last half century major advances have revolutionized understanding of the temperate zone including:

- measurement of contemporary processes gave important implications for interpretation of landscapes and landforms, exemplified by the contribution made by Professor Des Walling (Box 8.1). This is significant for a zone with some of the slowest rates of geomorphological processes, despite variations as a result of relief, tectonics and human activity.
- great advances made in understanding runoff production (see. p. 85ff, Chapter 4), imperative in enhancing understanding of the mechanics of temperate landscapes as the zone of rain and rivers.
- refinement of the Quaternary time scale (p. 136, Chapter 5) enabling knowledge gained from ice cores and deep sea cores to refine interpretations of the legacy of glaciations and periglacial conditions.
- assessment of the nature and magnitude of human impact: in most parts of the temperate zone the energy released and expended by human activity is large compared with the energy of environmental processes.
- realization that environmental management should include restoration to complement sustainability, conservation and preservation, especially as human impact has effected so much change.

How far does the temperate zone extend? Climates prevailing between the tropics of Cancer and Capricorn (23.5°) and latitude 66.5°, provide very broad limits, with much the greater land mass being in the northern hemisphere. Maritime or oceanic and continental climates are usually distinguished, although the continental climate is confined to the northern hemisphere because there are no large land masses in the southern hemisphere at suitable latitudes. Oceanic influence moderates the maritime climate, especially on the western edge of continents, giving relatively low ranges of temperature throughout the year. The seasonal climates usually do not have great extremes of temperature or precipitation, although the controlling continental and maritime air masses are not always easily predictable, thus accounting for considerable variability in temperate zone climates. Continentality increases inland, as the effect of land on heat receipt and loss increases, giving long hot, often humid, summers and severe winters often with prolonged snow cover. In North America, the north–south Rocky Mountains act as a climate barrier to the mild maritime air blowing from the west, extending the continental influence as far as the east coast of North America because the western cordillera cuts off maritime air from the Pacific. By contrast, in Eurasia, the Alps–Caucasus chains running east–west allow maritime air into the continent.

Inherited conditions, pertinent throughout the temperate zone, are reviewed (8.1) before considering contemporary processes (8.2), and some characteristics of five subzones (8.3).

8.1 THE INHERITED PALIMPSEST

Palimpsest is a parchment from which text has been partially or completely erased to make room for other text (see Chapter 7, p. 166); the word is therefore employed generally for something reused or altered but still bearing visible traces of the earlier form. Glaciologists have used the term where there are contradicting glacial flow indicators, usually consisting of smaller indicators (e.g. striae) overprinted upon larger features (e.g. stoss and lee topography, drumlins); historians use it to describe perceptions as layers of present experiences over faded pasts. The temperate zone is a palimpsest because its characteristics were partly removed during Quaternary phases when new features were created, subsequently human impact has caused other very significant changes, and finally present processes have not been sufficient to remove the character inherited from past conditions. It was for this reason that a major part of the zone was styled 'zone of former valley formation' by Budel (1977) who suggested that 95% of the total relief of mid-latitudes consists of forms inherited from pre-Holocene conditions. In fact Lewin (1987: 203) suggested that in this sense mid-latitude landforms are not temperate at all but derive from prior cold or even pre-Quaternary quasi-tropical conditions, and the great complexity of the zone has been attributed to the varying intensity of glacial and/or periglacial processes superimposed on older landforms which still include relicts of Tertiary time (Selby,1985).

This complexity indicates why it may not have been the most appropriate zone in which to develop geomorphological theories! Inherited impacts of the Quaternary (8.1.1) and of human activity (8.1.2) are succeeded by assessing the dominant characteristics upon which these inherited features have been overprinted (8.1.3).

8.1.1 QUATERNARY IMPACTS

Knowledge of Quaternary impacts is revolutionized against the background of the extent and chronology of Quaternary glaciations, compiled on a global scale (see, for example, Figure 5.5, Colour plate 3) with a series of digital maps now showing glacial limits in the GIS system ArcView (Ehlers and Gibbard, 2004). The broad components of these episodes in Quaternary history (Chapter 5) embrace the extent of glaciations, loading effects, sea level changes and impacts of other climate changes. Critical questions for the temperate zone include how far did glaciations extend in particular areas, how often, how much geomorphological work was accomplished, what landforms were produced by erosion, deposition and deglaciation, what landforms remained to be inherited by the present landscape, what other changes occurred and what happened in adjacent areas – topics all represented in the numerous papers on Quaternary morphogenesis of specific areas.

Knowledge of the extent and chronology of Quaternary glaciations has been reconstructed from glacial limits, end moraines, ice-dammed lakes, glacier-induced drainage diversions for Europe and parts of northwestern Siberia (Mangerud et al., 2004) and for North America (Ehlers and Gibbard, 2004). Impacts on landscape are reflected in individual landforms and sediments which can be visualized more broadly in terms of large scale patterns of erosion and deposition (Bennett and Glasser, 1996), expressed in terms of recurrent landform sediment assemblages – sometimes referred to as glacial landsystems (see Table 8.1). Different types of till deposits (see Table 8.2) up to 30 m thick in places, together with fluvioglacial, glaciolacustrine, and glacioaeolian deposits mantle many mid-latitude landscapes, together with the fluvial deposits in adjacent areas, all collectively representing the impact of past Quaternary environmental conditions.

Table 8.1   GLACIAL LAND SYSTEMS (DEVELOPED FROM THE PROPOSAL BY BENNETT AND GLASSER, 1996; SPECIFIC LANDFORMS ARE REFERRED TO IN TABLE 7.2 AND GLACIAL PROCESSES IN TABLE 4.12)

Glacial land system/landform sediment assemblage	General character and landforms
Erosional	
Selective linear erosion	Deep glacial troughs separated by upland areas of little or no glacial erosion in which preglacial landforms and sediments may survive.
Landscapes of little or no erosion	Areas of preglacial landforms and sediments. Occur beneath cold-based ice where glacial erosion largely ineffective. Associated with cold continental climates, areas of high relief or thin ice and ice divides.
Areas of areal scour	Associated either with warm-based ice or ice in thermal equilibrium. Intensity of scour varies from areas in which preglacial land surface may influence the morphology of individual landforms to areas in which no preglacial legacy survives. All rock surfaces show signs of glacial abrasion and plucking and assemblage of landforms including roches moutonnées, whalebacks, glacial troughs, rock basins and striated surfaces.
Landforms of local glaciation	Valley glaciers and cirque glaciers cut troughs and cirques, glaciation roches moutonnées, whalebacks, rock basins, striated surfaces and other landforms associated with warm-based ice.
Depositional	
Supraglacial landform-sediment assemblage	Associated with high englacial and supraglacial debris, giving large areas of kames and hummocky moraine.
Subglacial landform-sediment assemblage	Associated with warm-based ice and best developed in areas of soft deformable sediment, includes megaflutes, drumlins, rogens and megascale lineations. Deformation and lodgement tills interbedded with subglacially deposited units of sand and gravel. Subglacial eskers may occur.

Table 8.1 *(Continued)*

Glacial land system/landform sediment assemblage	General character and landforms
Maritime ice-marginal	At warm-based margins in maritime climates. Seasonal fluctuation of the ice margin common and meltwater production usually high, best developed where margin showing net retreat. Seasonal push moraines along the front, local areas of hummocky moraine, ablation moraine, outwash fans, outwash plains, eskers, sedimentary and landform assemblages vary according to locality.
Continental ice-marginal	At warm-based ice margins in continental climates. Seasonal fluctuations of the ice margin uncommon, meltwater production may be low, and lower mass balance gradients mean that ice velocities usually small. Ice marginal moraines occur only where ice margin advancing or stationary. If recession is steady then irregular spread of supraglacial meltout or moraine till results. Some small outwash fans may occur.
Surging landform-sediment assemblage	Valley glaciers and outlet glaciers of ice sheets that are prone to surging. Landform assemblage only described from active glacial environments. Large composite push moraine may be ice-cored, with stagnant ice behind. After ablation crevasse-squeezed ridges revealed as well as complex topography of hummocky moraine and kames formed by meltout of stagnant ice and by glaciofluvial deposition on melting ice. Presence of crevasse-squeezed ridges may be indicative of glacier surge.
Cold based or polar ice	Glaciofluvial processes largely absent. Sediments consist of basal and supraglacial meltout till and sublimation till. Ice-marginal sedimentation dominated by dump moraines or ablation moraines formed by concentration of supraglacial and englacial debiris at the ice margin by englacial thrusting and shearing at the ice margin. May have hummocky moraine surface.
Glaciofluvial	Along major meltwater discharge routes within ice sheets especially where geometry of ice margin or underlying relief concentrates or confines meltwater. Large outwash fans, meltwater channels, outwash fans, kame terraces, kames, eskers and braided eskers common.
Glaciolacustrine	Where ice margins dam water against reverse slopes or terminate in large proglacial lakes and have distinctive assemblage governed by topography and geometry of ice margin. Meltwater channels, lake shorelines, De Geer moraines, ice contact outwash fans, ice contact deltas, delta moraines, fluvial deltas, eskers and moraine banks.
Glaciomarine fjord	At glaciers that terminate in fjords or depressions submerged by the sea. Geometry of landforms and sediments restricted by relief of the fjord. Subaqueous

(Continued)

Table 8.1   *(Continued)*

Glacial land system/landform sediment assemblage	General character and landforms
	push moraines, outwash fans, moraine banks, ice-contact deltas and eskers may occur. Lateral moraines and meltwater channels may mark lateral glacier margins. Shorelines and deltas also possible.
Glaciomarine continental shelf	Where large parts of an ice sheet terminate on the continental shelf. Large moraine banks and delta moraines may develop. Sediments consist of sands and gravels, laminated silts and fine sands, and diamictons deposited by ice rafted debris or by sediment gravity flows.

Table 8.2   TYPES OF TILL (DEVELOPED FROM OWEN AND DERBYSHIRE, 2005)

Type of till	Genesis	Particle size
Glaciotectonite	Subglacially sheared sediment and bedrock	Poorly sorted
Comminution	Subglacially crushed and powdered local bedrock	Poorly sorted skewed towards fine
Lodgement	Subglacially plastered glacial debris on a rigid or semi-rigid bed	Poorly sorted with strong up-valley dip
Deformation	Subglacially deformed glacial sediment	Poorly sorted
Meltout	Glacial sediment deposited directly from melting ice	Poorly sorted but may be stratified
Sublimation	Glacial sediment deposited directly from sublimated ice	Poorly sorted with ice foliation preserved
Flow till	Sediment deposited off the ice by debris flow processes and may be classed as debris flow rather than till	Poorly sorted with downslope dip

Impacts of these glacial land systems have occurred on several occasions reflecting more than 20 glacial advances up to the maximum extent of Quaternary ice advance (see Figure 5.5, Colour plate 3) requiring features and deposits created from separate advances to be deciphered. Each deglaciation, especially the most recent, was the phase when the present drainage system was gradually re-established. The preserved record in landforms and sediments is used in a strategy called an inversion model (Kleman and Borgstrom, 1996) to extract 'hidden' information available to reconstruct palaeo-ice sheet flow patterns and mass distribution. In this way satellite imagery and ground survey data have been used to reconstruct the integrated pattern of the principal longitudinal and transverse features produced on a continent-wide scale by the last ice sheets in Europe and North America (Boulton et al., 1985). Whereas palaeoglaciological reconstructions of the dimensions, geometry and dynamics of former ice sheets were based mainly on glacial depositional evidence, the inversion model can now be used so that landforms of glacial erosion together with exposure-age dating techniques can be of great value in ice mass reconstruction in palaeoglaciology (Glasser and Bennett, 2004).

Several areas have distinctive glacial landform sediment assemblages including central Europe, Siberia and southern Canada, the central lowlands of the USA, and the Great Plains, where there are flat or gently undulating till plains with classic end moraines in the south and east, and drumlins, tunnel channels, ice thrust features and high relief hummocky moraine to the north (Mickelson, 1987). By contrast, erosional landscapes (see Table 8.1) are found in the Rocky Mountains, and in the southern hemisphere in New Zealand and Patagonia. Beyond these regions occur areas with periglacial landforms, as well as landscapes mantled by loess deposits. At the end of each glaciation the relatively rapid period of ice decay was when major characteristics of the present temperate zone appeared; recovery of vegetation from glaciations had occurred many times during the Quaternary and it was deglaciation immediately after the last glaciation that created some of our characteristic landform assemblages (see Table 5.6). In addition to the Great Lakes other dramatic changes occurred in palaeohydrological conditions. Collaborative investigations focused on the palaeohydrology of the temperate zone (Starkel et al., 1991) were subsequently extended to global investigations (Gregory et al., 1995), including reconstruction of palaeodischarges. With removal of water from melting glacial ice, often in time periods much shorter than required for the ice sheets and glaciers to build up, there were dramatically large discharges much greater than those of the present time. Because permafrost existed in marginal areas there were contrasting palaeohydrological conditions in non-glaciated areas as well – augmented by the gradual recovery from glaciation. Some palaeohydrological sequences are illustrated for specific basins below (see section 8.3).

8.1.2 HUMAN IMPACT

In addition to the shaping of the landforms of deglaciated areas, a vegetation sequence evolved during the Holocene as the environment recovered from major glaciations (p. 143, Chapter 5), a series of stages during which human impact gathered pace. The post-glacial vegetation sequence was, typically, a succession from tundra, to low scrub, coniferous forest and deciduous forest, so that many temperate areas should now be characterized by a natural vegetation cover of woodland, with perennial stream flow, and dominated by slopes that are usually gentle or moderate. Theoretically the north–south sequence of natural vegetation in the northern hemisphere should range from temperate boreal forest, comprising evergreen conifers, to deciduous temperate forests, as well as extensive areas of grassland occurring in the continental interiors, characterized by well-known names of prairie, steppe, or pampas and dry Mediterranean woodland. However any idea of natural vegetation for the temperate zone now has to be seen in the context of the interplay of the gradual recovery of temperate vegetation belts after the last glaciations/glacial phase, and the inexorable modification of the vegetation cover by human impact. This interplay means that it is now difficult to determine what is, or should be, the natural vegetation in any particular part of the temperate zone.

Although the broad outlines of human impact upon vegetation have been known for many years, it is the careful reconstructions by palaeoecologists, archaeologists and others that provided the details for many areas. The first inroads into the 'natural' vegetation were made by fire used by hunter–gatherers prior to the advent of agriculture in the Mesolithic (c.9000 years ago), continuing in the Neolithic (5000 years ago). Considerable deforestation had already occurred in Britain before the Romans arrived in the first century BC, and subsequently a great wave of deforestation in medieval times changed the appearance of western and central Europe. In Mediterranean areas deforestation occurred as early as 4600 years ago, removing the dry woodland natural vegetation of mixed evergreen and deciduous forest of oaks, pine, beech and cedars, and, in China, clearance of most of the temperate forest was completed even earlier. Riparian woodland persisted in many temperate areas despite deforestation, although it is estimated that in North America and Europe more than 80% of riparian river corridors have disappeared in the last 200 years (Hupp, 1999). In Mediterranean areas the interplay of human impact and climate fluctuations combined to affect sequences of soil erosion and deposition. When the first Europeans arrived in temperate North America they found woodland from the Atlantic coast to the Mississippi but then more deforestation occurred in the subsequent two centuries than in Europe in the previous 2000 years (Butzer, 1976). The grasslands of the North American prairies and the Eurasian steppe were equally influenced by population expansion during the eighteenth and nineteenth centuries, as the Russian steppes were colonized from the sixteenth century, although ploughing of the Eurasian grasslands on a large scale began only in the eighteenth century, reaching a peak under Soviet rule, with large-scale mechanization affecting most of the steppe. Significant cultivation of the North American prairies began about a century later. In the subsequent 150 years about two-thirds of the native vegetation of the Canadian prairies and about half that of the US prairies disappeared as the land was converted into huge cereal farms and cattle ranches. In these ways the temperate zone has become one of the most anthropogenically transformed parts of the earth with deforestation or ploughing of grassland, followed by conversion of land to grazing, arable or urban land.

Such changes in land cover and vegetation often instigated by human activity were superimposed on the way in which land-forming processes gradually adjusted from the last glaciation. A brilliant conception of the later stages of this sequence of change was provided by Wolman (1967) who demonstrated how sediment yield and river channel condition changed since 1700 in the north-east USA (see Figure 11.4). In temperate areas under a relatively stable climatic regime, all areas modified by human activity are at some stage along the time axis of this model – although the time axis from forest, through cropping to

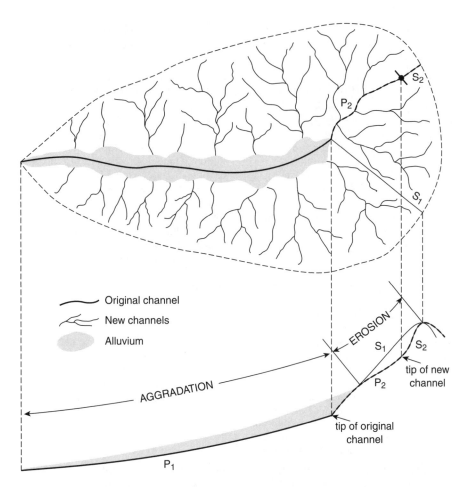

Figure 8.1 A change of land use can induce channel extension by gullying and downstream aggradation with floodplain growth (after Strahler, 1956)

urbanization, is just two centuries in the USA; in Old World temperate areas such as western Europe the same time axis covers at least two millennia. In addition the spatial implications of such adjustments have to be kept in mind and Strahler (1956) devised a model suggesting how changes in vegetation cover over the drainage basin could be associated with alterations in headwater channels leading to aggradation downstream and to changes of floodplain characteristics (see Figure 8.1). River responses to anthropogenic influences over the last 15,000 years, compiled from palaeohydrological research investigations in a variety of temperate areas (Gregory, 1995), illustrate the range of human activity throughout the temperate zone (see Table 8.3). Sufficient fluvial sediment sequences have now been investigated and analysed, especially in Mediterranean basins, to evaluate how environmental signals are autogenically processed and recorded by the fluvial sediment system over three nested time periods spanning the past 20,000 years (Macklin and Lewin, 2008) as shown in Table 8.4.

Table 8.3   EXAMPLES OF RIVER RESPONSE TO ANTHROPOGENIC CHANGE IN THE LAST 15,000 YEARS (BASED ON RESULTS FROM OVER 160 RESEARCH INVESTIGATIONS COLLECTED IN GREGORY, 1995 WHICH GIVES FURTHER DETAILS)

Age/Period	Human activities	Some river responses
Neolithic, Bronze Age, Iron Age	Deforestation/land clearance, change in farming techniques employed, intensification of agricultural activity, settlement expansion.	Greater runoff, increased rates of flood plain sedimentation in distinct phases, rise in water table, possible waterlogging of flood plain, onset of frequent flooding, new river channels developed, channel incision, increased soil erosion and sedimentation, accelerated erosion of surface soils.
Romano-British period, Anglo-Saxon period	Forest clearance for agriculture, intensification of agricultural activity, land drainage, settlement changes, changes of farming techniques, modification to waterways.	Increased runoff and erosion, extensive flooding, increased flood frequency and alluviation in channels and on flood plains, large-scale erosion from soils left bare in winter, increased runoff, channels confined.
Mediaeval times	Forest clearance for agriculture, intensification of agricultural activity, changes in farming techniques, shift to grassland, modifications to waterways, mining activities, land drainage.	Minor rise in water table of lakes, higher rates of sediment influx, some gullying on slopes, meander patterns changed to braided channels, flood peaks increased, waste products deposited in channel and on floodplain.
Seventeenth to nineteenth centuries	Forest clearance, intensification of agricultural activity, changes in farming techniques, modifications to waterways (including engineering for navigation), mining and settlement changes including expansion and impact of industrialization.	Aggradation, bars and islands appeared separated by multilinked channels, sediment influx substantially increased, coarse sediment mobilized, channel incision, increased size of channels, greater concentrations of metals in sediments. Increases in peak flows, increased rates of annual sediment accumulation, increased erosion from farmland, valley floor transformation, river width and depth increased, changes in sinuosity.
Nineteenth and twentieth centuries	Expansion of settlements, urbanization, changes in farming techniques, land drainage, modification to waterways through clearance, channelization, channel regulation, dikes, restoration of rivers.	Deposition of alluvium on to floodplain, larger and more frequent floods, excessive sedimentation, channel incision, reduced area of inundation in some areas, restoration schemes at the end of the twentieth century.

Table 8.4   RIVER RESPONSE AND RECORD OF LATE QUATERNARY
ENVIRONMENTAL CHANGES (AFTER MACKLIN AND LEWIN, 2008)

Period	Fluvial record preservation	Dating potential and resolution	Alluvial response	Sensitivity possible
Late Pleistocene	Episodic	Luminescence, U–series, ^{14}C., hundreds to thousands of years.	Aggradation phases not generally internally differentiated in the literature.	Dansgaard-Oescher, Heinrich events.
Holocene		^{14}C, luminescence, hundreds of years.	Alluvial changes, sediment size, frequency of palaeochannel fills, changes of degree for most part. Record confused by ongoing allogenic change.	Century-scale climate episodes; introduction of human effects.
Post-mediaeval		^{14}C, latterly unreliable, historical evidence may be annual/event.	Some system modification (channel metamorphosis, aggradation and incision effects).	Little Ice Age flood changes, human-induced flood sedimentation.
Instrumental/contemporary	Continuous	Records available	Erosion/sedimentation rates in flood sequence.	Flood events.

8.1.3 TEMPERATE ZONE FEATURES

Despite the Quaternary inheritance can we identify characteristic features – a zone which has inherited many landforms from past conditions, has relatively slow rates of geomorphological processes, but has been dramatically affected by human activity? Three major components (also see Figure 8.2) are:

- plateau areas where many erosion features were initially produced by pre-Quaternary erosion, often the remnants of landscapes evolved under Tertiary conditions which may have resulted in planation surfaces generated under tropical or subtropical climates although later affected by erosional and depositional processes of glacial or periglacial environments. Such remnants of Tertiary landscapes are evident in the Appalachians, in Wales and Scotland, and in the older massifs of Europe including the Massif Central of France, whereas mountain areas including the Alps, the Rocky Mountains and New Zealand are the consequence of more recent uplift (see later, p. 209)

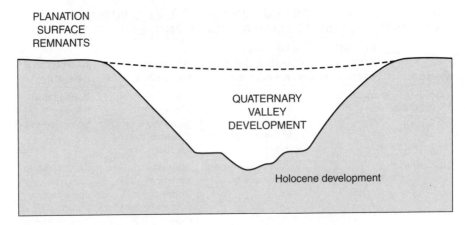

Figure 8.2   Three major components of temperate landscapes (Gregory, 2005a)

- major valley systems and plains, which are the product of the Quaternary erosional and depositional phases of the last two million years when the alternation of glaciations and interglacials was accompanied by major sea level fluctuations (see Figure 8.3);
- the detailed development of the environments along the present river courses and valley floors, which are the products of the last 10,000 years of the Holocene when climate changed from glacial to interglacial, the impact of human activity became increasingly pronounced and some changes in slope or river dynamics occurred within short periods of time.

These three components can be thought of as the broad outlines, the major and often distinctive landforms and landform assemblages, and the finishing touches! In many landscapes evidence for past development is no longer present as landforms – historical geology was suggested (Kimball, 1948) to be composed of two parts: stratigraphy dealing with what is there and denudation chronology concerned with 'what isn't' Part of the fascination of geomorphology is to deduce the what isn't that is the basis for reconstructing what happened in the past. Because of the limited evidence available Lewin (1980) drew the analogy of information gleaned through windows of limited and varying opacity and size. It is only through the surviving 'windows' in time that we glean evidence necessary to reconstruct past evolution of the temperate zone, much of it in deposits (see Table 8.5), so that perhaps in no other zone is the disjunction of modern processes and inherited landforms so apparent (Butzer, 1976: 355). Although as much as 95% of the total relief consists of forms inherited from pre-Holocene conditions (Budel, 1977), fluvial and coastal processes of the remaining 5% came to dominate the geomorphological enquiries of temperate zones, with slopes becoming central to early qualitative models of landscape development suggested in the Davisian normal cycle of erosion and in the Penck model of landscape development involving parallel recession of slopes (Chapter 2, p. 35).

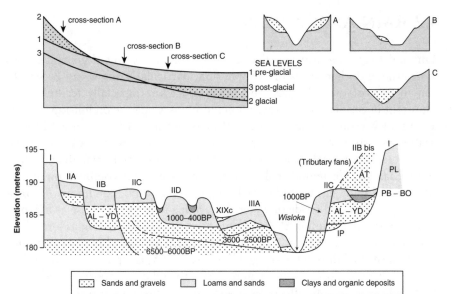

Figure 8.3 Stages of Quaternary valley development (above) and (below) an example of a valley cross-section, the Wisloka valley, southern Poland after Starkel, 1983)

Table 8.5 TYPES OF DEPOSIT THAT MAY OCCUR IN TEMPERATE AREAS (FOR AMPLIFICATION SEE GREGORY, 2005a)

Type of deposit	Characteristics
Relict from former conditions:	
Glacial	Till deposits, lake clays, gravels and sands; distribution and character not easily deduced; rapid variations in thickness.
Periglacial	Angular scree deposits, unsorted slope deposits, fine windblown material; distribution localized but character reflects locally available rock types, may be on slopes which are relict and unstable when modified.
Tropical	Clays which are remnants of deep weathering; often localized on plateau sites, to considerable depth in pockets.
Contemporary, but some may be inherited from past conditions:	
Weathering and colluvial	Medium grain size, not usually deep, may develop above fossilized deposits, some lateral variation due to subsurface drainage.
Fluvial	Range of grain sizes, often incorporate remains of fossil deposits, frequent vertical and lateral changes; small changes in river position can release new exposures of relict sediments.
Marine	Range of grain sizes present, may incorporate fossil deposits from cliff deposits or from off-shore sediments; deposits above sea level may mark former shorelines.
Aeolian	Fine silts in some areas including where vegetation removed; may be mantle over the surface not related to deposits of very different character beneath.

8.2 RAIN AND RIVERS

Rain and rivers is a neat appellation for the processes now dominant in this zone although other forms of precipitation, particularly snow, can be significant in four of the subzones, and other processes, especially tectonics, are important in certain areas. Gravitational processes operate on hillslopes in the context of variants of the hypothetical nine-unit model (see Figure 4.2) with weathering processes (p. 74, Chapter 4) including both mechanical and chemical processes potentially leading to a reduction in material strength and hence to slope failure. The temperate hillslope, composed of any combination of nine hypothetical land surface units (Dalrymple et al., 1969), conditions soil profile development and mode of water flow. Slope failures may arise as a consequence of the reduction of the material shear strength occurring as a result of high intensity precipitation, vegetation removal or ploughing for example. Water transferred through drainage basins (see p. 85, Chapter 4 and Table 1.6) follows several possible routes (Figure 4.5B); the relative importance of quick-flow and delayed flow affects hydrographs; the magnitude of delayed flow relates to groundwater recharge; and the extent of flow in the network of stream channels can change in short periods of time. Perennial stream channels are fed by groundwater or delayed or base flow during dry periods, whereas intermittent streams flow only when the water table is seasonally high, and ephemeral streams have water flow only during particularly heavy rainstorms. Under natural conditions the constantly changing stream channel network affects river hydrographs (see Figures 4.5 and 8.4), and the drainage network is most extensive just before times of peak stream flow, whereas when discharge is low it is the less extensive network of permanent streams that contributes to river discharge. These three modes of water flow in a drainage basin are members of a continuum. At the head of the ephemeral streams there may be natural pipes which can be up to 0.5 m in diameter and flow also occurs through the soil, as either matrix flow or diffuse flow. The continuum of types of water transfer through the basin, ranging from water flow through soil pores at one extreme, to open channel flow in a large river channel at the other, gives contrasts in the velocities of water flow (see Table 4.7). The drainage network density (Dd = total length of stream channels of all types/ basin area), its composition (in perennial, intermittent and ephemeral components), and dynamics (involving how it expands and contracts according to storm events and climate conditions) all vary throughout the subdivisions of the temperate zone with ephemeral elements more evident in the strongly seasonal climatic regimes.

Sediment and solutes derived from slopes and channels are integral elements of temperate drainage basin processes. However, in many areas the sources of sediment are limited so that the movement of sediment is less than would be expected and takes place at sub-capacity levels. Most sediment sources occur near stream channels so that as little as 10% of a drainage basin may contribute

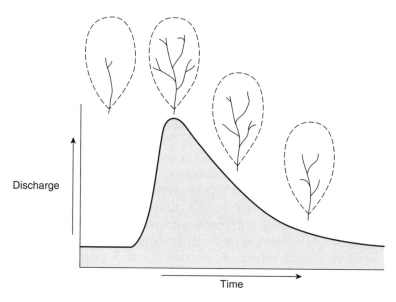

Figure 8.4   The relationship between extent of the drainage network and the river hydrograph. The network extends during storm events so that more rapid flow affects the form of the hydrograph

to sediment transfer, erosion rates tend to be fairly low, unless the erosion regime is very seasonal and characterized by high rainfall intensities sometimes greater than 25 mm. hour^{-1}.

Contemporary processes in the temperate zone are distinctive in at least two ways. First, through the influence of inherited conditions: for example, a considerable amount of sediment storage occurs on slopes, in flood plains and along valley floors and much can be released only if the hydrological system is changed substantially. In many temperate areas floodplains store the sediment released when the vegetation, usually forest, over the catchment areas was cleared by human activity which could have begun some 4000 years ago. In the Severn and Wye catchments of the UK it was suggested (Brown, 1987) that although one might expect much of the Severn's floodplain sediment, rapidly deposited around 2000–3000 years ago, to have been remobilized by increased fluvial action due to increases in flood magnitude and frequency, the channelization and regulation of British lowland channels prevented this by lengthening the residence time of sediment in floodplain storage. The quantities of stored sediment in lowland basins such as that of the Severn showed how human activity altered relationships between the erosion, transport and storage processes, so that as much sediment went into storage as left the Severn and Wye basins during the Holocene. The sediment storage component is not spatially uniform in such areas and it is likely that the majority of hillslope erosion went into proximal colluvial storage.

Second, human impacts, particularly deforestation and land use change, affected flow and sediment generation and the wood in river channels. The great phase of deforestation occurred in central Europe for 200 years after AD 1050, having a significant influence upon the supply of wood and upon river dynamics. When temperate areas were forested coarse woody debris accumulations were much more common (Gregory, 1995), so that as a driver of landform development wood may have been overlooked in the interpretation of palaeo-landscape change along river corridors (Francis et al., 2008). Riparian woodland has persisted despite deforestation, and although now occupying less than 2% of stream channel length, woody debris accumulations can be responsible for half the total flow resistance, can account for 4% of the vertical drop in a channel long profile, and for 70% of the sediment stored in the stream channel. Originally debris dams could be very densely distributed, with average spacing as little as every 2.8 m, significantly affecting stream channel processes.

The range of human impacts is illustrated by their major effects on river channels (see Table 8.6). However, effects of human activity are sometimes complex and the chronological sequence may be difficult to unravel. Deforestation in the Mediterranean basin produced accelerated erosion which has been reflected in a series of stages recorded in the alluvial chronology (see below, p. 208). Increase in sediment supply as a consequence of human activity is exemplified over the catchment area of the Huang He (Yellow River) basin of China, responsible for the largest sediment load of any world river, originally carrying nearly 10% of all the sediment transported to the oceans from the surface of the globe (Walling, 1981) and averaging 1.69 billion tonnes annual sediment production 1919–1996 (Changming, 2000). As a result of the high sediment load, the lower reaches of the river channel silted, many sections of the river bed rose, and the 'suspended river' meant that the river bed in the lower reaches became generally 3–5 m higher than the floodplain behind the levees and in some sections the height above the floodplain is 10 m (Changming, 2000). The river ceased flowing in its lower reaches on 21 occasions in the period from 1972 to 1998, and in the 1990s the river lower reaches dried up every year for increasingly longer periods. Problems arising from the drying up of the river include increased flood hazard (as a result of the suspended channel), serious ecological effects in the delta area, increased saline intrusions, and water supply problems. The progressive reduction in water discharge together with reduction in suspended sediment load of the Yellow River (see Figure 8.5A) has been attributed to climate change and direct human impact, especially increasing water abstraction, sediment trapping by an increasing number of both large and small reservoirs and an extensive programme of soil and water conservation. It has therefore been suggested that 30% of the decrease in the sediment load of the Lower Yellow River could be attributed to decreased annual precipitation over the basin, with the remaining 70% attributable to human activities (Walling, 2008).

The complexity of the sequence of alluvial sediment accretion is illustrated by the agricultural Coon Creek basin, Wisconsin where erosion accelerated by

Table 8.6 EXAMPLES OF HUMAN IMPACTS ON RIVER CHANNELS
(FOR FURTHER DETAILS SEE GREGORY, 2006)

Impacts	Some characteristics and examples
Dams	At least 45,000 large dams (>15 m high or 5–15 m high of reservoir volume >3 million m³); more than 400,000 km² reservoir inundated behind the world's large dams; fragmentation of nearly all rivers in North America due to presence of dams, with 80,000 dams >1.83 m and including all structures may be 2.5 million in the USA; more than 1500 large- and medium-sized dams in India and 100 barrages on all major river systems; Australia has 447 large dams and several million farmdams which modify river flows.
Channelization	In the USA 26,550 km of major works gives a channelized density of 0.003 km/km². In England and Wales 8504 km of major or capital works gives a density of 0.06 km/km², and there is also a further 35,500 km of river which is maintained.
Channel modification	Average of <10% of length of Alpine rivers is in a semi-natural condition ranging from 2.5% in Germany, 4.9% in Switzerland, 9% in Italy and 18% in France.
River diversions	By the end of the thirteenth century the 1780 km Beijing–Hangzhou Grand Canal built to link five river basins and transfer water from Yangtze to North China Plain.
Water extraction	Approximately 11% of freshwater runoff in USA and Canada withdrawn for human use. Water abstraction of about one-fifth of total water resources with 87% for irrigation agriculture in China.

Many reasons for specific changes to river channels (see Gregory, 2006: Table 3) include:

Changes of	Reasons for change identified
Channel cross-section/point	Dam construction, weirs, diversion of flow, including mill leats, HEP, abstraction of flow, return flow, drains, outfalls, bridge crossings, culverts under roads and crossings.
Channel reach	Desnagging and clearing, grazing, embankments and levee (bank protection and stabilization, resectioning, dredging, channel straightening, cutoffs), clearance of riparian vegetation, tree clearance, beaver removal, sediment removal, mining gravel extraction, sediment addition, mining spoil, boat waves, bank erosion, invasion by exotic vegetation species, afforestation, conservation measures, restoration and allied techniques.
Drainage basin network	Drainage schemes, agricultural drains, irrigation networks, ditches, stormwater drains.
Spatial, drainage basin	Deforestation, grazing, fire, burning, agricultural ploughing, land use, conservation measures, afforestation, building construction, urbanization.

human impact was subsequently reduced by conservation measures so that sediment budgets from 1938–1975 were a fraction of those of 1853–1938 (Figure 8.5B), and in 1975–1993 were only about 6% of the rate of the 1930s (Trimble, 1999). Much of the recent sediment yield derives from storage loss, especially stream bank erosion, and the lower main valley continues to aggrade at about 6% of the rate of the 1920s and 1930s, although changes, including fish shelter structures and protected cut banks, now inhibit natural stream

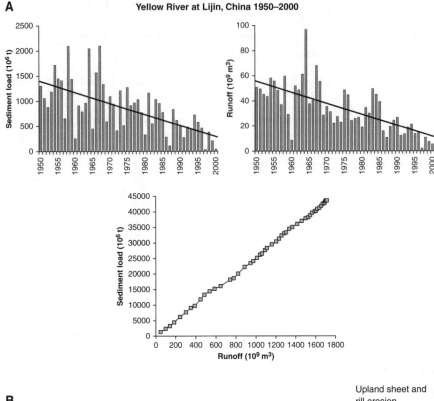

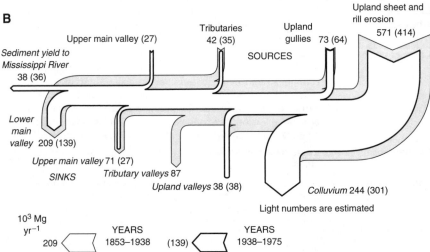

Figure 8.5 (A) Trends in the annual runoff and annual suspended sediment load for the Yellow River at Lijin, China (from Walling, 2008, Copyright © 2008 IAHS Press); (B) sediment storage for Coon Creek, Wisconsin (after Trimble, 1983)

Numbers in Figure 8.5B are annual averages of sediment supply in 10^3 Mg per year and show the importance of different sediment sources and also compare the storage before and after conservation measures (1853–1938) with the subsequent situation (1938–1975). The storage 1975–1993 is reduced significantly (Trimble, 1999) and the most recent situation is explained by Trimble (2008). See also Figure 2.3.

migration, bank erosion and downstream sediment transfer (Trimble, 2008). Human impacts inhibit natural processes in many areas: there are more than 75,000 dams in America with 137 of the large ones altering the flows of every large river, the geomorphological consequences including 32% larger low flow channels, 50% smaller high flow channels, 79% less active flood plain area, and 3.6 times more inactive flood plain area (Graf, 2006). Rivers in the central region of the USA have been most altered by flow regulation and changes in sediment supply associated with dams, diversions and channelization (Wohl, 2004) but there are now many cases where rivers are being restored to more natural conditions, for example by dam and weir removal (Graf, 2003) and other modifications of the river network can have substantial implications for parts of the temperate zone (e.g. Downs and Gregory, 2004), as introduced in Chapter 12 (p. 291).

8.3 SUBDIVISIONS OF THE TEMPERATE ZONE

Several areas can be distinguished (see Table 8.7), with the imprint of Quaternary conditions substantial in each zone, although earlier imprints were often obliterated or modified by the most recent. Deglaciation was often characterized by extensive areas of stagnant ice and by large discharges derived from melting ice or from drainage of glacial lakes. Reconstruction of the palaeohydrology has indicated how river systems developed after the last glacial maximum (LGM). Many river systems are now what Dury (1965) described as underfit, where meandering rivers occur in much larger meandering valleys (see Figure 8.6). Recovery from the LGM initially saw large discharges with braided rivers succeeded by meandering rivers which were subsequently characterized by aggradation as human activity released large amounts of sediment which accumulated in alluvial deposits along river valleys. The paucity of contemporary landforms together with lack of knowledge of contemporary processes instigated process monitoring and process investigations (e.g. Box 8.1). The sequence of Holocene change is outlined with particular reference to river processes and fluvial activity in the following subzones.

8.3.1 MARITIME ZONE

Western and central Europe have many landforms from Quaternary glaciations. Whereas the Scandinavian ice sheet covered the northern part of central Europe, including the site of present day Berlin during the LGM, southern zones were under continuous permafrost (Gregory and Benito, 2003). The morphology and landform distribution in areas glaciated by Scandinavian ice sheets are strongly controlled by the nature of the local rocks and sediments especially in the case of thrust moraines (Van der Wateren, 2003). Late glacial fluvial regimes were characterized by high flow seasonality, but subsequently as forest vegetation started to spread throughout the region during the Bolling interstadial (13,000–12,000 BP)

Table 8.7   BROAD SUBDIVISIONS OF THE TEMPERATE ZONE

Subzone	Climate and surface processes	Major hazards and problems which may occur
1 Maritime zone of middle latitudes (SE USA, NW & central Europe, China, SE Australia, New Zealand, Japan)	Maritime without severe winters. No large seasonal variations in temperature or humidity. Chemical erosion limited by moderate temperatures, some frost action but penetration rarely reaches bedrock. High angle slopes can be stable where still covered by forest. Many ancient deposits over landscape.	Accelerated erosion, avalanches, soil heave and collapse, floods, landslides on devegetated slopes, coastal erosion, flooding may increase downstream of vegetation changes.
2 Continental zone of middle latitudes (Midwest of USA, Russia)	Severe winters and seasonally distributed precipitation. Heavy showers and snowmelt can produce higher stream flow rates than in zone 1, mechanical processes more important as frost penetration is great and can reach bedrock. Chemical erosion limited by winter frost.	Drought, severe thunderstorms, hailstorms, snowstorms, landslides when vegetation removed. Downstream flooding increases when vegetation changed and other catchment characteristics altered.
3 Subdesert steppes and prairies (Great Plains, S Russia, Turkey, Mongolia)	Summer rainstorms, dry cold, severe winters. Transitional to temperate deserts with some frost action in winter. Wind action, occasional sheet wash and gullying.	Drought, tornadoes, soil erosion, deflation encouraged by removal of vegetation, gullying where land ploughed.
4 Mediterranean zone of middle latitudes (southern Europe, California, South Africa, SW Australia)	Seasonal precipitation, mild winters, warm/hot summers. Frost uncommon at low elevations. Alternation of wet and dry conditions can induce landslides. Seasonal streamflow regime can give high seasonal discharges which elevate coarse debris and rapid dissection and gullying where vegetation removed or degraded. Also intense seasonal precipitation events.	Soil erosion, floods, high spatial and temporal variability, high sediment yields along rivers, earthquakes, volcanic eruptions, landslides and sheet erosion where vegetation removed, increased flooding downstream, and gully development may occur.
5 Mountain areas (Rocky Mountains, Basin and Range Plateau, Alps, New Zealand)	Climatic conditions vary with altitude, include cryosphere conditions at highest levels (see p. 178–9)	Floods, shrinkage of glaciers, avalanches (see Table 7.6), other mass movements.

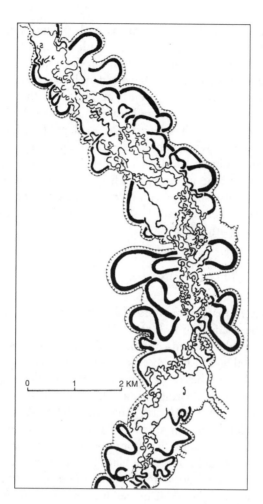

Figure 8.6 Palaeochannels on the Prosna valley floor, north-western Poland (after Rotnicki, 1991)

Several generations of channel are shown and the bankfull discharges through the largest channels were three times greater than those of the present channel.

fluvial systems shifted from braided to large palaeomeanders and subsequently to small palaeomeanders at the beginning of the Holocene.

The Vistula basin (198,000 km²), well documented from research led by Professor Leszek Starkel, drains from the Sudetes and Carpathians in the south through loess plateaus to areas of the North European plain that were glaciated by Scandinavian ice (Starkel, 1991). In the last 20,000 years the deglaciation of the northern part of the basin, the rise in sea level in the Baltic, and the extension of forest followed by deforestation from the Neolithic were significant controls. Evidence surviving includes the Vistula river terraces, the highest being the loess terrace (15–20 m high), the lowest the Holocene alluvial plain (8–15 m high), together with 2–5 alluvial fills associated with palaeochannels. Associated with the incision and decreasing sediment load, the change from braided to meandering channels occurred at c.13000 BP in the upper course and tributary valleys, and large meandering channels existed in the late glacial (see Figure 8.6), but became smaller in the Holocene. In the Holocene there were significant changes in flood frequency and straighter river channels but, during the last 2–3 millennia, aggradation has been more evident. In addition to tectonic influence, human impacts were visible at the end of the Atlantic phase – extensive aggradation occurred during late Roman times, building up of valley floors by channel deposits in mediaeval times, braiding of channels during the Little Ice Age, then regulation of Vistula upper and lower courses in the second part of the nineteenth century initiating a new phase of erosion, although the middle course preserved its braided pattern.

8.3.2 CONTINENTAL ZONE

In North America the maximum limits of ice sheets occurred 21,000–18,000 BP with a considerable readvance 15,000–14,000 BP. The Laurentide ice sheet created the spectacular glacial landscape of northern North America in which six land systems (see Table 8.8) have been recognized (Colgan et al., 2003). Large floods resulted from breaches of ice dams that blocked the large mid-continent drainage systems along the Cordilleran ice sheet margin and along the margins of the Laurentide ice sheet in North America. The Mississippi river drained almost the entire southern margin of the Laurentide ice sheet although the response is complicated because the enormous contributing area includes some western tributaries that did not receive glacial meltwater runoff (Knox, 1995), but four environmental episodes are recognized (Knox, 1999). The main river received great volumes of glacial meltwater and glacial sediment between 20,000 and 14,000 BP (Episode 1) when large bedload inputs to the system meant that the braided Mississippi profile was steeper than the present river and in Minnesota now stands as a terrace 45 m above the present floodplain, although 1000 km downstream it is only about 5 m above the modern floodplain. Downcutting after 14,000 BP (Episode 2) was facilitated by lower sediment loads and catastrophic discharges from rapid drainage of glacial lakes. For example proglacial Lake Wisconsin drained in about a week when the ice dam failed catastrophically with discharges between 3.6 and $5.3 \times 10^4 m^3.s^{-1}$ in the lower Wisconsin River – estimates which improve our understanding of the processes that have produced the morphology and behaviour of present-day upper Midwest river systems (Clayton and Knox, 2008). Episode 3 followed with Holocene floods when remobilization of colluvial and alluvial sediment from tributaries aggraded the Mississippi and buried much of the Holocene record. The fourth episode was initiated by the introduction of Euro-American agriculture in the early decades of the nineteenth century, with the replacement of prairie and forest vegetation with corn, wheat and hay crops and pasture, producing increased magnitudes of

Table 8.8   LAND SYSTEMS OF THE SOUTHERN LAURENTIDE ICE SHEET

Striking landforms in the area of the southern Laurentide ice sheet analysed in terms of glacial land systems are described in detail by Colgan et al., 2003.

Land system	Characteristics
Low relief till plains and low relief end moraines	Formed when ice at or near maximum extent and during recession, involves ice-marginal and subglacial zones.
Drumlins and high-relief hummocky end moraines	Reflects subglacial erosion and supraglacial erosion near an active ice margin.
Low relief aligned hummocks and ice thrust terrains	Reflects ice-lobe surges and widespread ice stagnation over large areas.
Bedrock-dominated glacial landscapes	High relief bedrock terrain draped by till, with glaciofluvial and glaciolacustrine sequences in most valleys, mostly reflecting glacial erosion of pre-existing high relief bedrock terrain advance to the last glacial maximum.
Glaciofluvial	Outwash and ice contact features.
Glaciolacustrine	Glacial lake plains.
Dissected terrain on older tills	Created before the last glacial maximum although has continued to develop to the present.

high frequency floods, five to six times greater than previous values in small tributary watersheds. Accelerated flooding was associated with greatly increased erosion and sedimentation rates so that long-term overbank sedimentation rates ranged from 1 to 3 cm.y^{-1} whereas the pre-agriculture long-term average Holocene rate was just 0.02 cm.y^{-1}. Since 1940 however improved land conservation practices have reduced the magnitudes and frequencies from the peak values attained in the late nineteenth and early twentieth centuries. In the Croatan area of North Carolina's lower coastal plain, culturally accelerated erosion, sedimentation and geomorphic change include historic alluvial sedimentation from 70 to more than 200 cm in the past 300 years, demonstrating how impacts of human activity can be difficult to recognize because they are intertwined with other phenomena (Phillips, 1997).

8.3.3 SUBDESERT STEPPES AND PRAIRIES

Although not extensively glaciated, their landscapes, including the Great Plains, southern Russia, and Turkey, often have rolling plains, and include the results of proglacial drainage, loess deposits and periglacial features. For example the Dnieper valley, Belarus (Kalicki and Sanko, 1998) which now has a snow regime with a long lasting high water level period (March–May) and some summer rainfall floods, has always flowed to the Black Sea in the Quaternary but at the LGM was affected by ice-dammed lakes at the ice front. During each period of global deglaciation a vast and interconnected system of proglacial lakes developed across Canada and the adjacent USA, similar to systems developed in Europe and Asia (Teller, 2003). The Lakes along the southern margin of the Laurentide ice sheet included glacial Lake Agassiz and a number of others, many drained catastrophically (see Teller, 1995). The glacial Great Lakes system to the east occupied previously existing basins and were ice-marginal for part of their history. Just as the central grasslands of the USA were changed significantly to become the most altered and endangered ecosystem of the US with 95% of the original tallgrass prairie replaced by other vegetation or land use (Wohl, 2004), so major changes elsewhere created new environmental systems.

8.3.4 MEDITERRANEAN ZONE

At the LGM the region in Europe was deglaciated except for the Pyrenees, Alps and Pindus mountains, although at that time the cooler climate, with lower total precipitation and development of steppe vegetation, favoured aggradation along many Mediterranean rivers (Macklin et al., 2002). In the western Mediterranean, fluvial activity occurred during detectable phases, and a general trend of river channel metamorphosis from braiding to meandering occurred throughout the Holocene in many areas of the northern Mediterranean region (Gregory and Benito, 2003). Soil erosion by water is now one of the most important land degradation processes in Mediterranean environments (Poesen and Hooke, 1997), slopes tend to be steep sometimes characterized by distinctive landforms such as flatirons, and there is extensive development of erosional plains. In the Mediterranean basin there is a well documented history of erosion and deposition against a background of human activity, inspiring vegetation change and more recent conservation measures so that this is therefore a classic area for geoarchaeology (e.g. Brown, 2003). Gully and channel erosion may be the dominant sediment sources in a variety of Mediterranean environments, albeit affected by land use changes. There are many local names for characteristic valleys and gullies including calanchi landscape (badlands) made up of many singular landforms, each of which (calanco) corresponds to a hydrographic unit (Morretti and Rodolfi, 2000). Extreme conditions associated with floods in the region, the variability of flows and of flood zones, the mobility of the channels and the high sediment loads create particular challenges for channel management, especially as trends in land use and channel management can exacerbate these problems. Hence a holistic approach to management of the fluvial system is recommended (Poesen and Hooke, 1997). In addition tectonic events include earthquakes which can trigger landslides and generally energize slope and fluvial processes.

In the Iberian peninsula fluvial archives, travertine and slope deposits provide sensitive resolution records of environmental changes during the last 170,000 years, showing that large scale tectonics triggered the general downcutting trend, whereas the main aggradation and incision phases occurred during periods of major sea level changes (Schulte et al., 2008). Some basins show evidence of flooding at particular periods and radiocarbon dating of slackwater flood deposits reveals six periods of flood clusters at dates of 10,750–10,240; 9550–9130; 4820–4440; 2865–2350; 960–790; and 520–290 cal BP (Benito et al., 2008).

Figure 8.7 Val d'Herens, south west Switzerland – the Arolla glacier occurs above the valley train

8.3.5 MOUNTAIN AREAS

Of the mountain zones (see Figure 6.1) mentioned in Chapter 7 (p. 178), the Alps (see Figure 8.7), the Carpathians, the Pindus Mountains and the Scandinavian mountains in Europe and the Rockies in North America together with the Alps of New Zealand have landforms influenced by mountain conditions together with those inherited from local glaciations. Large floods resulted from breaches of ice dams that blocked the large mid-continent drainage systems along the Cordilleran ice sheet margin, and include the cataclysmic outburst of glacial lake Missoula (see p. 138) which released 2100 km^3 of water at a peak discharge of c. $17 \times 10^6.m^3.s^{-1}$ producing a suite of erosional and depositional flood features over 40,000 km^2 of Montana, eastern Washington and Oregon.

8.4 CONCLUSION

The temperate zone embraces a range of environments, showing clear evidence of previous stages of development in some inherited landforms, all combined in its complexity. The major stages of development include the pre-Quaternary evidence of development of extensive planation surfaces, some fashioned under tropical conditions, and now evident in landscapes as accordant summits of upland areas such as the Appalachians, Wales or parts of Australia; the imprint of the succession of Quaternary stages with the legacy of glacial land systems in some areas but of periglacial or extraglacial systems beyond the former ice margins; complemented by the consequences of climate change in the Holocene and the imprints of human activity which have been so substantial in some areas. Each of these stages fostered a particular approach to geomorphology, including those described as denudation chronology, Quaternary geomorphology, or process geomorphology – such approaches were conducted separately and sometimes jealously guarded so that many authors have not thought it necessary to recognize temperate environments in the way that cold, arid or tropical environments have been the subject for co-ordinated study. However such separate investigations are being succeeded, or at least complemented by attempts to fuse previously separate research areas. Ways of bringing timeless and timebound investigations together include:

- Studies of contemporary glacial and periglacial processes assist interpretation of Quaternary evidence – many glacial and periglacial landforms were first investigated in Quaternary glacial system assemblages.
- Elaboration of the stages of Holocene development (e.g. Oldfield, 2005) – by linking investigations of Quaternary sediments to studies of present processes, the signals for recent change, for example in flood history, have been identified.
- Using knowledge from contemporary process domains to extrapolate back into the past as palaeohydrology has utilized an approach retrospective from current processes (e.g. Gregory et al., 1995) illustrated by interpreting palaeomeanders and meandering valleys. The inversion model (p. 190) enables interpretation of glacial erosion patterns.
- Refined understanding of long-term landscape development by cosmogenic dating of erosion rates (see Table 5.3).
- Extending investigations of present processes using new **dating methods** such as fallout of the artificial radionuclide ^{137}Cs resulting from atmospheric testing of nuclear weapons during the mid-twentieth century thus enabling quantitative analysis of changing rates of overbank sedimentation over 100 years (e.g. Walling and He, 1999).
- Recognizing the advantages of multidisciplinary investigations, for example in geoarchaeology (e.g. Brown, 1997). Reconstruction of environmental change in temperate landscapes depends upon an increasingly multidisciplinary approach, endeavouring to understand how contemporary process–response systems are at variance with those of the past as a consequence of human activity.
- Awareness of the need to envisage and design future environments requiring knowledge of the future of the past (e.g. Oldfield, 1987) in the context of restoration and management approaches considered in Chapter 12.

In the complex palimpsests that compose temperate landscapes we are able to detect more and more signals to reconstruct past land systems; these should be the foundation for determining sustainable land systems of the future.

BOX 8.1

PROFESSOR D.E. WALLING

Professor Des Walling appears as a traditional academic scientist but also an exceptional one: one website is headed by the statement 'What is it that has made Professor Walling so well known that the World beats a path to his door?' His internationally recognized research contributions were initiated by process investigations in small instrumented catchments, later extended to larger basins with results subsequently applied to periods of change, as he developed the potential for using fallout radionuclides (for example, ^{137}Cs and ^{210}Pb) to trace the mobilization and transport of fine sediment in drainage basins. His research over 40 years, underpinned by an intensive network of long-term measuring sites in the Exe basin, has coupled empirical investigations with laboratory analysis of river water quality and sediment properties to provide the basis for wide-ranging analysis of the behaviour of drainage basin sediment systems. He has illuminated understanding of the magnitude and frequency characteristics of fluvial sediment transport, and advanced the design of effective monitoring and load estimation strategies, with important results for sediment delivery processes and sediment budgeting, soil erosion and sediment sources, floodplain sedimentation, and sediment fingerprinting.

His published output is outstanding, as an author or editor of 30 books and of more than 460 scientific papers and contributions to edited volumes. More than 60 PhD students have been supervised by him at the University of Exeter where he graduated and obtained his PhD and was appointed to the academic staff, becoming Reardon Smith Professor of Geography in 1998. His work has been acclaimed by many awards not confined to one discipline: the Back Award of the Royal Geographical Society (1985) and its Victoria Medal (2000), the Vollenweider Award of Environment Canada (1990), the President's Prize of the British Hydrological Society (1995), the Linton Award of the British Society for Geomorphology (2007), the International Hydrology Prize awarded jointly by IAHS, UNESCO and WMO (2007), the Chien Ning Award of the World Association for Sedimentation and Erosion Research and the Chien Ning

(Continued)

(*Continued*)

Foundation (2007) and the prestigious Hydrologic Sciences Award of the American Geophysical Union (2008). He is a past President of the International Commission on Continental Erosion (ICCE) and the International Association of Sediment Water Science (IASWS); he is currently President of the World Association for Sediment and Erosion Research (WASER) and Honorary President of the International Commission on Continental Erosion.

Des Walling demonstrates the value of fundamental empirical field research linked to meticulous interpretation at national, international and global scales. Applications of his work to catchment management and the design of effective sediment management strategies led to involvement with sediment-related issues. He is an extremely approachable, modest and dedicated scientist who has built up a great fund of knowledge and expertise, much pertinent to temperate environments, and he has brought great credit to geomorphology by achieving multidisciplinary recognition.

FURTHER READING

Gregory, K.J. (2005) Temperate environments. In P.G. Fookes, E.M. Lee and G. Milligan (eds), *Geomorphology for Engineers*. Whittles Publishing, Dunbeath, Caithness, pp. 400–418.

The millennium ecosystem assessment at: http://www.millenniumassessment.org/en/Synthesis.aspx.

Turner, B.L., Clark, W.C., Kates, R.W., Richards, J.F., Mathews, J.T. and Meyer, W.B. (eds) (1990) *The Earth as Transformed by Human Action*. Cambridge University Press, Cambridge.

TOPICS

1 It is suggested that internet searches give relatively few results for the temperate zone. Can you improve on this?

2 Elaborate Table 8.1 with examples for particular locations.

3 Illustrate ways in which investigation of the temperate zone is becoming less timebound or timeless (see section 8.4) by examples of each strand or of other linking developments.

9

ARID ENVIRONMENTS

Arid environments can conjure up an impression of deserts with sand dunes and camels! However just 25% to 30% of desert regions are covered by sand deposits; arid environments technically also include large expanses of polar desert in the Arctic and Antarctic (see Chapter 7), excluded from this chapter which focuses upon temperate and tropical deserts. Other popular images of the desert are as the home of Western films, where bleached bones can lie by alkali pools, or as static and unchanging places – whereas in fact the Sahara desert was almost completely covered by grass and shrubs as recently as 5000–6000 years ago. During the Quaternary there were times when deserts were even more extensive (Stokes et al., 1997), so that it is suggested that, at the last glacial maximum, active dunefields occupied 50% of the land area between 30°N and 30°S compared with just 10% now (Sarnthein, 1978). At other times the reduced extent of deserts is evidenced by palaeolake shorelines, fluvial deposits and cave speleothems. Deserts, of arid and semi-arid drylands, cover about one-third of the land surface of the Earth, although globally they can cover between 20–50% of the land surface, according to definition and mapping criteria (Tooth, 2007).

Arising from the *International Year of Deserts and Desertification in 2006* a UNEP contribution (see http://www.unep.org/geo/GDOutlook/index.asp, accessed 12 January 2009) suggests three ways in which deserts have been defined. First, by the Aridity Index (AI), which is a numerical indicator for the degree of dryness of climate, calculated in several ways, one of which (UNEP, 1992) is the ratio between mean annual precipitation (P) and mean annual potential evapotranspiration (PET) as AI = P/PET (see Table 9.1). Aridity is greatest in the Saharan and Chilean–Peruvian deserts, and it is generally lower in the Thar and North American deserts. Second, bio-ecological criteria identify the ecoregions of the world, using desert vegetation characterized by xerophilous

Table 9.1 DRYLAND TYPES ACCORDING TO ARIDITY INDEX (AI)
DEFINED AS THE RATIO BETWEEN MEAN ANNUAL PRECIPITATION (P)
AND MEAN ANNUAL POTENTIAL EVAPOTRANSPIRATION
(PET): AI = P/PET (UNEP, 1992)

Classification	Aridity Index (AI)	Mean annual precipitation	Percent of global land area
Hyper-arid	AI < 0.05	<50 mm	7.5%
Arid	0.05 < AI < 0.20	50–250 mm	12.1%
Semi-arid	0.20 < AI < 0.50	250–500 mm	17.7%
Dry subhumid	0.50 < AI < 0.65		9.9%

life-forms and desert-adapted plants. Third, a land-cover index from AVHRR satellite imagery can classify land-cover categories including uniform regions with extremely low vegetation cover as the Desert Biome.

How many deserts occur and where are they? Major deserts of the world, many known by characteristic names, are listed in Table 9.2, which also indicates landscape characteristics. The range of desert types explains why different classifications have been used. A climate classification, according to the amount of precipitation received, used by Peveril Meigs (1953) to distinguish extremely arid (at least 12 consecutive months without rainfall), arid (less than 250 mm annual rainfall) and semi-arid (250–500 mm including grasslands generally referred to as steppes), was used for the UNESCO (1979) classification of hyper-arid and arid (see Table 9.1). Further subdivision distinguishes deserts with cold winters (e.g. central Asia, parts of North America), coastal deserts where fog can be prevalent (Atacama, Chile, Oman, Namib) and hot deserts (Sahara, Australian) where frosts are rare. Further variety is introduced by rain shadow deserts (Gobi) and monsoon deserts (Rajasthan Desert of India, Thar Desert of Pakistan). A structural classification differentiates those underlain by stable cratons of ancient igneous and metamorphic rock denuded to become platforms of low relief, subsequently overlain by thin mantles of little deformed sedimentary strata (Sahara, Arabian, Kalahari, Namib, central Australia), from basin and range deserts where recent earth movements have produced fold mountains or horst and graben topography (west of USA and most Asian deserts). Ecological classifications distinguish hot and dry (e.g. Chihuahuan, Sonoran, Mojave, Australian); semi-arid (e.g. sage brush of the Great Basin); and coastal (e.g. Atacama) deserts. In addition to polar deserts (see Chapter 7) classifications can also recognize palaeodeserts (e.g. the Nebraska Sand Hills – 57,000 km^2 of inactive dunes, some up to 120 m high, now stabilized by vegetation) and extraterrestrial deserts (Mars and Venus are planets on which aeolian features have been identified). Figure 9.1 indicates the UNEP (1992) world distribution of drylands, including dry subhumid areas, thus emphasizing the transitional nature of many zones (Chapter 6, p. 160).

Whichever classification is employed, there is a gradation from sand deserts (erg) through stony deserts (reg), rocky deserts (hammada) to those characteristic of basin and range deserts.

Table 9.2 MAJOR WORLD DESERTS AND THEIR CHARACTERISTICS

Areas are approximate, reflecting transitions indicated in Table 9.1. Some indications of early exploration of specific deserts are given in Graf (1988: 4–24) and a good review of their characteristics is given in Thomas (1997: 467–573).

Desert	Area km²	Characteristics
Africa		
Sahara	9,100,000	Includes sandstone plateau, Tibesti and Hoggar massifs, broad closed basins including Chad, deflational regs, occasional yardangs, and ergs and sand seas in northern Sahara.
Kalahari	260,000	Gently undulating sand plain. Dune desert and sand seas in the arid south-west, parallel lines of dunes, concentrated in the west. Four major national parks and many conservation areas help protect the delicate ecosystems.
Namib	135,000	One of the oldest deserts, up to 55 million years old, five major dunefields separated by gravel plains with inselbergs, sand dunes in the south up to 305 m high in dune seas.
Eurasia		
Gobi	1,300,000	Cold rain shadow desert (temperatures can reach 40°C in summer and – 40°C in winter) blocked by Himalayas. On plateau 900–1500 m above sea level, bare rock and gravel plains, sandy desert at lower levels in west, as well as alkaline basins. Expanding due to desertification. Great Gobi National Park established in 1975 and UNESCO designated the Great Gobi as Biosphere reserve in 1991.
Rub'-al-Khali	650,000	One of the largest sand deserts with dunes up to 350 m high. Very dry, no significant perennial rivers.
Kara Kum and	350,000	Two great ergs/sand seas comprise the Turkestan desert.
Kyzyl Kum	300,000	Kyzyl Kum extensive plains as well as sand dunes have areas covered with clay coatings (takir), some large closed basins, agriculture along rivers and round oases.
Takla Makan	270,000	One of largest sandy deserts, surrounded by mountains fronted by alluvial fans (see Figure 9.4, Colour plate 8), more pluvial in Quaternary.
Dasht-e-Kavir	260,000	On Iranian plateau, almost flat pans and playas with high salt content, known as Great Salt desert. Dasht-e-Lut to south is mainly sand and desert pavement.
Syrian	260,000	Plains on major basalt lava flows, low hills, much desert pavement, many fluvial deposits associated with wadi systems.
Thar or Great Indian desert	214,000	Vast tract of sand hills with closed basins and salt with deposits, has a short, intense monsoon season with dry winds. Crossed by snow fed Indus and tributaries. Parts may be Holocene, some relict dunes in south east, and population considerable in Rajasthan desert area.

(Continued)

Table 9.2   *(Continued)*

Desert	Area km²	Characteristics
Australasia		
Great Victoria	647,000	Sandhills, grasslands and salt lakes, with areas of stabilized sands.
Great Sandy	400,000	Higher rainfall than most deserts, some stabilized sands.
Simpson	145,000	Sand sea, has the longest parallel sand dunes in the world, extremely hot, underlain by Great Artesian basin. Dunes affected by glacial cycles, sand movement greater in the past at times of last glaciations. Lake Eyre basin.
North America		
Great Basin	492,000	Elevated, cold winters, valleys and mountain slopes with extensive areas of scrub and grassland, just 1% of the area is sand dunes. Lakes Bonneville and Lahontan were Quaternary lakes and about 100 basins contained pluvial lakes during late Quaternary.
North Mexico, Chihuahuan Desert	450,000	At high elevation (700–1600 m), hard winter freezes are common. Many closed basins with endoreic drainage (bolsons) with alluvial and playa deposits. Many species of low shrubs, leaf succulents and small cacti.
Sonoran	310,000	One of the hottest deserts in North America, relatively lush with variety of plants including saguaro (*Cereus giganteus*) cactus. Block-faulted mountain ranges, pediments, alluvial plains. Sonoran National monument (c.2000 km²) designated in 2001. Some irrigated cultivated areas.
Mojave	65,000	Fault block mountains, endoreic basins, landscape development since Miocene, evidence of former pluvial conditions. Winter rainy season, hard freezes common.
South America		
Patagonian	673,000	Rain shadow in lee of Andes, cold winter (temperature rarely exceeds 12°C and averages 3°C), stony desert, piedmont gravel plains and eastward-sloping plateaus with sandstone canyons, fluvial deposits from the Andes, many endoreic basins.
Atacama	140,000	The driest desert (possibly no significant precipitation between 1570 and 1971), rain shadow, composed of salt encrusted playas (salars), sand and lava flows, few areas of sand dunes, cultivation in far north enabled by drip irrigation systems supplied by groundwater.

9.1 SAND DESERTS AND SAND SEAS

Wind and water often work in partnership because water is required for weathering and can also transport silts and sands that can be moulded or transported by aeolian processes. Sand deserts cover 25% of desert surfaces, representing 5% of the

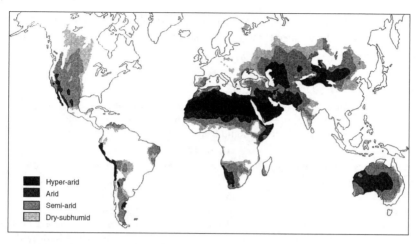

Figure 9.1   Global distribution of drylands (after World Atlas of Desertification, UNEP, 1992)
Reprinted with permission.

global land surface, but up to 25% of Africa. *Erg*, originally derived from an Arabic word meaning dunefield, is now used for desert areas that extend 125 km² with sand covering more than 20% of the surface, whereas smaller areas are known as dunefields. The largest sand sea is the Rub' al Khali in Arabia with others in the north Sahara, Kara Kum, Taklimakan and Kalahari. When accumulated in subsiding structural and topographic basins, sand seas can be up to 1000 m thick, but elsewhere, such as the ergs of linear dunes in the Simpson and Great Sandy Deserts of Australia, they may just be individual dunes superposed on the alluvial plain. Of the order of one million years can be required to build ergs with very large dunes such as those on the Arabian Peninsula (Wilson, 1973). Sand seas and dunefields generally occur in areas down-wind of copious sources of dry, loose sediment, such as dry riverbeds and deltas, floodplains, glacial outwash plains, dry lakes and beaches, in areas too dry to support extensive vegetative cover, where there is wind energy (erosivity) to entrain and transport the sediment. Particular sand seas often have dunes of a single type, so that ergs of linear dunes, crescentic dunes, star dunes, or parabolic dunes tend to have broadly consistent orientations and sizes. Large sand seas may contain many dune types and the corridors between the dunes can have hard rock or pebbly surfaces in which dry lakes have accumulated, retained by the disruption of surface drainage by the dunes. Dunes range from less than 1 m to over 200 m high, and McKee (1979) distinguished simple dunes, compound dunes that include several of the same type merged, and complex that include several dune types. Types of dune (see Table 9.3), primarily resulting from differences in formative wind regime, are still the basis for controversy about their origin, development and rate of movement. Sediment type and sediment supply affect dune types and when vegetation exerts a significant influence; nebkhas are small dunes formed around plants, parabolic dunes (hairpin dunes) are crescentic with horns pointing away from the direction of dune movement, developed where vegetation provides resistance to sand movement as in the Thar desert, and blow-outs may form as depressions within a dune complex. Phylogenetic dunes, which can form with vegetation cover up to about 35% of the surface, can also occur in dunefields of coastal areas.

Table 9.3 TYPES OF SAND DUNE

Mobile dunes can include some structures which are compound and composed of more than one type. Although convenient to classify, other varieties can be recognized and dunes may occur in a continuum ranging from transverse, through linear to star dunes.

Type	Morphology	Typical dimensions	Characteristics
RIPPLES	Small scale ridge of sand.	0.1–5 cm high, 0.005–10 cm apart.	Impact ripples may result from saltation.
DUNES	Mounds or ridges of wind-blown sand.	0.1–15 m high, 3–600 m apart.	Created as wave patterns developed where air flow interacts with ground surface creating turbulence.
Linear	Long slightly sinuous ridges generally symmetrical in cross-section.	Usually c.20 m high and up to 1 km apart. Can be up to 200 m high.	Most common form of desert dune, occurs in extensive dunefields, forms parallel to sand drift direction. Most common on desert plains, e.g. Namib, Australia.
Seif dunes	Sharp crested large, elongated dune or chain of sand dunes.	Up to 2 km long in North Africa.	Oriented parallel to the prevailing wind movement, can be c. 10 m.yr^{-1} (Thomas, 1997: 401)
Star dunes	A large, isolated sand dune with a base in plan view resembling a star, has 3 or more sharp-crested ridges radiating outwards from a central peak.	Can be 100 m above the surrounding plain, sometimes up to 300–400 m high.	Develop in multidirectional, complex wind regimes, tend to occur at the depositional centres of sand seas or where air flow modified by topographic barriers. Include greater volumes of sand than other dune types, movement up to 110–200 m.yr^{-1} have been reported (Thomas 1997: 402).
Rhourd	Pyramid dune similar to a star dune.	Can be 150 m high and 1 to 2 km across.	A pyramid-shaped sand dune, formed by the intersection of other dunes, developed where two draa chains cross.
Transverse or crescentic	Where there is ample sand, barchans may join up to form transverse dunes.		Aligned normal to sand drift direction. Develop where air has wave motion downwind from obstacle, ridge or other disturbing influence. Well developed in the Chinese deserts including the Taklimakan.

Table 9.3 *(Continued)*

Type	Morphology	Typical dimensions	Characteristics
Akle	Overlapping networks of bulbous advancing ridges and curved re-entrants/ network of sinuous ridges.		From French *ecaille* (fish scale).
Zibar	Low transverse regular ridges without slip faces, no distinct dune forms.	Up to 5 m high, spacing can range from 50–400 m with local relief < 10 m.	Composed of coarse sand, lacking slip face development, occur on the edges and between larger dunes in sand seas and cover quite large parts of the Sahara and Arabia, commonly occurs on sand sheets, in interdune areas, or in corridors between larger dunes.
Barchans	A crescent-shaped dune with tips extending leeward (downwind), making this side concave and the windward (upwind) side convex. Tend to be arranged in chains in the dominant wind direction.	Height typically up to 15 m, total length c. 200 m, but can be much greater.	Can move 50 m per year. Tend to form either where transport rates are high or where sand supply limited. The 'barchanoid' type consists of three intergrading subtypes: the barchan, the barchanoid ridge, and the transverse dune (Bullard and Nash, 2000). Maximum height usually one tenth of width.
Dome dunes	Low circular or oval mound lacking a slip face.		Coarse sand, no slip faces. May result where barchans are re-moulded.
MEGADUNES/ DRAAS	Very large whaleback dunes which can occur around hill massifs.	Height of 20–400 m and wavelength of up to 5500 m.	May be interrupted by large nodal star dunes (rhourds). Move slowly and may be compound or complex dune forms. The Draa Malichigdane in Mauritania extends for 100 km downwind of its parent hill.

Thomas and Wiggs (2008) recognize that since the work of Bagnold (see Box 4.2), two scales of aeolian research have focused on small-scale investigations and on landform/landscape analyses. However, against the background of field studies, wind tunnel studies and mathematical modelling, new complex systems models suggest the need to return to a larger-scale perspective where dunes are not just considered as individual elements, but as integral parts of a dunefield (Wiggs, 2001). Desert winds can carry more sediment ($m^3.yr^{-1}$) than any other geomorphological agent, with hundreds of km^2 of dunes progressing at 15 $m.yr^{-1}$ being not unusual, and the Sahara probably produces between 130 and 1400 million tonnes of dust per year (see Goudie, 2006: table 5.2) while the river Niger carries a mere 15 million $tonnes.yr^{-1}$ of sediment (Cooke et al., 1993). Although it is difficult to generalize about measurements of sand transport, in high energy environments sand drift can be 25–40 $m^3.m^{-1}$ width. yr^{-1}, and 15–25 in intermediate energy environments (Thomas, 1997: 378). Although the location and general extent of individual sand seas and dunefields are depicted in many publications, a global **database** of geographically accurate maps of individual desert and other inland dunefields and sand seas, constructed using GIS methods (see http://www.dees.dri.edu/Projects/Dune_Atlas), is providing a significant advance (Lancaster, 2008). Recent studies using OSL dating have revealed how regional and multi-regional chronologies of periods of dune formation demonstrate responses to Quaternary changes in climate and sea level, manifested by changes in sediment supply and availability, and dune mobility. Anchored dunes, attached to objects that cannot be moved by wind, are distinguished from stabilized dunes, which have been immobilized by cementation or vegetation.

Recent attention given to aeolian dust (McTainsh and Strong, 2007) has shown that dust transport systems operate on very large spatial and temporal scales, involving much larger quantities of sediment than previously realized. Goudie (2008; also see Box 9.1) has demonstrated the importance of deflation, especially in hyper-arid areas with centripetal drainage; the existence of deflation hot spots such as the Bodélé Depression; and that wind activity was greater during glacial phases than at present as indicated from analysis of ice and ocean cores and loess deposits. In addition, desertification, although not used as a term until 1949 by Aubréville (Plit, 2006), was recognized in the 1930s when parts of the Great Plains in the United States turned into the 'Dust Bowl' as a result of overgrazing, drought and poor farming practices.

Beyond the sand seas, wind can be very active, especially in hyper-arid areas, where there is deep stone-free or clayey sediment as in ancient Quaternary or Holocene deposits, bedrock outcrops or old lake beds that wind can erode unimpeded. Wind abrasion eroding materials and deflation entraining sediment can result in ventifacts, which are rocks shaped by aeolian abrasion; yardangs, which are aerodynamic stream-lined, wind-abraded ridges with their long axes parallel to the wind, varying in height from several metres to 200 m and in length up to several kilometres, with their morphology often betraying the character of the underlying rock; pans and wind-eroded basins; and plains of deflation which may include ridges topped by meandering or braided river deposits now elevated because the marginal finer deposits have been deflated (Cooke et al., 1993: 290–306). Such features testify to ways in which climates affecting deserts have changed over the Quaternary, prompting intriguing names such as suspendritic drainage or perched wadis (Cooke et al., 1993: 305).

9.2 ROCKY DESERTS AND BASIN AND RANGE ENVIRONMENTS

Many types of desert landscape exist other than sand seas, ranging from those within sand deserts, such as corridors between dune ridges, open plains, or enclosed basins between dune complexes, to the landforms that occur on shield deserts and in mountain and basin environments. Where sand seas are absent, shield deserts have many features inherited from former more pluvial conditions, including river gorges, duricrusts, remnants of deep weathering and several landforms associated with seasonal tropical landscapes (p. 242). The crystalline basement rocks underlying the shield deserts of Australia are exposed in inselbergs, although in the central Sahara, in the Ahaggar and Tibesti mountains, basement rocks are complemented by surface outcrops of more recent volcanic rocks. Mountain-and-basin deserts occur in areas affected by recent tectonics where up-faulted mountains alternate with down-faulted basins, exemplified by the American deserts and the deserts of central Asia. Distinctive characteristics of scenery in these areas are often graphically described by local names (arroyo, wadis, buttes, mesas, badlands, playa, bajada). However a geographical bias in our knowledge and in ongoing research persists, for example because our understanding of the rivers in the North American, Middle Eastern and Australian arid zones greatly exceeds that of many other arid parts of the world, such as in Africa, South America, and Asia (Tooth, 2008).

Absence of significant vegetation cover often endows a striking landscape appearance for which an idealized profile (Lee and Fookes, 2005) is reminiscent of the nine-unit hypothetical land surface model (see Figure 4.2). The slope components (see Figure 9.2) are in four main zones (see Table 9.4): backslopes and uplands (I), footslopes and fans (II), plains (III) and enclosed basins (IV). This sequence can be repeated several times across desert areas so that the backslopes may be the lower member of another sequence. Uplands or desert mountains make up about 40% of the surface area of drylands in the Sahara, Libyan desert, Arabia and the south west of the USA (Fookes, 1976). The plain and backslope can be capped by duricrust (see p. 239) providing a protective capping for prominent features. Desert surfaces consisting of bedrock or fragmented material, resulting either from weathering or debris transport and deposition, may include hamada, reg and desert pavement (see Table 9.4). Desert pavement of abundant stone fragments of pebble size, with stones often embedded in loamy soil material, can result from several processes acting together, including deflation and sheet flow removing fines, and wetting and drying and heating and cooling bringing larger stones to the surface, with additions of dust very important in explaining the sedimentology of these features. Coarser fragments may be covered by desert varnish which is a dark coating several millimetres thick, highly enriched in iron, manganese and clay minerals, sometimes very old, indicating the remarkable stability of

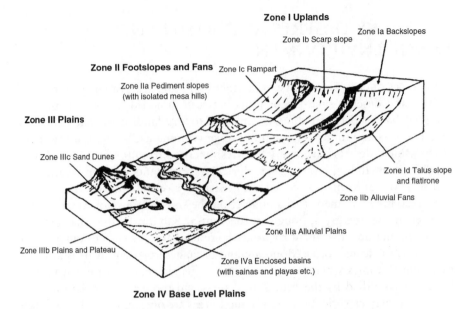

Figure 9.2   Slope components (based on Lee and Fookes, 2005) –
the four zones are explained in the text

Reprinted with permission from *Geomorphology for Engineers*, P.G. Fookes,
E.M. Lee and G. Miligan (eds), Whittles Publishing.

some desert surfaces: several theories for its origin include accumulation by
micro-colonial lichens and bacteria. Leaching of cations from desert varnish
has now been shown to be useful as a novel dating technique. Upland margins
are composed of free faces and scarp slopes together with talus slopes and
ramparts. Bare rock exposures, especially in the free faces, are more common
in dryland areas than elsewhere because the rate of debris production by
weathering is usually less than the rate of removal. However the morphology
of the scarp slope or free face depends upon the lithology of the rock beneath,
with steep cliffs frequent below a duricrust caprock or on igneous rocks, sandstones and limestones (see Figure 9.3), with slope angles characteristic of particular lithologies. In general the stronger the caprock the larger the free face
because more undermining is required before failure. Conversely the thinner
the bedding and the closer the jointing the smaller the face and the cliff fails
more easily. Rocks dipping away from the caprock can accentuate slumping
and sliding of the caprock, and thin caprocks may give low cliffs which retreat
relatively rapidly. If impermeable rocks overlay permeable ones, seepage erosion can give cave formation and collapse of the rock above. Considerable variations in rates of retreat of cliffs can be up to one or two mm.yr^{-1}, and show
significant changes over periods of time, with a decline in hillslope angle being
consistent with the Davisian model and parallel recession of the slope with
extending pediments giving a pediplain (p. 85).

Table 9.4 COMPONENTS OF DRYLAND SLOPE PROFILES (ALSO SEE FIGURE 9.2)

Zone	Component	Characteristics
I Uplands and backslopes – low relief plain at the top of cliffs or scarp slopes.	Hamada	Flat bare rock surface, may have cobbles and boulders reflecting character of underlying rock or that of adjacent scarp.
	Reg	Stony desert or sheets of gravel, usually deposited by streamflow or sheet floods or resulting from deflation of fines.
	Desert pavement	Accumulation of stones and pebbles, may be tightly packed crust or cemented by saline deposits, overlying a soil in which there are fewer pebbles.
	Free face, scarp slope	Steepest slope element is often exposed bedrock.
	Talus slope	Scree or apron of material derived from free face above.
	Rampart	Inclined rock surface at the base of the scarp or free face, often mantled with rockfall debris.
	Badlands	Intricately dissected, high drainage densities, including gullies and sharp-edged ridges, usually developed by fluvial erosion in weak rocks such as mudstones, where there is sufficient storm rainfall but insufficient soil and vegetation cover.
II Footslopes – terrain surrounding uplands composed of pediments and alluvial fans.	Pediment	Relatively gently sloping erosional surface of low relief cut in bedrock, may be partly covered by a thin veneer of alluvium which may contain buried palaeochannels.
	Alluvial fan	Fan-shaped deposit, surface forms a segment of a cone radiating downslope from the point where the stream leaves the source area, usually has distributaries.
	Bajada	Coalesced fans form a broad, gently inclined, unconsolidated depositional piedmont, often encircling desert basins.
III Plains – zone of deposition of sediments, alluvial and aeolian, derived from uplands and footslopes.	Alluvial plains	Deposits and morphology associated with alluvial channels, often multithread.
	Entrenched channels, arroyos	Trench with roughly rectangular cross-section in valley floor alluvium with major stream channel.
	Gullies	Deep V-shaped channels on valley sides or valley floors usually with ephemeral streams, initiated by climate change or human activity, generally is an obstacle to wheeled vehicles and too deep (> 0.5 m) to be obliterated by ordinary tillage.
	Sand dunes	Thin sheets or dunes may occur on plain.

(Continued)

Table 9.4 *(Continued)*

Zone	Component	Characteristics
IV Enclosed basins – almost flat, vegetation-free ephemeral lakes (playas, salar or pans).	Playa	Usually dry and nearly level lake plain that occupies the lowest parts of closed depressions, periodically inundated by water after storm events.

Figure 9.3 Desert landscape, Zabriskie Point, Death Valley
Compare with zones I and II in Figure 9.2. Tertiary lake sediments have been eroded to badland topography and the dark caprock on top of some badland ridges is late Tertiary volcanic lava and explains the height of Manly Beacon on the right.
Source: David Miller, USGS

Below the free face the talus slope accumulates as fragments fall from the cliff above, possibly increasing progressively to cover the free face. The rampart, with a slope angle characteristic for the debris accumulating and where removal just balances production, is sometimes referred to as a Richter slope, defined as a straight rock slope unit with an angle usually 32°–36° at the maximum for the stability of the thin talus cover above the bedrock. Below the free face mass movement processes can include debris flows and landslides, and some areas have intricately dissected landscapes formed by dense drainage networks eroding poorly consolidated sediments such as mudstones that lack protective soil and vegetation cover. These are termed badlands – a term probably derived from French explorers' reaction to areas in North America's western plains as *mauvaises terres à traverser*. Although not restricted to deserts, badlands form where weak unconsolidated materials are subject to periodic high rainfall intensity and runoff, and are typically steep and rounded slopes, intensely dissected by rills and gullies, sometimes with hoodoos (see p. 74). Erosion rates of badlands vary but 3–20 mm.yr^{-1} in Chaco canyon, New Mexico are nearly two orders of magnitude greater than they were 5000 years ago (Wells et al., 1983).

Footslopes (see Table 9.4) typically have gently sloping rock pediments and/or alluvial fans. Pediments are low angle surfaces cut across a range of hard rock lithologies, acting as a transfer zone for erosion debris from mountains to basin storage sites, that may be thinly veneered with alluvium and/or with an in-situ weathering mantle. Their angle can be 11° near the scarp foot, where the sharp piedmont junction angle is sometimes marked by a 'scarp foot nick', but as little as 0.2° at their lower end. Regolith thickness can be 2–4 m (Strudley and Murray, 2007) and they are often associated with alluvial fans at the base of mountain fronts. They may be diversified by duricrusts and irregular topography punctuated by inselbergs and networks of bedrock-lined and alluvial wadis. In addition to extending from the base of mountains as apron pediments covering a large portion of the landscape, terrace pediments can develop along major river courses. Pediment domes occur where there are no intervening mountain masses as in central Australia and South Africa (Graf, 1988: 179). Pediment origin is not undisputed but, since McGee in 1897 suggested that they were planed off by sheet floods in the American South-West, they have also been attributed to some combination of lateral planation by mountain front streams and rill erosion of weathered materials, progressive stripping of tropical deeply weathered material, and parallel recession of the mountain front leaving the ever expanding pediment at the base.

Alluvial fans that appear as a segment of a cone of poorly sorted coarse sediment (boulders, cobbles, gravels and some sands) radiating downslope from the point where the stream leaves the source area, also serve as a transfer system for materials eroded from mountain masses being transported to deposition in basins. They vary in size from a few metres in length to more than 20 km (see Figure 9.4, Colour plate 8), many large fans are thicker than 300 m, their debris decreases in size down fan and they are especially evident in tectonic areas. Two groups of processes are responsible: first, debris flows transporting boulders or cobbles or muds supplied by landslide processes including rockfalls, rock avalanches, and slides; second, sheet floods produced by short duration intense rainstorms depositing coarse sediments but with some fines. Therefore there are two types of fan: firstly, those developed by debris flows, from catchments underlain by bedrock that weathers to produce boulders, cobbles, silt but little sand, which may have effects on secondary processes including gullies, rills and boulder lags on the fan surface; secondly, those developed by flash floods, which have a prominent sand skirt around their distal margin, a smooth surface with average slope angles 2–8°, often being associated with catchments that are underlain by fractured or jointed granite, gneiss or friable sandstones. Fans frequently comprise complex sequences of coalescing or segmented sections, often of several dates. Alluvial fans occur in many areas, Surrell (1841) possibly being the first to describe them in the Alps, and Drew (1873) the first to apply the term alluvial fan to cone-shaped deposits of stream sediments at the mountain-basin boundary where streams issue from narrow valleys. In the hyper-arid conditions of the Atacama desert cosmogenic nuclides have been used to estimate limits on the surface exposure duration and erosion rates of alluvial fans and bedrock surfaces (Nishiizumi et al., 2005) demonstrating that maximum erosion rates for cobbles on alluvial surfaces are uniformly < 0.1 m.Ma^{-1}, so that some landforms in the Atacama Desert are remarkably stable, still bearing considerable resemblance to how they appeared in the Miocene. Where many adjacent fans coalesce, particularly if there is ample material or a prolonged period of deposition, the alluvial apron or depositional piedmont that can form along the mountain front is termed a bajada (Blackwelder, 1931).

Plains, below the pediments and alluvial fans, are the zone of net deposition of sediments supplied from erosion of the uplands and footslopes comprising alluvial and wind blown sediments. Where alluvial they can have a complex network of braided stream channels with extensive floodplains. In some cases erosion of the valley floor alluvial deposits, by channel erosion, results in the development of incised channels and the conversion of the original flood plain into terrace remnants. Such entrenched channels (Graf, 1988: 218) occur in all dryland areas but in the American South-West arroyo refers to a trench with a roughly rectangular cross-section, excavated in valley bottom alluvium with a major stream channel on the floor of the trench. Arroyos may be accompanied by discontinuous and continuous gullies as a result of channel erosion due to land management, climate change or internal (intrinsic) geomorphic adjustments. Some segments of plains may be surface pavement of closely packed stones (reg or sarir in the Sahara, gobi in Asia, gibber plain in Australia) often in close association with boulder-strewn rock outcrops with a scatter of stones (hamada). Basins include broad, low-lying depressions bounded by steep scarps or terrace sequences such as the Quattara, Siwa and Jaghbub depressions in the eastern Sahara. Although they have been thought of as the result of extensive and prolonged wind deflation a variety of factors probably contributed, so that the Quattara depression (down to 134 m below sea level) may have been excavated as a stream valley, subsequently modified by solution weathering of adjacent limestone plateaux, and further extended by mass movement, deflation and surface water erosion (Graf, 1988). Playas or pans, denoted by a variety of local names including sebkha in North Africa and Arabia, are closed depressions which occupy regional or topographic lows, lack surface outflows, are generally above the water table but ephemerally occupied by surface waters, usually very flat, where evaporation is greater than inputs, and are largely vegetation free (Shaw and Thomas, 1997). They are net accumulation zones of both fine-grained and non-clastic sediments receiving inputs from aeolian and fluvial processes, capillary rise or throughflow. Several types (Lee and Fookes, 2005:

436) include some with saline, soft, permeable and loosely compacted surfaces possessing microrelief up to 15 cm, others with thin salt crusts of halite or gypsum of thicker salt pavements with surfaces that can be wet, soft and sticky, and sometimes mounds developed around saline springs rising up to 30 m high.

Several themes have prevailed in the interpretation of the range of dryland landforms. Whereas wind was originally thought to have been very influential, the effects of water, including the significance of high magnitude low frequency storm events, have subsequently been given greater attention. Water flow occurs occasionally along networks of ephemeral, intermittent, and occasional perennial streams (see Chapter 4, p. 86) which are distinctive in several ways. First, the ephemeral channels are very extensive whereas intermittent are restricted and perennial seldom occur except where they are allogenically maintained by flow from elsewhere as in the case of the Colorado and the Indus; second, the absence of vegetation enables runoff contributions by overland flow or sheet flow as well as pipeflow and throughflow during storms; and third, streams lose water through the channel bed, described as transmission losses or effluent flow. Flood events, although rare, can be very effective, so that Schick (1977) estimated that for a 10 year period in the Nahal Yael catchment, Sinai, 99% of the erosive work occurred in just 5 days with only 7 runoff events responsible for 20% of mean annual runoff. However up to 60–70% of runoff from small catchments may not reach a main channel because of infiltration losses, the lag time from peak precipitation to peak discharge can be extremely short, and very large sediment yields are facilitated by the widespread availability of coarse sand and loose sediment on slopes and by the paucity of vegetation. Often described as flash floods, flows can be augmented if temporary dams, created across main stream channels by debris flows or floods from tributaries, are later breached during storms producing flood waves or 'walls of water' downstream. The 12 largest floods ever recorded in the USA have all occurred in arid or semi-arid areas (Costa, 1987) and such events can dramatically alter channel morphology.

Further themes include the way in which processes and landforms are very evidently the result of energy or force derived from inputs of water from the atmosphere interacting with landscape resistance based upon earth structure and materials (Graf, 1988), as well as the way in which some landforms are inherited from more pluvial conditions, and have evolved over considerable periods of time, so that some landscapes entombed by sand have 'erosional paralysis' (Oberlander, 1997). Thus the interrelated Quaternary fluvial and aeolian activity related to climate change on Cooper Creek in the Lake Eyre basin in southwestern Queensland (Maroulis et al., 2007) shows how the extensive muddy floodplain is characterized by buried sandy palaeochannels, now almost entirely invisible but stratigraphically connected to source-bordering dunes that emerge as distinctive sandy islands through the floodplain surface. Whereas the channels once determined the location of source-bordering dunes, in an interesting role reversal the remnant dunes now determine the position of many contemporary flood-channels and waterholes by deflection and confinement of overbank flows. A recent trend has been to view arid landscapes as the integrated product of spatial and temporal interactions both within *and* between different subcomponents of the geomorphological system (Tooth, 2008), requiring awareness of atmospheric-oceanic coupling (e.g., Heinrich events, El Niño-Southern Oscillation dynamics), the geosphere (e.g., deep crustal and mantle processes), and the biosphere (e.g., the interface with human activities).

9.3 SENSITIVITY AND CHANGE

Sensitivity (p. 132–3) is the likelihood, extent and rapidity with which a given landform changes in response to a single unusual process event or to a reinforcing sequence of events. Landscape stability can reflect the temporal and spatial distributions of the resisting and disturbing forces, thus described by the landscape change safety factor as the ratio of the magnitude of barriers of change to the magnitude of the disturbing forces. The sensitivity of dryland environments can be greater than that of other zones. Paucity of vegetation cover and resistance means that they can be particularly vulnerable to climate change as they are marginal environments where small shifts in climatic variables such as rainfall amount or intensity can have direct and indirect effects on geomorphological processes (Tooth, 2007), by increasing the disturbing forces or decreasing the magnitude of barriers to change.

This has already been significant in the Quaternary. Cold, dry conditions that dominated in hot deserts in Africa, Asia and Australia during the last glacial maximum of the Quaternary were succeeded by a spectacular transition to warmer and wetter early Holocene conditions with rising lake and groundwater levels peaking about 8000–9000 years BP, although the Caspian and western North American lakes were high in late glacial time from 15,000 to 11,000 BP (Baker et al., 1995). During the Holocene anthropogenic impacts have triggered change, related to the development of arroyos or the stability of sand seas. Past variations in sand transport occurred in the Taklimakan sand sea with aeolian sand activity high during the 1960s and mid-1980s, but reduced from the mid-1980s to the late 1990s (Wang et al., 2004). Desertification refers to the degradation of land in arid, semi-arid and dry subhumid areas, caused primarily by human activities and climatic variations (FAO, 2005) affecting 70% of all drylands, and one-quarter of the total land area of the world. The year 2006 was declared the International Year of Deserts and Desertification (IYDD) by the United Nations General Assembly, and the 'Global Deserts Outlook' (Ezcurra, 2006) was launched, presenting an overview of the environmental status of the world's deserts in the context of global climate change and other impacts and pressures. It is thought that desertification in China is probably controlled by climate change and geomorphological processes, exacerbated by human impacts (Wang et al., 2008).

Sensitivity can also apply to future changes. A model using an improved dune mobility index indicates how Kalahari dunes might react to global warming, reactivating the dunefields of the mega-Kalahari, partly due to erodibility increasing as precipitation declines, and partly as wind energy increases, especially towards the end of the dry season when surfaces are least vegetated (Thomas et al., 2005). This study suggests that by the end of the twenty-first century, all the dunes in the Kalahari region are likely to be on the move. Such scenarios are not easy to construct because of the complexity of the variables involved and the non-linear response of land surface systems to forcing factors, but extending the study to model seasonal dunefield activity 2070–2099 with global warming (Thomas and Wiggs, 2008) is an excellent example of scenarios for future geomorphological change that Goudie (2006) has suggested should be developed as a major research priority (see Chapter 12). Although uncertainties exist in the prediction of arid ecosystem responses to elevated CO_2 and global warming (Lioubimtseva, 2004), other scenarios developed include use of global climate change models to deduce that even moderate decreases in precipitation could significantly decrease perennial drainage and surface water access across a large area of South Africa by the end of this century (de Wit and Stankiewicz, 2006). Focus needs to be given to the way in which predicted shifts in the balance between aeolian and fluvial activity might impact on landscape change and human activities, possibly requiring a more integrated approach such as the Drylands Development Paradigm (DDP) proposed to deal with the inherent complexity of desertification and dryland development, by identifying and synthesizing the factors important to research, management and policy communities (Reynolds et al., 2007).

Equilibrium that is rare in dryland fluvial systems is even more unusual in systems designed for river, urban and agricultural management (Graf, 1988: 293) – for example in Californian deserts it is suggested (Wiltshire, 1980) that management agencies have not fully utilized the technical understanding of desert systems that is available. Further scope exists for geomorphological analyses using **non-linear methods,** not simply translated from temperate or other zones which are greener and less sensitive to adjustment.

BOX 9.1

PROFESSOR ANDREW GOUDIE

Professor Andrew Goudie has undertaken research on deserts for nearly 40 years, most recently on dust storms but also on duricrusts, rock and salt weathering, dunes and sand movement. In addition to numerous contributions to the field of desert geomorphology, he has authored significant text books which have profoundly influenced the way in which we consider the land surface of the Earth. Although some of these, such as *Deserts in a Warmer World* (1994), *Great*

(Continued)

(Continued)

Warm Deserts of the World: Landscapes and Evolution (2002) and *Desert Dust in the Global System* (2006), have focused on arid environments he has also authored and edited other key works including *Environmental Change* (1977, 3rd edn 1992), *The Human Impact on the Natural Environment: Past, Present, and Future* (1980, 6th edn 2006), *The Nature of the Environment* (1984, 4th edn 2001), *The Dictionary of Physical Geography* (1985, 3rd edn 2000), *Geomorphological Techniques* (1990), *The Changing Earth: Rates of Geomorphological Processes* (1995), *Encyclopedia of Global Change: Environmental Change and Human Society* (2002), and *Encyclopedia of Geomorphology* (2004). His contributions have therefore significantly advanced geomorphology not only by research on deserts but also by communicating understanding about the land surface of the Earth to the specialist and also to the more general reader.

　　Professor Goudie graduated at the University of Cambridge (1967, PhD 1972), has been on the geography staff at the University of Oxford since 1970, being a Professor since 1984, head of the school of Geography 1984–94 and Master of St Cross College since 2003. He was awarded the DSc degree by the University in 2002. His many services to societies include President of the International Association of Geomorphology (2005–9). His research contributions and publications have been recognized by many awards including the Founders Medal of the RGS (1991), the Mungo Park medal of the RSGS (1991), the Geological Society of America's Farouk El-Baz Prize for desert research in 2007, and the Linton award of the British Society for Geomorphology in 2009.

　　Andrew Goudie, always guaranteed to give lively and stimulating lectures, demonstrates how a geomorphologist needs to contribute not only through research papers but also in writing innovative books to advance understanding of the land surface of the Earth.

FURTHER READING

Comprehensive references are:

Thomas, D.S.G. (ed.) (1997) *Arid Zone Geomorphology: Process, Form and Change in Drylands*. Wiley, Chichester.

Parsons, A.J. and Abrahams, A.D. (2009) *Geomorphology of Desert Environments*, 2nd edn. Springer, New York.

A stimulating book on fluvial processes is:

Graf, W.L. (1988) *Fluvial Processes in Dryland Rivers*. Springer Verlag, Berlin.

Application to management problems is given in:

Lee, M. and Fookes, P. (2005) Hot drylands. In P.G. Fookes, E.M. Lee and G. Milligan (eds), *Geomorphology for Engineers.* Whittles Publishing, CRC Press, Dunbeath, pp. 419–453.

TOPIC

1 Consider why it is difficult to characterize the landforms characteristic of specific deserts.

10

HUMID AND
SEASONALLY HUMID
TROPICS

The tropical zone, between the Tropic of Cancer (23.3°N) and the Tropic of Capricorn (23.3°S), covering about a third of the Earth's land surface, is often thought of as dense, humid forest frequently incorrectly termed jungle, but can actually be divided into three major zones on the basis of climate: wet zones, the humid tropics, often characterized as hot wetlands (Douglas, 2005) usually nearer the equator and having rainfall in the range of 2000 to 13,000 mm.yr^{-1}, which include montane, superwet and wet climates; the monsoon zone, located mainly around the Indian Ocean, of two seasons with pronounced differences in rainfall; the wet–dry zone, the seasonal tropics often referred to as 'savanna', where annual rainfall totals range from below 800 mm but can approach 1600 mm per year, where the dry season is more than 4 months and located in intertropical and subtropical areas not influenced by monsoons. Savanna environments are transitional between humid tropics and the dryland zones (Chapter 9).

Several themes characterize the geomorphology of these tropical zones.

- the length of time available for landform evolution and landscape change. No major interruptions have occurred, sometimes for as long as 200 million years and many humid tropical land surfaces are old with deep weathering profiles, so that the largest areas of savanna occur on extensive ancient plateau surfaces of the southern Gondwana continents, characterized by leached, nutrient-poor oxisol and ultisol soils, often underlain by deep saprolites possibly capped by duricrusts (Thomas, 1994).
- dense vegetation cover disguises landforms and the character of the land surface in many areas, although if removed the consequences for earth surface processes can be very dramatic.

- very high energy environments for some parts of tropical landscapes, even where buffered by vegetation prior to its removal, so that major mass movements and rockfalls can happen frequently as a consequence of the amount, intensity and magnitude of rain events and earthquakes, which also affect some of the world's largest river systems within the zone.
- a tendency for geomorphologists to apply uniformitarian principles based on a temperate normality to all water-worn landscapes of the tropics (Thomas, 2006). Thus the tropics were not perceived as different from other places (Gupta, 1993), although research in the humid to seasonal tropics tended to concentrate primarily on the most characteristic landform assemblage of this zone, namely that of stepped, largely undissected etchplains, often dotted with inselbergs and cutting across ancient basement rocks (Wirthmann, 1987). Recent geomorphological expectations regarding the geomorphology of the humid tropics have altered following research in the latter part of the twentieth century (Thomas, 2008b).
- the complexity of Quaternary change is now known to have contributed to the present landscape, including the consequences of millennial scale 'rapid' change during the Quaternary as recorded in GRIP and GISP2 ice-core records, and also found in tropical oceans (Thomas, 2004). Whereas slope failures, floods, colluvial/alluvial sedimentation could reflect short-term changes in the record, the reorganization of slope and fluvial systems involves significant time lags or delays, often on a millennial time scale, associated with vegetation changes, and variations in landscape sensitivity that may typify regional patterns of change. Thus the legacy of Quaternary changes persists and large sediment stores remain in present landscapes, often in areas of potential sensitivity to erosion (Thomas and Thorp, 1995).
- seismically active zones occur in large areas of the tropics: tectonic activity can be associated with 38% of the rainforest areas of the South-East Asia–Australasia area, 14% of American tropics and 1% of African tropics – even beyond these areas tectonics including volcanism can be important (Douglas and Spencer, 1985: 321).
- profound, rapid landscape transformation with widespread changes of land cover, development of urbanization and intervention in fluvial systems is occurring in many regions within the tropics.

Investigation of tropical geomorphology has depended upon detailed investigations undertaken by many individuals, including Professor M.F. Thomas (see Box 10.1).

10.1 DEEP WEATHERING

Despite many recent changes and Quaternary shifts of landscape-forming processes, the time available for weathering has been considerable. Extensive weathering profiles developed because many intertropical areas have been dryland for long periods of time – in places as much as 200 million years – with humid tropical conditions more widespread during the late Mesozoic and Palaeogene. Therefore the challenge has been to relate the deep weathering profiles, the landscapes of stepped plains, often cutting across ancient basement rocks and dotted with inselbergs, with the range of landforms encountered throughout tropical landscapes. Some explanations were conceived for the particular characteristics of the savanna or for humid tropical forest environments but more recent understanding of climatic history has underlined the need for an interpretation which applies flexibly to the range of present tropical environments, also allowing for the tectonic activity significant in some areas.

Depths of weathering up to 100 m and an earlier distinction of four weathering zones has been refined (Fookes, 1997) by recognizing six weathering grades (see Figure 10.1 and Table 10.1). These range from Grade I which is fresh rock with no indication of rock material weathering, rising upwards through grades of faintly, slightly, moderately, highly, and completely weathered to Grade VI at the ground surface which is residual soil where all rock material is converted to

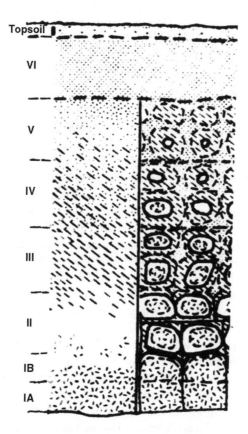

Figure 10.1   Idealized weathering profile showing six weathering grades (Fookes, 1997) with (right) and without (left) corestones.
Reprinted with permission from The Geological Society.

Table 10.1   TROPICAL WEATHERING GRADES (DEVELOPED FROM FOOKES, 1997)

Grade (see Figure 10.1)	Description
Humus, topsoil	
VI Residual soil	All rock material converted to soil; structure and material fabric completely destroyed. Large change in volume but no significant transport of soil.
V Completely weathered	All rock material is decomposed and /or disintegrated to soil. Original mass structure largely intact.
IV Highly weathered	More than half of the rock material is decomposed or disintegrated to soil. Fresh or discoloured rock is present either as a discontinuous framework or as corestones.
III Moderately weathered	Less than half of the rock material is decomposed or disintegrated to soil. Fresh or discoloured rock is present either as a continuous framework or as corestones.
II Slightly weathered	Discoloration indicates weathering of rock material and discontinuity surfaces. All the rock material may be discoloured by weathering.
IB Faintly weathered	Discoloration possible on major discontinuity surfaces.
IA Fresh	No visible sign of rock material weathering; some slight discoloration possible on major discontinuity surfaces.

soil with the mass structure and the material fabric completely destroyed. In Hong Kong there are depths of up to 60 m of silty-sand residual soil corresponding to Grades IV–VI with large corestones in the matrix or exposed on the surface (Douglas, 2005). Weathering of primary minerals is more complete than in temperate climates and occurs at greater depths. Organic matter tends to remain near the ground surface, is subject to rapid biodegradation and recycling, with geochemical processes (e.g. neutral or slightly acid hydrolysis) resulting in higher concentrations of free oxides than in temperate region soils. Such freed iron and aluminium oxides as remain in the soils produce the characteristically bright soil colours. Many of the processes, which are biogeochemical, have been somewhat neglected in geomorphology (Thomas, 1994: 14).

Several phases of weathering processes have been recognized, characterized by increasing degree of weathering of primary minerals, an increasing loss of combined silica, and increasingly marked dominance of neoformed clays (Duchaufour, 1977). In phase 1 (fersiallitization) there is a dominance of 2:1 (expanding lattice) clays rich in silica, there are considerable amounts of free iron oxides that are more or less rubified (red or ochreous in colour) and the exchange complex is almost saturated by the upward movement of bases during the dry season; illuvial horizons are less well developed. Phase 2 weathering represents the final stage of pedogenesis in climates that are less hot or marked by a dry season, and weathering products correspond more or less to ultisols

(USA) or acrisols (FAO). Phase 3, ferrallitization, is characterized by complete weathering of primary minerals except quartz, and the clays are all neoformed, directly precipitated in the soil from solution and consist solely of kaolinite.

Although deep weathering can be the norm in many areas, it varies in depth and character according to rock type, availability of moisture, and the age of the land surface. Thus the growth of the Antarctic ice sheet during the Miocene appears to have been associated with drier climates in many parts of the tropics and subtropics in the southern hemisphere so that the weathering systems in South America were effectively 'switched off' at that time (Thomas, 2005). Lower rainfall in the Miocene may have induced erosion and redistribution of the deep saprolites that had previously been produced, but subsequently, in the later Neogene and throughout the Quaternary, the renewal of weathering and saprolite formation occurred. Climate fluctuations of the Quaternary, including more than 20 cycles of alternating humidity and aridity, were responsible for changes in the vegetation distribution and density. Hence weathering systems fluctuated in their effectiveness, responding to regional climatic and hydrological controls, sometimes over extremely long time periods.

Understanding the landscape of tropical environments depends upon the balance between the processes of weathering and accumulation of residual materials, their removal and deposition. Deep weathering profiles directed attention to the two surfaces where change potentially occurs – or what Budel (1957) referred to as a double surface of levelling 'doppelten einebnungsflachen'. Deep weathering continues to extend the basal weathering front or surface – that is the line at variable depth below the surface which marks the contact of sound rock beneath from weathered rock above. The shape of this weathering front, often at depths as much as 30–50 m below the surface, depends upon rock composition and structure, the hydrological conditions, and the recent history of the area. In addition to change occurring at this lower surface, there is also denudation on the upper land surface, by weathering, erosion and deposition, especially by water in fluvial processes.

Relating changes at these two levels it is possible to envisage one scenario where deep weathering increases because the lowering of the basal weathering front occurs more rapidly than lowering of the land surface. The alternative, when lowering of the surface is more rapid than that of the basal weathering front, can culminate in a situation when the deep weathering layer is comparatively thin so that the land surface becomes dominated by surface processes. The basal weathering front, especially in areas of granitic rocks, may be diversified according to joint spacing, by corestones and what appear to be buried inselbergs. Thus if stripping of the deep weathered material occurs, not unlikely in view of the climate fluctuations now known to have occurred during the Quaternary, this basal weathering front may become exposed to create a landscape of plains and inselbergs. Such plains have been described as etchplains, that originated sub-surface essentially etched from the rock by deep weathering followed by subsequent removal of weathered products. A range of etchplain types have been identified by Thomas (1965, 1994) according to the degree of completeness of the stripping of the

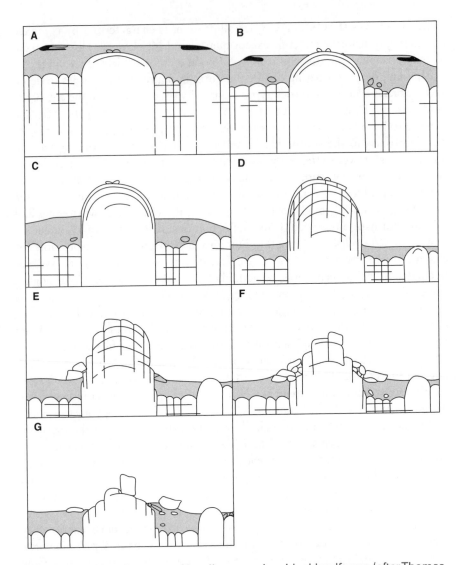

Figure 10.2   Development of inselbergs and residual landforms (after Thomas, 1974, 1994)

The dome exposed in stage C gradually disintegrates (D to G) to form a kopje (see Table 10.2).

weathered material and the exposure of the weathered front (see Figure 10.2). Just as there are a series of potential stages or alternatives in the ways in which the deep weathered material is stripped away, so there are also different potential hillslope profiles which can be created (Douglas, 2005; Fookes, 1997). The nature of water action, and in some cases the character of the networks of channels and streams, can exert a significant influence upon the morphology that is produced. The two-stage concept of landform development, with significance for the origin of a wide range of landforms, the role of water and weathering, and the age of landforms and landscapes, is one of the most fruitful developed (Twidale, 2002).

During landscape evolution alterations in the weathering profile, especially associated with moisture levels, can mean that the horizons rich in iron and aluminium oxides become indurated and hardened, thus coming to function as caprocks which have a significant effect upon slope morphology. The origin of such deposits and the terms used to describe them has produced debate amongst pedologists, geologists, geomorphologists and others and includes the creation of laterite and duricrusts. Laterite, a term first used in India for vesicular mottled red and cream clay which was dug out of the ground, shaped into bricks and dried in the sun, has been defined geomorphologically as 'a highly weathered material rich in secondary oxides of iron, aluminium or both ... is either hard or capable of hardening on exposure to wetting and drying' (Thomas, 1974: 49). Duricrust is the term for a hard crust or nodular layer first formed in saprolites (weathered or partially weathered bedrock in situ) or sediments by precipitation of hydrated oxides of aluminium, iron, silicon or of calcium carbonate or sulphate. The variety of names employed relates to the dominant oxides: aluminium in alucrete and bauxite; iron in laterite and ferricrete; silica in silcrete; or calcium carbonate in calcrete. Alucrete/bauxite tend to occur in the most humid areas progressing sequentially to calcrete in the more arid locations.

Any location in the tropics demonstrates the results of the history of weathering and erosion so that landforms and surface processes are significantly affected by inherited characteristics. In addition, a contrast exists between the stable old, cratonic surfaces and those areas influenced by recent tectonic mountain building – the tectogene tropics. Such contrasts affect the three main subdivisions of the humid and seasonal tropics – rainforest, savanna and monsoon environments.

10.2 RAINFOREST LANDSCAPES

Tropical rainforests, also referred to as lowland equatorial evergreen forests, occur principally in South America (especially the Amazon basin), Africa (especially the Congo basin), and South-East Asia, but also in Australia, central America and on Pacific islands. Annual rainfall usually ranges from 1750 to 2000 mm and mean monthly temperatures exceed 18°C throughout the year. Covering 7% of the Earth's land surface area, these forests contain half of all living animal and plant species, with some 82% of the world's known biodiversity. They are often imagined, incorrectly, to be jungle – which is the result of disturbance. Under undisturbed conditions rainforest is structured into five well-defined storeys, of which the top three are forest trees, the fourth is shrubs, and the fifth is ground cover. At least seven major types of tropical evergreen forests have been recognized including those associated with mountain, alluvial and swamp areas. Early explorers imagined that the immense trees up to 50 m high and the abundance of foliage signified very fertile soils but by the mid-twentieth century this perception had been revised, so that tropical soils are now known to be as varied as those in other parts of the world with some alluvial and volcanic soils highly fertile but others having limitations for plant growth.

Different forest types blanket the environments and disguise landforms beneath, often regulating the efficiency of the surface processes. High rainfall amounts and intensities of >100 mm per hour are known meaning that potentially high rates of erosion can occur, so that erosivity $(kJ.m^2)$, which is the potential for soil to be eroded, can be as much as 25 times greater than in humid temperate areas (Douglas, 2005). Erosivity can be measured in terms of drop size diameter mm; drop number; rainfall intensity; and energy load $W.m^{-2}$ (Kowal and Kassam, 1976). The character of the forest affects the amount of precipitation intercepted; running down trees as stemflow, and lost by evapotranspiration, with some 11–25% of precipitation being intercepted, and 40–65% possibly transpired. In the Luquillo Experimental Forest in eastern Puerto Rico measured interception loss was 50% of gross precipitation as a result of a high rate of evaporation from a wet canopy (Schellekensa et al., 1999) whereas, in a small watershed (1.3 km² area) near Manaus in the tropical rain forest of the central Amazon region, a mean precipitation of 2209 mm.yr^{-1} was recorded, 67.6% of which was lost to the atmosphere through evapotranspiration (Leopoldo et al., 1995).

Humid rainforest environments potentially have high energy processes but conditions at any time vary according to depth of weathering, stability of the weathered mantle, surface slope angle, tectonic influence, and rate of water transmission. Vegetation removal can lead to enormous increases in erosion. Weathering profiles can be up to 40 m deep in stable areas, although in tectonically active areas mass movements are sufficiently frequent to limit depths to as little as 1 m. Mass movements can occur after sudden wetting of the weathered material giving potential for metastable collapse of residual soils which can lead to a combination of erosion by gullying and landslide activity. Rainforest disturbance regimes (Spencer and Douglas, 1985: 20) vary in their location according to tree falls, landslides and earthquakes of which any may act as triggers to instigate change in slope activity. In Malaysia, residual granite and sedimentary rock soils that occur over more than 80% of the country's land area can collapse when wetted. The weathering environment can be viewed in terms of transport-limited and weathering-limited erosion regimes (Johnsson, 2000). If slope erosion processes operate more rapidly than weathering processes then erosion is limited by the rate at which the rock is weathered. Conversely if the rate of weathering exceeds the rate at which detachment and transport processes operate then erosion is limited by the efficiency of the transport process and a deeper weathering profile develops. Landslides are necessarily one of the most dominant geomorphic processes affecting humid tropical environments. Studies of eight small drainage basins by Ng (2006) in Hong Kong recorded 451 landslides, demonstrating how landslide locations are related to a headward progression of the drainage network.

Spatial variations occur with soil terrain types and with the pattern of water drainage. Thus overland flow, with associated surface sediment transport, may be a potentially important mechanism of runoff generation in forested Acrisol but not in Ferralsol landscapes (Elsenbeer, 2001). In the north-western region of the main island of New Caledonia (south-west Pacific) there is a succession of hard rock protrusions and weathering troughs, whose depth varies greatly (Beauvais et al., 2007). Saprolite- and ferricrete-mantled source areas can be distinguished with the former resulting from a regolith erosion process by shallow landslides, but the latter from a secondary ferruginization process of reworked lateritic debris. The deepest troughs underlie saprolite-mantled source areas above channel heads, which are characterized by a low permeability saprolite, relatively high slope gradient, and lower area/slope ratios. Such source areas generate fairly high runoff, sustaining rivers and creeks with relatively high erosion power, whereas the ferricrete-mantled source areas are characterized by higher permeability and area/slope ratios, leading to lower runoff and less erosion but further chemical rock weathering. The ferricrete of those source areas acts as a protective hard-cover against mechanical erosion of the underlying regolith. This ferricrete reworks, at least partly, allochtonous lateritic materials inherited from a previous disaggregated ferricrete that suggests past erosion processes driven by alterations in hydro-climatic conditions.

A precipitation event activates flowpaths that generate a stream storm hydro-graph, with more rapid overland flow from some areas, saturated overland flow from others, and throughflow or interflow elsewhere. It is suggested that this is a pulse-driven system (Douglas, 2005), with pulses triggered by storm rain events, capable of instigating small mass movements, stream rise and sediment slugs, with tropical storms giving tree fall which can produce debris dams across stream channels. Terrain units can be recognized according to the processes that are most dominant, possibly including runoff and surface erosion on upper slopes; net sediment transportation and mass movement on mid-slopes; with fluvial effects, including possible debris flow development in drainage lines, and deposition on lower slopes. However, many geomorphological processes may be active within the same terrain so that, on hillsides of moderate gradient and medium to low energy, fluvial valleys may at varying times be affected by erosion, transport or deposition in response to rainfall, landslides, flooding or drought. Landslide activity can be widespread – especially when vegetation is removed or modified. In Hong Kong some 200–300 slope failures are reported each year, usually triggered by heavy rainfall and occurring on man-made slopes (Douglas, 2005; Evans and King, 1997). There can also be a severe soil erosion hazard after vegetation is removed prior to building construction. Even in humid tropical systems, there can be large intra-seasonal and inter-annual variability of rainfall amounts which can significantly affect all components of the water balance, although a strong memory effect in the groundwater system can buffer seasonal climate anomalies and the deep unsaturated zone can play a key role in reducing most of the intra-seasonal variability, also affecting groundwater recharge (Tomasella et al., 2007).

Rainforest environments therefore vary quite significantly in the geomorphological processes which occur. Further variations arise where distinctive rock types occur and tropical karst, on calcareous rock outcrops, includes polygonal or cockpit karst (landscape pitted with smooth-sided and soil-covered depressions) and tower karst (landscape of upstanding monoliths across surface of low relief, see Figure 3.4). In addition karst-like landforms can occur on quartzose rocks in areas which include the Guyana shield and dolines and towers in Venezuela (Reading et al., 1995). Distinctive areas of tropical coasts can include coral, algal mats and mangroves; the latter occupy over 75% of the coastal fringe of intertropical and subtropical areas, together with extensive mudflats affording protection from tropical storms and also assisting as land builders encouraging sedimentation. Significant effects can arise in the variations in climate from year to year and in 2005 parts of the Amazon basin experienced the worst drought in 100 years: the rainforest could be vulnerable to several years of drought conditions. Perhaps most dramatic are the effects experienced when forest is cleared (for example see Figure 10.7): in four small montane, humid-tropical watersheds in Puerto Rico affected by human disturbance, the two with primary forest cover had 43–71% of landslide-caused erosion associated with road construction and maintenance (Larsen, 1997).

10.3 SAVANNA LANDSCAPES

Although derived from an American Indian word meaning grassland without trees, since the late 1800s the term savanna has been used for land with both grass and trees. These landscapes, often perceived as extensive plains featuring elephant, giraffe, zebra and other animals, are transitional from rainforest to semi-desert, having a range of vegetation types including open deciduous woodland, woodland/grassland mosaics, and open grassland. Many have particular regional names such as cerrado (small trees and grasses) and caatinga (thorn woodland) in Brazil, but more usually are characterized by the dominant species such as *Brachystegia* woodland in Africa and *Eucalyptus* savanna in northern Australia. Although temperatures usually remain high throughout the year, mean monthly temperatures range from 10°–20°C in the dry season and 20°–30°C in the wet season. Annual precipitation of 500 to 1600 mm is markedly seasonal with the dry season varying between 4 and 11 months. Variations in precipitation often occur from year to year and fires in the dry season are thought to regulate the extent of tree growth. The most extensive areas of savanna are found on ancient plateau land surfaces of the southern Gondwana continents, characterized by leached, nutrient-poor oxisol and ultisol soils, often underlain by deep saprolites possibly capped by duricrusts. Therefore geomorphological research has concentrated on the most distinctive landforms, namely the stepped, largely undissected etchplains, often punctuated by inselbergs cutting across ancient basement rocks. Because the extensive plateau surfaces were thought to have been produced by denudation in the late Mesozoic and early Cainozoic, such ancient landscapes have necessarily been affected by climate changes, including both wetter and drier periods in the past. In addition there are some areas of more recent volcanic rocks including the Deccan plateau in India and the East African rift zone.

Although this zone is transitional between rainforest and desert and has been affected by swings of climate for more than 30 million years, the dramatic landforms and scenery including pediments (see Table 9.4) have invited many deductions about their origins, sometimes concluding that a particular combination of land-forms is characteristic of this particular morphoclimatic zone (e.g. Budel, 1982; Tricart, 1965). However this is now thought to be misleading (Thomas, 2005) because similar landforms are also found in adjacent wetter and drier zones. On ancient shields broad valley floors, wide interfluvial areas and footslopes, with isolated and residual hills (see Figure 10.3), compose striking savanna landscapes which have attracted speculation about their origin since the earliest explorers and travellers, subsequently leading to proposals for the development of the land-scape as a whole. Inselbergs, residual hills that abruptly interrupt tropical plains, commanded much attention, including speculation about their origin and devel-opment in theories which sought to link the several identified types of residual hill. Figure 10.4 schematically demonstrates how various forms of residual hill or inselbergs can be interrelated. In Figure 10.2A the weathered material occurs above the basal weathering front which varies in depth according to joint spac-ing. Subsequent stages depend upon the relative change of the surface and the basal weathering front: in B the top of a dome appears, becomes more extensive in C/D, begins to collapse in E, F and G. Such a sequence is compatible with understanding of the development of the weathering profile (see Figure 10.1 and Table 10.1) and landforms, with names often derived from particular countries associated with each of the stages.

Figure 10.3 Slopes of inselberg Domboshawa, Zimbabwe

This granite residual is protected by National Monuments of Zimbabwe and includes depressions containing mixed deciduous woodland trees.

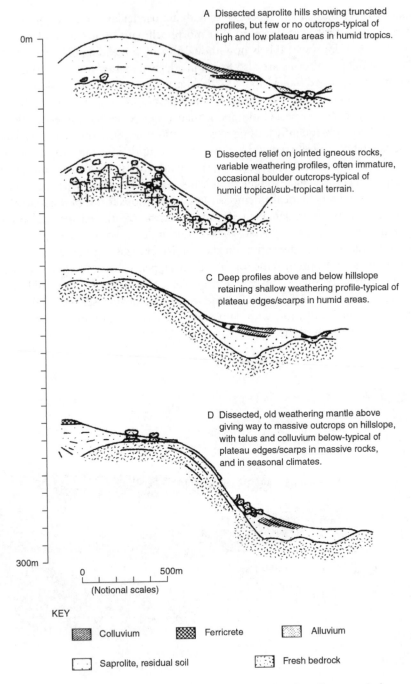

A Dissected saprolite hills showing truncated profiles, but few or no outcrops-typical of high and low plateau areas in humid tropics.

B Dissected relief on jointed igneous rocks, variable weathering profiles, often immature, occasional boulder outcrops-typical of humid tropical/sub-tropical terrain.

C Deep profiles above and below hillslope retaining shallow weathering profile-typical of plateau edges/scarps in humid areas.

D Dissected, old weathering mantle above giving way to massive outcrops on hillslope, with talus and colluvium below-typical of plateau edges/scarps in massive rocks, and in seasonal climates.

0m

300m

0   500m

(Notional scales)

KEY

▨ Colluvium   ▩ Ferricrete   ▦ Alluvium

☐ Saprolite, residual soil   ☐ Fresh bedrock

Figure 10.4 Schematic sequence of residual hill or inselberg and slope development (from Fookes, 1997)

Reprinted with permission from The Geological Society.

Table 10.2   LANDFORMS IN SAVANNA LANDSCAPES

The list cannot be comprehensive, in view of the range of local landforms, they are not exclusive to savanna landscapes, but they give some idea of the way in which savanna landscapes may be composed (see Thomas, 1994, 2005).

Landform	Definition	Development
Inselberg	German for 'island mountain'. Prominent rocky hills, usually domed in profile and often granite. May be subdivided into bornhardts and tors, but also include hills of sedimentary rock capped by duricrust.	Prominent, isolated, residual knob, hill, or small mountain, usually smoothed and rounded, rising abruptly from an extensive lowland erosion surface in a hot dry region; generally bare and rocky although the lower slopes are commonly buried by colluvium (see Figure 10.2).
Bornhardt	Broadly rounded dome, bare rock surfaces and steep sides, with shape controlled by jointing and sheeting. May have steep rise from pediments or from deeply weathered material with 'scarp foot nick'.	
Castle kopje	Afrikaans for a small isolated hill. Scattered steep-sided hills called kopjes, or koppies.	Sub-surface weathering followed by second stage of stripping of surrounding weathered material (see Figure 10.2).
Tors Blockfields	Piles of smaller blocks. Rock cores residual from weathering.	
Caprock-protected hills, plateaus and benchlands in valleys	Duricrust exposed and hardened horizons rich in aluminium, iron, silicon or calcium hydrated oxides or other compounds. Hills represent inversion of relief.	May have developed below the surface, sometimes affected by fluctuating water table, then exposed by fluvial dissection of landscape, and hardened.
Ruware	Low rock pavements, may grade upwards into domes or large tors. May be incipient domes or tors.	See stage B in Figure 10.2.
Karst and pseudo karst (silica) features	Cavernous macro-voids, surface features ranging from minor forms such as rillenkarren (surface channels or corrugations) through sinkholes to macro-tower karst.	

(Continued)

Table 10.2   *(Continued)*

Landform	Definition	Development
Breakaways	Low cliffs where exposure of duricrusts results from loss of topsoil and dissection.	
Lateritic pseudo-karst	Water excavating cracks through duricrust can lead to collapse giving karst-like depressions.	
Surface gravels and stone lines	Removal of finer material may lead to concentrations of coarse material of duricrusts.	Some rolled pebbles of fluvial origin derive from time when deposits were in weathering profiles prior to dissection and hardening.
Landslide debris	Including boulder accumulations from rock falls or slides, earth flow debris from slumps and translational slides.	
Coarse talus and debris fan deposits		See Figure 10.4.
Scarp foot depressions or nicks	Marked linear depressions at the foot of escarpments or inselbergs.	Controversy about origin, may be because several origins; stream action, possibly diverted by fans, or due to differential weathering at the foot of the scarp.
Pediment	Plane or gently concave erosional surface of low relief cut in bedrock, may be partly covered by a thin veneer of alluvium, usually undissected by gullies or permanent stream channels.	See Table 9.4.
Dambos, fadamas	Shallow, seasonally waterlogged valleys.	Related to dynamic drainage network and are associated with or alternative to intermittent and ephemeral channels.
Alluvial deposits	Floodplain facies from meandering streams, deep channel fills, flood sediments, clay fills and clay pans.	
Gully	Often develop in deep colluvium.	See Table 9.4.
Termite mounds	Termitaria can reach 10 m high and 30 m in diameter, densities of 2–5 per hectare and very large mounds can be 700 years old.	

Figure 10.5 Zimbabwe, west of Harare, immediately after rainstorm showing overland flow in a dambo and then into gullied section

Specific landforms associated with inselbergs include the effects of protective cappings by duricrusts. On hill slopes shallow debris slides and flows may occur because deep-seated landslides may be limited as a result of restricted amounts of groundwater in such seasonal climates (Thomas, 2005). Rainfall often occurs in intense downpours, with as much as 100 mm in a single storm and intensities over 15–30 minutes exceeding 100 mm.hr^{-1}, potentially triggering small slides and mudflows or initiating overland flow (see Figure 10.5). Landforms related to water flow in savanna landscapes arise from the seasonal climate and intense storms can engender surface flow and quickflow giving extensive drainage networks that may have major ephemeral and intermittent components, including shallow, seasonally waterlogged, valleys known as dambos in Africa (see von der Heyden, 2004), especially in Zimbabwe and Zambia – in the latter country they have been estimated to occupy about 10% of the surface area. In the savanna of Nigeria, Guinea, Sahel and the Sudan, low-lying areas including streams channels and streamless depressions which are waterlogged or flooded in the wet season are known as fadamas. Such seasonal wetlands are valuable for grazing and agriculture but in many areas their susceptibility to change and particularly to the incidence of gullying has been demonstrated. In Zimbabwe accelerated gullying is deduced to be caused by breaching of the dambo clay by subsurface water movement (McFarlane and Whitlow, 1991); and the interfluve deforestation and cultivation of shallow-rooted crops that conserve water in Malawi, raising the groundwater level, also appears to have occurred in Zimbabwe promoting gullying. Hence if dambos are to be conserved and satisfactorily utilized, revision of land and water policy is required.

Figure 10.6   Gullies near St Michael's Mission, central Zimbabwe

The gullies are produced after deep cracks are enlarged into pipes in sodium rich clays in gully fill sediments. Recent gullying appears to be more extensive and intensive than earlier cycles of gullying in the late Quaternary (see Shakesby and Whitlow, 1991).

Gully initiation can result from alteration of the surface flow in the extensive drainage systems including dambos, or from decrease of the vegetation resistance, as a result of grazing or cultivation. In some locations (see Figure 10.6) there is sedimentary evidence of earlier phases of gullying but contemporary gully development can also be instigated by growth of subsurface pipes especially in sodic clays; bank erosion aggravated by game animals such as buffalo; or by roadside culverts and ditches which lead road drainage into valley heads without associated conservation measures. Landforms encountered (see Table 10.2) are not exclusive to savanna landscapes, particularly in view of the length of time available for landscape development and the way in which a succession of wet and arid phases has affected the generation of the land surface features, but human impacts have been significant.

10.4 MONSOON ENVIRONMENTS

Monsoon refers to wind that changes direction with season, blowing towards the sea in winter and towards the land in summer. The largest monsoon-dominated region is in India and South-East Asia, comprising the south-west (Indian) and the south-east (east Asian) monsoons, but monsoon areas also occur in East Africa, the Guinea coast of West Africa and in northern Australia. The monsoon system is a thermodynamic atmospheric circulation characterized by strong seasonality of wind direction, temperature and precipitation (Kale et al., 2003; Ramage, 1971). Monsoon characteristics in India include the season of the north-east monsoon from January to May and the season of the south-west monsoon from June to December, with the bulk of the rains in the first four months. Multi-proxy records from China and the Indian and north Pacific oceans indicate that the Indian as well as the east Asian monsoon systems were established about 8 million years ago, that there is evidence of a stronger monsoon system 3.5 and 2.6 million years ago, and that, since the onset of glaciations in the northern hemisphere some 2.5 million years ago, the strength of the Asian monsoon has varied on both long and short time scales (Kale et al., 2003). The monsoon can lead to destructive flooding all over India and South-East Asia from April to September, environments being distinguished by their particular seasonal hydrological regime and the fact that they embrace mountain areas and some major river basins that in turn can also be susceptible to the effects of tropical cyclones.

Areas affected by the Asian monsoons include large river basins, the Indus, Ganges, Brahmaputra, Irrawaddy and Mekong, all extensively influenced by human activity where fluvial processes predominate. Kale (2002) demonstrates the large seasonal fluctuations in flow and sediment load characteristic of the rivers of India where large floods are the major formative events. Whereas Himalayan rivers occupy highly dynamic environments with extreme variability in discharge and sediment load with earthquakes and landslides having a great impact, Peninsular rivers contrast because adjustments are less frequent and of a much smaller magnitude. Almost all the geomorphic work by the rivers is carried out during the monsoon season, especially by floods: in the Ganga-Brahmaputra Plains annual floods appear to be geomorphologically more effective than the occasional large floods whereas the geomorphic effects of floods are modest on the more stable rivers of the Indian Peninsula (Kale, 2003). The major river basins (see Table 10.3) in monsoon areas, accounting for over 7% of world runoff, have been managed and extensively modified and their delta areas are sensitive not only to the influence of tropical cyclones but also to the consequences of global warming. It has been suggested that the summer monsoon initiates, amplifies and terminates climatic cycles in the northern hemisphere, influencing conditions producing the greenhouse effect injecting a large amount of water vapour into the atmosphere and by affecting snow accumulation rates. This underlines why understanding the linkage between monsoons and global climates is important in reconstructing global Quaternary climate and hydrologic change (Kale et al., 2003). As significant rising trends in the frequency and magnitude of extreme rain events during the monsoon seasons have been demonstrated from 1951 to 2000 (Goswami et al., 2006), a substantial increase in hazards related to heavy rain is expected over central India in the future.

Table 10.3 SOME MAJOR DRAINAGE BASINS IN THE MONSOON
TROPICS

Compiled from various sources; some statistics quote length of basin, others quote
main river length.

River	Drainage area (km²)	Length (km)	Runoff (km³) (world %)	Notes
Mekong	795,000	2600	538.3 (1.4%)	Extends from Tibetan plateau to delta where it joins the South China Sea. Tectonic controls on northern part of basin and tectonic influence and bedrock confined river reaches are evident at many localities throughout the lower course. 21% eroded area, 69% of original forest lost, 4 large dams, delta 94 km² × 10³.
Indus	963,000	3180	269.1 (0.7)	From glaciers on Tibetan plateau through deep gorges to Punjab and Sindh plains where it is very braided. 4% eroded area, 90% of original forest lost, 10 large dams, delta 30 km² × 10³.
Ganges	489,000		439.6 (1.1)	10% eroded area, 85% of original forest lost, 6 large dams, delta 106 km² × 10³.
Irrawaddy	431,000	2170	443.3 (1.1)	9% eroded area, 61% of original forest lost, delta 21 km² × 10³.
Brahmaputra	935,000		475.5 (1.2)	Floods result from melting snow in spring and monsoon rains June to October, 11% eroded area, 85% of original forest lost. Merges with Ganges to form delta which is the largest in the world; 106 km² × 10³.

Table 10.3 *(Continued)*

River	Drainage area (km²)	Length (km)	Runoff (km³) (world %)	Notes
Yangtze	1,959,000	6380	(1.7)	Flood season May to August, 27% eroded area, 17 large dams, discharge to sea greatly reduced as result of water use.
Xi Jiang/ Pearl River	409,458	2200		22% eroded area, 80% of original forest lost, 7 large dams.

10.5 EVOLVING TROPICAL LANDSCAPES

Theories of landscape evolution fostered by the time available for tropical landscape development have been modified as knowledge has increased about the impact of alternating phases of pluvial and arid landscape-forming processes. Although some, such as the Davisian model, were translated from temperate environments, others such as Lester King (1962) suggested that tropical areas, and then the world's landscapes, could be interpreted in the light of the desert model. More flexibly the suggestion by Budel of double surface of levelling (Thomas, 1974: 231) enabled earlier models to be accommodated within this idea. An etchplain, as the culmination of landscape evolution by the surface lowering of previously weathered material over long periods of time, in terrain where resistant areas of unweathered rock may persist or become exposed as the surface is lowered (Douglas and Spencer, 1985: 327), is the basis for five types of landform (Thomas, 1994), namely mantled, partly dissected, partly stripped, stripped or etchsurface, and complex etchsurface.

The tropics were affected by Quaternary climate changes. Although the database concerning late Quaternary environmental change in the humid tropics depends upon records from scattered sites, Thomas and Thorp (1995) conclude that available evidence indicates that prolonged aridity affected all but a few favoured core areas of equatorial climate after 20,000 BP, lasting 5–7 ka. Dry conditions at the LGM were marked by semi-arid landforms and reduced stream activity. Large palaeofloods occurred after 13,000 BP but dry conditions returned 12,800–11,600 cal BP (Younger Dryas), before the early Holocene pluvial led to abundant sedimentation lasting nearly 2000 years after 10,500 BP, in Africa, Amazonia and Australasia (Thomas, 2008a, 2008b). Re-establishment of the lowland rainforests was delayed until after 9000 BP in Africa, Australia and Brazil, and several wet–dry oscillations followed in the mid-Holocene period. In many areas of the humid tropics there has already been substantial human impact: rainforests once covered

14% of the Earth's land surface but coverage has been reduced to 6% and this could be seriously further reduced in the next 40 years. The Amazon rainforest, which accounts for 54% of the total rainforests left on Earth, has been reduced by 10% since 1970 with suggestions that up to 40% may have been removed after the next two decades (see Figure 10.7). In the Cherrapunji of the Meghalaya region under natural conditions the effects of many extreme rainfalls each year were buffered by the dense vegetation cover but, after deforestation and extensive land use, the fertile soil was removed and either the exposed bedrock or an armoured debris top layer protects the surface from further degradation, just facilitating rapid overland flow, so that a new 'sterile' system has been formed (Soja and Starkel, 2007).

Figure 10.7   Amazon rainforest clearance

Above shows burning and deforestation of Amazon forest for grazing land. Below compares the extent of deforestation of Rondonia to the outline of Florida. This GOES image shows deforestation in grey and white

Sources: NASA and NOAA.

Many regions within the tropics are undergoing rapid development involving extensive land cover transformation, modification of fluvial systems and urbanization. In the future many areas could be susceptible to the impact of climate change and global warming. Sea level rise estimates vary (Houghton, 1997) but for 2100 could be from about 15 cm to 1 m, and under the IPCC business as usual scenario could be 12 cm by 2030 with a further 18 cm by 2100. Potential impact of 12 cm by 2030 and up to 0.5 m by 2100 could be significant because half the world population lives in coastal zones, with the tropics particularly vulnerable. According to Gupta and Ahmad (1999) many of the cities of tropical areas were established in hazardous or environmentally sensitive areas, often developing across inappropriate terrain conditions such as floodplains, coastal swamps, steep slopes or sand dunes. Those located near active plate margins and tropical cyclone belts are especially vulnerable. Bangladesh, in the complex delta region of the Ganges, Brahmaputra and Meghna rivers, has about 7% of the country's habitable land less than 1 m above present sea level and 25% below the 3 m contour. Taking subsidence and removal of groundwater into account gives some estimates of 1 m sea level rise by 2050 with nearly 2 m by 2100 (Houghton, 1997). Despite the large uncertainty in these estimates, the resulting loss of land could be complemented by the effects of storm surges, thus compounding the vulnerability; in these areas, and elsewhere in the tropics, there is an urgent need for geomorphologists to consider sustainable alternatives.

BOX 10.1

PROFESSOR MICHAEL F. THOMAS (1933–)

Professor M.F. Thomas is well known for many publications on the seasonal and humid tropics and for his book on tropical geomorphology (Thomas, 1994) – inevitably many of his publications have been referred to extensively in this chapter. His career has demonstrated the importance of fieldwork in a range of tropical environments dating from his first overseas University appointment at the University of Ibadan, Nigeria (1960–64). He obtained his Geography degree at the University of Reading in 1955, his PhD from the University of London (1967) and has been at the School of Biological & Environmental Sciences, University of Stirling since 1980, where he is now Professor Emeritus, having been elected Fellow of the Royal Society of Edinburgh (FRSE) in 1988. His fieldwork experience has been gained throughout the tropics providing an excellent foundation for his current research interests in tropical and applied geomorphology with particular reference to tropical weathering and landform development, and the impact of Quaternary environmental change on landscape stability and stream sedimentation. In view of his experience he was

(Continued)

(Continued)

the ideal contributor for a chapter on Geomorphology in the tropics 1895–1965 (Thomas, 2008a). He was a Leverhulme Emeritus Fellow 2002–03, received the Centenary Medal of the Royal Scottish Geographical Society in 2000 and the David Linton award from the British Society for Geomorphology in 2001.

FURTHER READING

Excellent coverage is given in:

Thomas, M.F. (1994) *Geomorphology in the Tropics.* Wiley, Chichester.

Change in Monsoon Asia is outlined in:

Kale, V.S., Gupta, A. and Singhvi, A.K. (2003) Late Pleistocene–Holocene palaeohydrology of Monsoon Asia. In K.J. Gregory and G. Benito (eds), *Palaeohydrology: Understanding Global Change.* Wiley, Chichester, pp. 213–232.

TOPIC

1   Follow up explanations for, and controversies surrounding, the detailed origin of features in Table 10.2 and any others that you can identify.

11

URBAN LANDSCAPES

Many geomorphology books ignore urban areas, so why include a chapter on urban landscapes which cover 2% of the Earth's land surface where landforms are largely obliterated by human activity? Bunge (1973) argued that physical geography (including geomorphology) is much needed in an urban setting because 'Cities are a karst topography with sewers performing precisely the function of limestone caves in Yugoslavia, which causes a parched physical environment, especially in city centres' (see Figure 11.1). However it is not just because of the proportion of world population that resides in cities but also because urban environments can be seen as assemblages of land forms (see section 11.1), just like karst; and also because of the way in which landforms and environment have been modified and created (section 11.2), and because knowledge of the way in which land-forming processes have been changed (section 11.3) enables us to manage them better in the future (section 11.4), with the theme of **design with nature**, as pursued by Ian McHarg (see Box 11.1), being pervasive.

According to the World Health Organization (WHO, 1990) an urban area is a 'man-made environment encroaching and replacing a natural setting and having a relatively high concentration of people whose economic activity is largely non-agricultural'; whereas the US Census bureau definition is more specific as areas with more than 386 people per km^2. The 2% of the Earth's land surface that is built up has included >50% of the world's population since 2000, compared with just 2% in 1850, and it is estimated that urban areas could include 60% by 2025. Some countries, such as Australia, New Zealand, Singapore and the UK, already have populations greater than 80% urban. In 2000 there were 60 cities of 5 million or more inhabitants, all consuming quantities of energy (electricity, transport, fuel), of food, materials and land, generating waste, and requiring investment in buildings and infrastructure. Most of the world's anticipated population increase to 8.1 billion by 2030 is expected to be in urban areas. The environmental impact of cities can be expressed as **'ecological footprint'** (see http://www.gdrc.org/uem/footprints)

Figure 11.1 Toronto from Sears Tower – the six perceptions (pp. 256–7) can be applied to this scene

which is the amount of land required to sustain them. If the world population consumed resources at the same rate as does a typical resident of Los Angeles, it would require at least three planet Earths to provide all the energy required.

11.1 PERCEPTION OF URBAN ENVIRONMENTS

Although the urban environment is created by human beings, the human mind perceives what has been created in at least six ways pertinent to the land surface of the Earth:

- The *location* of cities, encompassing the way in which physical environment of the urban area is accommodated in its geologic, topographic, hydrologic and ecologic setting. Many urban areas were initially founded according to the availability of water supply or where valleys, and therefore routes, converged. Geomorphologically this is analogous to the location of a system of landforms or a land system.

- as *a physical system*, the city is the angular, human-made and natural environment represented in three dimensions, equivalent to a major land system.

- made up of *network components* a city is analogous to the geometric composition of landform. An approach identifying the salient elements of the city (Lynch, 1960) based on paths, edges, districts, nodes and landmarks, is still used by planners and designers in analysis of existing conditions and developing plans for the future (e.g. Zube, 1999).

- The *fabric* of urban areas can be compared to landform assemblages, just as Bunge (1973) used the analogy with karst areas. Remote sensing characterizes urban environments on the basis of their high spectral and spatial heterogeneity so that most urban pixels in moderate-resolution imagery contain multiple land-cover materials. Powell and Roberts (2008) show how urban land cover can be generalized as a combination of vegetation, impervious surfaces and soil, in addition to water, so that multiple end member spectral mixture analysis (MESMA) can differentiate inter- and intra-urban variability, introducing the possibility of distinguishing urbanizing areas through time and across regions. Remote sensing enables categorization of urban forms in a range of different ways, including subdivisions in terms of age or function of urban zones, or classification according to type of building which can be houses, industries, transport, or open spaces.

- *processes*, can be a focus achieved, for example, by concentrating upon the constant movement of energy through the city system (Douglas, 1983: 9). Mean energy flows for cities at 0.45 cal.m^3.day^{-1} compare with 1.54 for tides and 0.04 for floods (Alexander, 1979). Other environmental processes can be investigated with regard to their manifestation under urban conditions.

- in terms of a *particular attribute or quality* such as scenic attributes. Thus the aesthetic quality of commercial and residential streets has involved consideration of variables such as naturalism, complexity, orderliness and openness (Nasar, 1988), an approach reminiscent of the way in which landforms contribute to scenery and scenic quality.

A city or urban area can therefore be viewed as composed of particular elements, just as a landform is composed of flats and slopes and landforms are built up into landform assemblages and landform systems. Viewing the physical character of urban areas alongside landforms echoes the way in which McHarg (Box 11.1) viewed such areas and advocated urban planning and design.

11.2 CREATION OF URBAN LAND SYSTEMS

The creation of urban systems involves transfer of materials in or out of the area, involving excavation or building foundations, using a range of materials (brick, stone, glass, concrete, asphalt) to construct the fabric of the urban area. These two ways of transforming the original landscape have been seen as equivalent to a geological and geomorphological agent; nearly a century ago Sherlock (1922: 333) suggested that 'Man is many more times more powerful as an agent of denudation, than all the atmospheric denuding forces combined'. Globally Douglas and Lawson (2000) suggest that the deliberate movement of 57,000 Mt.yr^{-1} of materials through mineral extraction processes exceeds the annual transport of sediment to the oceans by rivers by almost a factor of three, and in Britain the deliberate materials shift is nearly 14 times larger than the shift caused by natural processes. Analysis of direct and indirect modifications of landforms and transfer of earth materials in four study areas in Spain and Argentina showed that human activity is presently the main contributor to landform modification and earth material transport, so that mobilization rates due to construction and mining seem to be 2–4 orders of magnitude greater than natural denudation rates (Rivas et al., 2006). Global erosion rates and sediment transport are significantly less than construction plus mining mobilization. Therefore, urban geomorphic and related changes may represent manifestations of a global geomorphic change, with the 'human geomorphic footprint' expressing the new landform creation rate and mobilization rate (Rivas et al., 2006).

In addition much of the solid waste produced by cities is accumulated in landfill sites some within urban areas and some now undergoing restoration. About 57% of the solid waste generated in the United States is still dumped in landfills. Although landfill may initially use existing quarries for infilling, when such opportunities are exhausted the landfill has to be placed in other areas or in the sea. Once landfill has been accumulated it may be the location for use as parks, golf courses and nature reserves: a site in Cambridge, MA, was used to extract clay to manufacture bricks from 1847 until 1952, and it was then used as the city landfill site but was closed to active dumping in the early 1970s, being subsequently used as the site for Danehy Park, a 20 ha recreational facility. In the late 1970s and early 1980s, during the extension of the transport system, some additional 1–12 m of fill was placed above the 20 ha landfill so that over 1.5 million cubic metres of material were added to the site surmounted by an additional 0.3–0.7 m of sand and loam during the development of the park. Some settlement has occurred as is typical of landfill sites and is being monitored, but this proved to be an award-winning project including an artificial wetland, soccer pitches, and many trees exemplifying the way in which environments, including new landforms, can be created. Where space is limited, landfill is sometimes placed offshore and in Japan coastal areas are used for landfill sites with natural marine clay layers acting as bottom liners to prevent pollutant migration. Highly urbanized and industrialized Singapore (island land area of 697 km^2, population 4.2 million) required a solution to the six-fold increase in waste between 1970 and 2000. One project was to join two islands 8 km off the coast of Singapore by a 7 km perimeter bund embankment to create the world's first offshore landfill site (the Semakau landfill). This will be sufficient until 2040 and may be the site for an eco park.

It is not just the movement of waste material which occurs in urban areas but in some cases it is the creating or reclaiming of new land from the sea. In Dubai, in the United Arab Emirates (UAE), construction of unique human-made islands in the shape of date palm trees began in 2001. Palm Jumeirah island is primarily a residential area, with hotels, beaches and marinas (see Figure 11.2, Colour plate 8). These are the largest land reclamation projects in the world and will constitute the world's largest artificial islands, with the first two comprising some 100 million m^3 of rock and sand.

11.3 URBAN EARTH SURFACE PROCESSES

Earth surface processes are modified both directly and indirectly by urbanization. Direct modifications occur when existing processes (see Table 11.1) are changed as a result of building activity and urbanization, whereas indirect changes arise when the creation of urban climates or new ecosystems have feedback effects on processes.

Direct effects result from the way in which the fabric of urban areas impedes, controls or changes the character of earth surface processes. Impervious area, possibly

Table 11.1 EFFECTS OF URBANIZATION ON HYDROLOGICAL CHARACTERISTICS AND RELATED PROCESSES OF URBAN AREAS

Hydrology	Other characteristics
Increased: Magnitude and frequency of floods; annual surface runoff volume; stream velocities; sediment pulses; pollutant runoff; nutrient enrichment and bacterial contamination; toxics, trace metals, hydrocarbons; water temperature; debris and trash dams.	Temperature (0.5° to 4.0°C giving heat island); rainfall (5–15% more especially in downwind areas, due to localized pollutants acting as condensation nuclei); cloudiness (+5–10%); atmospheric instability (by 10–20%, caused by surface and near surface heating; increased turbulence from rougher city surface); thunderstorms (10–15% more frequent); PE and transpiration rates; air pollution (by 10 fold); dust (+1000% relative to rural, SO_2 + 500%, CO+2500); impervious areas; soil compaction; subsidence (draining of aquifers shrinks building foundations); soil pollution (with waste material and industrial pollutants); soil erosion (during building construction).
Decreased: Baseflow; infiltration; bank erosion (when sediment sources protected); pool riffle structure (removed in channelized reaches); groundwater reserves (pumping for water supply); aquatic life in rivers.	Solar radiation (up to 20% less); relative humidity (5–10% lower); wind speeds (20–30% lower); soil erosion (in built-up urban areas).
Other changes: Channel capacity; sedimentation (aggradation if sediment supply large; scour if sediment transport lower).	Thermal circulation; fog (+30 to +100%); topography and landforms (including accumulation of materials, extraction of building materials); erosion rates (greatly increased during building constriction, but may be decreased when urbanized).

Table 11.2 PERCENTAGE IMPERVIOUS AREA ACCORDING TO TYPE OF URBAN LAND USE (INDICATIVE FIGURES FROM BRABEC ET AL., 2002)

Type of urban space usage	Indicative percentage impervious area
Agricultural land/open space	up to 5%
Public and quasi-public	50–75%
Parks	up to 11%
Golf courses	up to 20%
Low-density	up to 19%
Medium-density	up to 42%
'Suburban' density	up to 25%
High-density	up to 60%
Commercial and industrial	up to 90%, usually greater than 60%
Highways	up to 100%

the most significant human impact, is composed of the buildings, roads, sidewalks/pavements and parking lots/car parks which are covered by impervious materials such as stone, brick, asphalt and concrete. Such surfaces are not rendered completely impervious and the percentage of impervious area varies according to the type of land use (see Table 11.2). Although it is tempting to assume that much of the urban area is completely impervious, there is a range from up to 19% impervious in low density housing areas to as much as 60% in high density housing areas, with commercial and industrial areas up to 90%, usually greater than 60%, and roads and highways up to 100% impervious. These figures, derived by combining figures from a number of different areas (Brabec et al., 2002), demonstrate the range that can exist (see Table 11.2). It has been suggested that, in the USA, impervious urban areas total some 110,000 km² (*Eos*, 15 June 2004), equivalent to an area nearly as great as the state of Ohio. The importance of impervious areas means that projects are under way to refine knowledge of their extent, including mapping impervious surfaces for the entire United States as one of the major components of the circa 2000 national land cover database, and also to demonstrate how their recognition can be ascertained by remote sensing. Thus a two-year mapping project detailed the extent of impervious surface area (ISA) for the state of Pennsylvania, by using satellite measurements to derive fractional vegetative cover maps (data for 2000 is available at the Pennsylvania Spatial Data Access, PASDA, website at www.pasda.psu.edu/access/newdata.shtml).

Impervious areas affect earth surface processes in three ways: they alter existing processes, they preclude others, and they also introduce new ones.

11.3.1 ALTERATIONS OF EXISTING PROCESSES BY IMPERVIOUS AREAS

Alterations arise as impervious areas reduce infiltration so that less water reaches the soil and groundwater recharge decreases, so that more of the precipitation

received flows across impervious surfaces as overland flow. A greater proportion of precipitation is discharged from urban areas, with more rapid runoff
shown by stream hydrographs that are much more peaked than those of comparable rural areas (see Figure 11.3). Increased amounts of surface runoff are
augmented by the surface water runoff system which collects water from road
and roof surfaces. Stream discharges from urban areas tend to demonstrate
higher peak flows and lower base flows, and the flood frequencies of rivers
draining urban areas also differ significantly from those before the urban area
existed. Urban areas also generate distinctive water quality characteristics,
including water temperatures often higher than those of rural areas, higher
solute concentrations reflecting additional sources including pollutants, and
suspended sediment concentrations, high during building activity but lower
after urbanization when sources are no longer exposed (see Table 11.1). Urban
ecology is changed as a result of exterminations and introductions: although
once thought of as ecological deserts, urban environments are now known to
support a variety of plant and animal species.

In addition to pollution in urban areas, problems arise from increased flooding, both within and downstream from urban areas, from sediment transport
leading to erosion rates up to 40,000 times greater than pre-disturbance rates
(Harbor, 1999) and sediment yield increased by up to 300 times by building
activity (Chin, 2006). Hydrographs from urban areas have peak discharges typically 2 to 4 times greater as a result of urban development, lag times decreased
to less than one-half former values, so that modelling urban hydrology is important in the design of the infrastructure of urban areas. It is not possible to prevent urban flooding entirely but structures and drainage systems can be designed
for precipitation events of specific characteristics and recurrence intervals.
However in Sydney, Australia, a flood of 17 August 1998, following up to
249 mm rain in 3.5 hours, saw close to a 100% rainfall-runoff relationship,
resulting in widespread erosion, especially where urban development had
encroached on channels (Reinfelds and Nanson, 2001). Planning policies in place
prior to the event were insufficient to prevent damage to some recent developments because in Sydney, as elsewhere, the stormwater drainage systems were
designed for different conditions (see Riley et al., 1986; Warner, 2000).

Many changes to stream and river channels in, and downstream from, urban
areas occur. This was envisaged in a perceptive diagram by Wolman (1967)
showing how sediment yield altered with land use changes and was associated
with changes in channel condition (see Figure 11.4). This general model relates
to the sequence of land use change from forest to farmland to building activity
to urbanization, because any area in the world is somewhere along that progression. The time scale in Figure 11.4, conceived with reference to the eastern part
of the USA, would be adjusted to more than 1000 years for the Old World.
Although this model provides a broad context for the change likely to occur,
individual research investigations have provided clear evidence for the existence
of larger channels in urbanizing rivers: it has been found that the amount of
channel change varies substantially with increases of channel cross-sections up

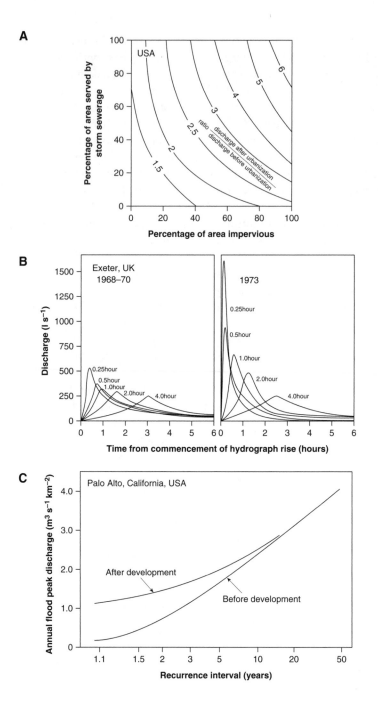

Figure 11.3   Hydrological effects of urbanization (from Walling, 1987)

(A) The relation between degree of urbanization and magnitude of the mean annual flood for a 2.59 km² basin; (B) comparison of unit hydrographs for natural and developed (1973) conditions for an Exeter catchment undergoing building activity; (C) compares hypothetical flood frequency curves for a basin under natural and urbanized conditions.

to 15 times, but usually between 2 and 4 times, and sometimes with no change or even a decrease, especially where large amounts of sediment are available. Change in channel cross-section can be expressed as the channel change ratio (cross-section after change/cross-section before change) but in some cases there can be changes of other characteristics such as channel sinuosity or channel slope. There are therefore significant differences between urban areas, and there are also contrasts in the time taken for channel adjustment to occur, and reaction and relaxation times (see Figure 5.2) for these adjustments can be several years to half a century for the former and several decades for the latter.

As urban areas are associated with flooding and drainage problems, urban drainage systems have been developed for 5000 years (Chocat et al., 2007) with some of the earliest urban drainage structures built during the Mesopotamian Empire (Wolfe, 2000). However drainage systems can further increase the amount of surface runoff and its speed of concentration, illustrated in the Italian city of Siena, where the Piazza del Campo (see Figure 11.5) evolved into its present fan shape, like a scallop shell, in the Middle Ages.

In addition to drainage and flood protection, the need for sanitation and for the provision of safe drinking water in urban areas meant that urban water management became a subject in its own right. Whereas the streams and ditches draining through cities were initially used for disposing of both storm and waste water, such streams became increasingly noxious sources for diseases including cholera and typhoid. Hence the sewer was invented as a way of covering up stream channels and conveying the water and its contents away from urban areas. Combined sewer and stormwater drainage systems were developed in Victorian times in the UK where many cities still have these underground drainage lines and sewers constructed more than a century ago. One consequence is that, as cities continue to grow, their demands exceed the capacities of the original stormwater and sewer systems, but the inherited systems are not easy to modify. Water supplied is often all of one quality despite its requirement for many purposes other than drinking: a combined sewer and stormwater system means that purification is an expensive process because all the water goes to the treatment works. Public hygiene and flood protection are major objectives of urban drainage, supplemented more recently by environmental concerns. Globally, however, still only about 15% of wastewater is treated and there is a great range of provision in urban areas across the world according to the ways in which development has, or has not, taken place.

Pollution in urban areas can arise from sewage, but also from oil and heavy metals from vehicles together with many forms of waste and debris. Existing problems (Chocat et al., 2007) centre upon the increasing quantities of water draining from impervious areas with associated flooding, low flows, river channel erosion, and degradation of biological habitats, together with the deteriorating quality of water drained from urban areas which may include heavy metals, nutrients, contaminated sediments, complex organics and pathogens. Such problems have been supplemented more recently by concerns for landscape aesthetics, ecology and beneficial uses. These occur in a context which is often characterized by an ageing infrastructure with impaired performance of wastewater

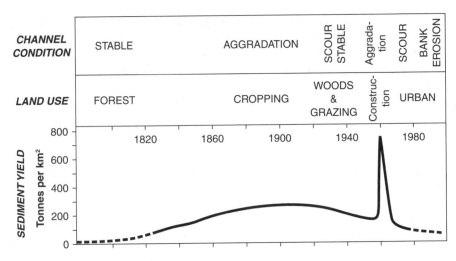

Figure 11.4   Change of sediment yield and channel change in urban areas (modified from Wolman, 1967)

Figure 11.5   Piazza del Campo, Siena

This was paved with red brick as early as 1349, divided into nine sections in honour of the Council of nine rulers of the city at the time. Each section was divided from the next by a shallow channel of travertine carrying drainage directed towards the *gavinone*, the central drain at the apex of the scallop shape.

treatment plants, and with the size of stormwater drainage pipes exceeded more frequently than their design should allow (a graphic example of culvert failure in Freeport, Maine, USA photographed by Kevyn Fowler during a storm is provided at http://www.wmtw.com/video/17144859/index.html, accessed 25 May 2009).

Other earth surface processes are also increased or decreased in the urban area, perhaps most dramatically in permafrost areas: if precautionary measures are not taken, urban development affects the heat balance of the surface layers, potentially leading to thawing of the permafrost and thence to the development of thermokarst features (see p. 176). This sensitivity is exacerbated by global warming, explaining why some of the areas most sensitive to climate change impacts are to be found in high latitudes. Yakutsk in Eastern Siberia, covering an area of about 135 km^2 with a population of c.240,000, is one of the largest world cities built on continuous permafrost. This city, with January temperatures averaging −42.8°C, is the location for the *Permafrost Research Institute* founded in 1941. Most buildings in Yakutsk are built on underground concrete piles, varying in depth according to the size of the building. Some ancient wooden houses, not built on piles, are heavily deformed as a result of adjustment of underlying permafrost.

Landslides can be responsible for hazards in urban areas (Alexander, 1989) and their incidence may be increased by the way in which impervious areas affect soil moisture and groundwater or by the loading effect of buildings on slope materials. Especially in tropical areas if the deeply weathered layers receive more water as a result of urban development, then mass movements are increased. However in some areas like Hong Kong the mass movements were more numerous where forest remained and water continued to infiltrate, whereas cleared areas allowed the water to flow over the slopes during storms of 1966 (So, 1971). Especially on steep slopes, land cleared for building activity can provide very large sediment yields either by sheet erosion or by gullying, often in illegal settlement areas where environmental considerations are less observed. In São Luís, Brazil, deforestation during irregular, unplanned and unauthorized urban settlement expansion (therefore lacking sanitation, rain pipes and paved roads), prompted land degradation and initiated gully formation which might be controlled by palm-mat geotextiles (see Figure 12.4; Guerra et al., 2009).

11.3.2 PROCESSES PRECLUDED BY IMPERVIOUS AREAS

Processes precluded in urban areas are where the surface is covered in impervious materials so that some forms of weathering are not possible, and slope and channel processes are inhibited. Although in such cases they may be perceived as non-existent, some weathering or slope activity may occur. This is analogous to flood plains where flooding is perceived to be precluded as a result of flood-prevention measures although experience has shown that the flood hazard may actually have increased.

11.3.3 NEW PROCESSES INTRODUCED BY IMPERVIOUS AREAS

New processes can be introduced by urban conditions. Withdrawal of groundwater or of minerals, especially oil, from below the surface may lead to subsidence on the surface. Subsidence may also occur as a result of compaction of sediments due to the pressure of buildings – the leaning tower of Pisa is a classic example. Subsidence recorded can be quite substantial with up to 10 m in Los Angeles above the Wilmington oil field; over 60 km² of Tokyo lies below sea level as a result of removal of groundwater from alluvial deposits below the central part of the city (Douglas, 1983); and the subsidence of Venice (Figure 4.12) is due to the extraction of groundwater and/or extraction of natural gas. Salt weathering, caused by salt crystallization, salt hydration or the thermal expansion of salts, can be significant below urban areas such as Bahrain (Cooke et al., 1982), because sodium salt concentrations in groundwater can affect concrete in buildings and foundations.

Urban areas induce urban climates because cities disrupt the climatic properties of the surface and of the atmosphere, altering the exchanges and budgets of heat, mass and momentum with a range of consequences. Localized effects of urban climate tend to be merged above roof level, forming the urban boundary layer (UBL), and below roof level is the urban canopy layer (UCL). As more heat is used to warm the air and the ground in urban environments, the relative warmth of the city provides its urban heat island and a large city is typically 1.3°C warmer annually than the surrounding area, although the heat island varies diurnally, with up to 10°C difference near midnight. The existence of urban climates not only affects earth surface processes, but cities influencing local and regional climates with the potential consequence of more intense storms can in turn be responsible for more frequent flood events.

11.4 MANAGEMENT OF URBAN ENVIRONMENTS

A range of ways, many inherited from different past conditions, have been developed to deal with water disposal and distribution, and with sediment problems as indicated in Table 11.3. During the course of further urban expansion the initial response to such problems was that, as Luna Leopold (Box 4.1) indicated in 1977, technology would fix it. One such response is channelization which includes processes of river channel engineering for the purposes of flood control, drainage improvement, maintenance of navigation, reduction of bank erosion or relocation for highway construction (Brookes, 1988: 5), often implemented to remove, or at least moderate, the flood hazard or to inhibit channel erosion. However channelization could be responsible for transferring the flood hazard from the urban area to another area downstream, and channelization dramatically altered the landscape from one with stream and river channels to one dominated by stark concrete channels (see Figure 11.6).

With greater awareness of the benefits of a more sustainable holistic approach (Downs and Gregory, 2004), alternative ways have been sought for managing hydrological processes in urban areas. Although not applicable everywhere, at least five types of approach are available (Chin and Gregory, 2009):

- review the characteristics of the urbanizing area before finalizing the nature, density and potential extent of new urbanization, in some cases deciding to locate urban expansion in an alternative area. So-called smart growth approaches have been conceived to minimize the runoff impact of urbaniza-tion (Tang et al., 2005). However in many cases the location of the urban expansion often cannot be altered.

- retain precipitation to reduce the speed of water movement through the urban area. For example, rainwater harvesting collects the rain which falls onto roofs of buildings and stores it in tanks until required for use. The water can be pumped to the point of use, thus reducing the demand for mains water. In this process, a volume of water is kept out of the stormwater management sys-tem, thereby helping to reduce flooding risks (Freerain Rainwater Management Solutions, 2006).

- delay runoff by slowly releasing rainwater from roofs of buildings using narrow down pipes, or by collecting surface water in gardens followed by its slow release or by other measures to delay runoff (see Table 11.3). For years, towns and cities have interrupted the transfer of water from roofs of buildings to the stormwater drainage system, for example by downpipes on the side of buildings releasing water to flow over pavements, sidewalks and roads before joining the stormwater drainage system. Such overland flow is slower than flow in pipes and may produce short-duration, localized flooding but reduces the rate of water transfer through the urban area. Indeed, in some areas where water collected on roof surfaces was used for domestic water supply, it was only with the installation of a piped water supply that the roof storage was no longer collected, resulting in a sudden increase of urban drainage into local rivers, as experienced in Armidale, NSW in the 1960s (Gregory 1977, 2002). Where appropriate, constructing a detention pond, balancing pond, or settling basin to collect water and sediment from peak flows enables the water to be released slowly – thus protecting downstream reaches from the highest peak flows. Alternatively reservoirs above the urban area can be maintained at 50% capacity so that they absorb storm runoff and so avoid rapid transmission of flood flows to urban areas downstream.

- methods of managing feedback consequences in the urban area, as alterna-tives to channelization which involved enlarging natural stream channels by widening or deepening, with other techniques have been developed (see Table 11.3) to smooth banks and straighten the course of the stream (Riley, 1998).

- mitigation of downstream consequences, often involving channelization in the past, but now resolved by other techniques where appropriate and feasible.

Table 11.3 SOME MANAGEMENT METHODS AVAILABLE FOR CONTROLLING HYDROLOGY OF URBAN AREAS (SEE CHIN AND GREGORY, 2009)

Management objective	Techniques included
Land-use planning – *determine where to locate urbanization*	Integrated basin management – *locate urban areas according to their potential impacts on the basin* Smart growth – *minimize impact of urban sprawl on runoff and systems affected by runoff processes*
Retention of precipitation – *reduce runoff production*	Rainwater harvesting (rain from roofs to tank storage) Road surface detention Disconnect roof areas from stormwater drain systems Rain gardens on housing plots – *encourage infiltration and pollutant removal* Reduce impervious area – *allow more infiltration* Flat roofed houses and roof detention
Delay of runoff – *reduce rate at which urban runoff is transmitted and conveyed*	Underground storage reservoirs – *slow release of stormwater* Collection of water on roof gardens, brown roof, green roof Downpipes on to pavements and roads – *not directly connected to stormwater drainage system* Soakaways Filter drains – *linear trenches of permeable material* Minimize connections between impervious surfaces Permeable pavement Detention ponds, balancing ponds Infiltration basins, bioretention areas, infiltration trenches Water conservation structures Sustainable urban drainage systems Low impact development techniques
Management of effects in the urban area – *mitigate likely consequences of urban drainage*	Separation of foul water and stormwater systems Restoration of baseflows – *groundwater cultivation by construction that facilitates infiltration* Reduce channel velocities and accommodate or delay pollutant loads Permeable revetment Swales – *shallow vegetated channels* Excavation of pools, plunge pools Channel restoration or rehabilitation Increase residence time in channels Set backs from the channel Filter strips – *drain water from impermeable areas and filter out silt* Sediment traps in channels Preservation of wetlands, floodplains, tree cover – *increases infiltration and reduces storm runoff* Daylighting – *excavation of culverted or buried streams*
Planning for downstream consequences – *minimize downstream effects*	Total catchment management – *zoning and ordinances to preserve open spaces* Channel management including channelization where necessary Protection of stream corridor Education – *to preclude or restrict dumping of debris in channels*

Figure 11.6 Channelization in the Los Angeles river (above) and Barranco del Rev Arena, south-west Tenerife (below)

Urban stormwater management systems now reflect the fact that urban drainage has moved away from the conventional thinking of designing for flooding towards balancing the impact of urban drainage on flood control, quality management and amenity (CIRIA, 2001). Sustainable Drainage Systems (SUDS) may be preferred to conventional drainage methods because they aim to manage runoff flow rates, reduce the impact of urbanization on flooding, protect or enhance water quality, serve the needs of the local community in environmentally friendly ways, provide habitat for wildlife in urban watercourses and, where appropriate, encourage natural groundwater recharge (Herrington Consulting, 2006). Developing the approach of Ian McHarg (McHarg, 1969, 1992; also see Box 11.1), low-impact development techniques (LID) also enable environmentally friendly land use planning by including a suite of landscaping and design procedures that attempt to maintain the natural, pre-developed ability of a site to manage rainfall. This approach also embraces the idea that stormwater is not merely a waste product to be disposed of; rather, it can be a resource (Massachusetts Government, 2006).

Management is now undertaken with greater reference to environmental considerations, sustainability and adopting a **holistic view**. For example the European Commission (18 February 2004) revealed plans to improve environmental aspects of towns and cities with a new EU wide strategy aiming to provide a 'best practice' style approach with successful projects implemented on a widespread basis across the Union. McHarg (Box 11.1) noted that geomorphology had not been considered as much as it could have been in urban planning (McHarg, 1996), thus raising the question why is it important? Ways in which geomorphology can be relevant are:

- *Contributing to the choice of areas for urban development according to environmental suitability, an approach involving urban land evaluation.* The essence of McHarg's approach was that in planning urban areas there should be an attempt to design with nature and therefore to embrace 'natural' conditions as much as possible. Land suitability analysis considers the spatial distribution of environmental characteristics in relation to their suitability for particular urban land uses. Thus some slope angles or some surface materials are inappropriate for certain types of urban land use. Using what McHarg characterized as a 'layer cake approach' it is possible to superimpose characteristics to determine the most suitable conditions for particular types of urban land use.
- *Recognizing environmental problems that may arise as a consequence of urban development.* Problems occur because of the way in which geomorphological processes – weathering (e.g. salt weathering, duricrusts), fluvial (e.g. flooding problems, sediment, erosion, alluvial fans), slope stability (landslides, subsidence, hydrocompaction), and aeolian (wind erosion, loess) – are affected by urban environments. Problems that may arise in the areas considered in the previous four chapters are shown in Table 11.4. Urbanization can affect ecosystem processes, both intentionally and unintentionally, as illustrated for Indian Bend Wash, Arizona where the alterations of land cover, stream channel structure, and hydrology involved the creation of new channels and artificial lakes (Roach et al., 2008).
- *Identifying when environmental problems of urban areas may produce urban hazards.* In addition to recognizing potential hazards it is also necessary to establish their likely frequency. Many aspects of urban design relate to design frequencies, in the way that the flood hazard can be expressed in terms of recurrence interval. Culverts, channelized rivers, or flood protection schemes may be designed to accommodate the 100 year event. However three problems which may arise are: (1) the 100 year event may

Table 11.4 ENVIRONMENTAL PROBLEMS THAT MAY ARISE IN THE FOUR AREAS CONSIDERED IN CHAPTERS 7–10

Area	Geomorphological problems	Associated problems
Arctic, Antarctic and high latitudes	Permafrost, active layer	Special types of construction for buildings and infrastructure.
Temperate and Mediterranean	Mass movement including landslides, flooding, coastal erosion	Hard engineering solutions sometimes necessary because softer solutions not always feasible.
Arid	Wind erosion, dune migration, flash floods, salt weathering	Devise management methods to cope with extreme events.
Humid and seasonally humid tropics	Deep weathering, frequent storm events, rapid surface erosion, frequent landslides; sea level rise	Unplanned urban expansion occurs rapidly and subject to **environmental hazards**.

occur next year and there is no guarantee that it will not occur again in the subsequent century; (2) there is a tendency for the public to assume that a 100 year design discharge provides insurance for 100 years; (3) a 100 year event selected several decades ago may become less appropriate as continuing urbanization increases the frequency of flooding. Many developing countries, located almost entirely within the tropics, have rapid urbanization with cities in hazardous or environmentally sensitive areas, requiring the interfacing of geomorphology with engineering practices and urban planning, particularly for cities like Kingston or Bangkok where hazards are so acute and widespread that a practical solution is difficult to achieve (Gupta and Ahmed, 1999). In addition to individual hazards the spatial distribution of hazards can be important, as demonstrated for Fountain Hills, Arizona (Chin and Gregory, 2005) where problems associated with stream channel change due to urbanization were identified (see Table 11.5).

- *Devising ways to avoid, mitigate or at least minimize environmental problems and hazards and their impacts, employing new innovative techniques and embracing a holistic view wherever possible.* A range of ways is now available for management of the physical aspects of urban environments (see Table 11.6) which can include avoiding new impervious surfaces that block natural water infiltration into the soil by making more use of pervious concrete or porous pavement, employing material that offers the inherent durability and low life-cycle costs of a typical concrete pavement while retaining stormwater runoff and recharging local groundwater systems. In addition maintaining natural areas within urban environments is essential for both survival of resident plants and animals and for wellbeing of inhabitants. Wise urban planning should be not only functionally diverse and aesthetically appealing, but also include self-perpetuating ecosystems that require minimal maintenance and are sustainable.

Table 11.5   STREAM CHANNEL HAZARDS (THIS EXAMPLE RELATES TO CHANNEL TYPES IN NORTHERN FOUNTAIN HILLS, ARIZONA – SEE CHIN AND GREGORY, 2005)

Hazard associated with	Manifestation
Channel system	Compartmented by road and rail network
Flood frequency	Increase
Drainage	Temporary floods
Bank erosion and scour	Along channel
	Downstream from crossings
	Below culverts
	Behind revetment
Aggradation and urban debris accumulation	Along channel
	Above crossings
	Buried structures
	Contracted bridge openings
Blockage	Due to culvert size or bridge opening
Quarrying of channel sediments	Dredging

Table 11.6   SOME METHODS USED IN MANAGING URBAN ENVIRONMENTS (SEE ALSO TABLE 11.3)

	Morphology and soils	Hydrology	Urban climate and ecology
Mitigation and minimization of impacts	Minimize loading effects of buildings Minimize exposure of bare soil (temporary ground cover – geotextiles, mulches, plastic sheeting) Control water and sediment on building sites Landslide protection measures Insulating procedures in permafrost areas Protection against salt weathering in drylands	Flow velocity reduction Channelization Floodplain levees land use zoning Diversion channels for water from construction sites Recharge of aquifers Land use regulation Flood insurance	Control programmes for atmospheric emissions Reduce pollution (to encourage organisms to return) Remove undesirable species Reduce use of fertilizers and pesticides
Design aspects	Slope stabilization and design Integrated planning for river corridors	Stream restoration (e.g. Urban Streams Restoration Act in California 1984) SUDS sustainable	Tree planting (to augment tree biomass and diversity and influence atmospheric environment)

Table 11.6   *(Continued)*

	Morphology and soils	Hydrology	Urban climate and ecology
	Include soil as integral component of park planning and management	drainage systems – militate against flooding and pollution Minimize connections between impervious surfaces Local storage such as rain gardens in each garden/ backyard to reduce runoff	Street tree planting Create urban nature reserves Establish conservation areas and natural reserves Botanical gardens and parks golf courses, public gardens, backyards Window boxes

- *Employing experience gained from research investigations to devise recommendations for urban management.* Research results on the impacts of urbanization on channel processes and morphology were the basis for a protocol developed by Chin and Gregory (2009) which could be considered during the management of river channels in urban systems.
- *Contribute to restoration schemes when these are implemented.* The objectives of restoring a channelized river (Downs and Gregory, 2004: 240–1) can range from making the river look as natural as possible (naturalization), restoring it to some former condition, or assisting it to adapt to a new environment (rehabilitation). Whichever approach is adopted it is necessary to consider what is a natural river (Wohl and Merritts, 2007) according to the particular geomorphological environment. It has been suggested that so far success rates for restoration schemes are low, that concerns are growing that conventional approaches to river restoration may be fundamentally flawed, and that particular issues arise in the case of urban environments (Clifford, 2007). Although the need to rehabilitate urban rivers has been questioned (Findlay and Taylor, 2006), in many areas the legacy and consequences of harsh engineering can be softened by a more holistic approach. Some restoration has involved daylighting, which is the resurrection of streams that were in underground culverts, and re-creation of streams on the surface. In Sutcliffe Park in south-east London, UK, the Quaggy river was placed in an underground culvert in the 1930s but, as part of a flood management programme, it was daylighted in 2004 (see Figure 11.7A) and by 2007 the environs of the river had been restored (see Figure 11.7B).
- *Establishing the environmental characteristics that people want, by investigating public perception and then suggesting appropriate initiatives.* Whereas nature used to be seen as a machine that could be engineered to provide the maximum output of desired products, this was superseded in the mid–late twentieth century by a more sympathetic attitude. This is exemplified by the suggestion for rivers that 'Where cities once exploited, abused and then ignored rivers in their midst, they are now coming to recognise, restore and

Figure 11.7   Daylighting Sutcliffe Park, south-east London, UK

The river Quaggy was placed in an underground channel in the 1930s; the park was originally opened in 1937. Downstream flooding problems led to the daylighting of the river Quaggy in 2004 (top) in a re-landscaped park designated as a Local Nature Reserve in 2006 and shown here (below) in 2007, two years after the re-landscaping.

appreciate them' (Bolling, 1994: 207). In attempting to make urban environments more natural do we fake nature or endeavour to recreate it? We should certainly obtain views of the public as to the environment that they prefer. Although inappropriate for the Australian urban environment, many gardens in Australia originally imitated the British style and, not until a new Nationalism was born in the 1950s, were lawns replaced by bush-floor effect with promotion of indigenous plants and maintenance of the bush garden ethos. Thus analysis of the way in which residents perceive their front and back gardens (yards) demonstrates how the residential landscape reflects expressions of self, status and conceptions of place that combine to create little understood 'dreamscapes' (Larsen and Harlan, 2006). Is this the manifestation of individual creation of environment, and sometimes landforms, in the urban environment?

These are some of the ways in which geomorphology can be helpful in the management of urban areas although any geomorphological input is often combined with other earth and environmental sciences. This multidisciplinary approach and the above themes are pertinent to management of geomorphological resources considered in the next chapter.

BOX 11.1

IAN MCHARG (1920–2001)

Ian McHarg, a landscape architect, published his book *Design with Nature* in 1969 pioneering an approach of ecological planning for urban areas, including ideas of significance well beyond landscape architecture itself. In this and his other multidisciplinary work he recognized that geomorphology was not involved as much as it should have been: in his autobiography (1996: 91), commenting that 'geography and the environmental sciences were conspicuously absent', and that there was 'a general conclusion that geomorphology was the integrative device for physical processes and ecology was the culminating integrator for the biophysical. These contributed to understanding process, meaning and form' (1996: 331).

Described as one of the great cultural figures in twentieth century planning and design, Ian McHarg was born in Clydebank, Dunbartonshire, Scotland, the son of a Church Minister in the depressed, near industrial, poverty-stricken Glasgow. His early experience, together with awareness of the rural countryside of the Firth Valley and the Western Highlands, had an early formative influence, probably providing the foundation for his later conviction that cities need to accommodate the qualities of the natural environment more effectively. After war service, he travelled to the USA, studying for degrees in landscape architecture and city

(Continued)

(Continued)

planning in the Harvard School of Architecture. Subsequently appointed at the University of Pennsylvania, he became Professor 1954–86, and he co-founded the Department of Landscape Architecture and Regional Planning at the university, serving as its chairman from 1955 to 1986. He brought together a unique faculty of geologists and hydrologists, ecologists, cultural anthropologists and even epidemiologists, to teach in the department alongside the design professionals. In 1960 to put his ideas into practice he co-founded the Philadelphia firm of Wallace, McHarg with city planner David Wallace; it later became Wallace, McHarg, Roberts and Todd and continues as WRT. He remained with the firm until 1981, creating a succession of planning and design projects for urban, metropolitan and rural regions which are still highly regarded models for ecological planning and design.

McHarg was one of the first to recognize the lack of knowledge of the environment involved in planning, design and engineering. His ideas were first developed in respect of the city where it was suggested (McHarg and Steiner, 1998) that an immeasurable improvement could be ensured in the aspect of nature in the city, in addition to the specific benefits of a planned watershed. His ideas contrasted with the French style of garden design which he saw as the subjugation of nature: he offered an alternative design with nature philosophy. His brilliantly titled 1969 book *Design with Nature* appeared after Rachel Carson's *Silent Spring* in 1962 but communicated ideas, intent and language that may have influenced the 1969 National Environmental Policy Act and other legislative instruments in the USA. The ecological planning method developed by him explored the physical, biological and social processes that shape particular places. As each layer of information was superimposed on top of the previous one, the primary patterns of the landscape were identified to guide the form of development – an approach that was a foundation for Geographical Information Systems now widely used as planning tools.

His lasting contribution was his approach, and which was sympathetic to the environment and was multidisciplinary, contrasting with the fragmentation of disciplines and lack of communication between them. His aim was to design and implement a balanced and self-renewing environment. His internationally acclaimed contribution was recognized by the Harvard Lifetime Achievement Award, the Thomas Jefferson Foundation Medal in Architecture, and 12 other international medals and awards, including the very prestigious Japan Prize in City and Regional Planning, and he was the only landscape architect to receive the US National Medal of Art. In 2001 Carol Franklin wrote in his obituary in *The Independent* that he was a charismatic and powerful public figure who was brilliant, poetic, funny, irrepressible and sometimes abrasive, able to go right to the heart of the matter, once telling the Japanese in a speech, 'Everything we [the West] have done badly you have done faster and worse', or announcing to a gathering of *Fortune* 500 executives that 'the time has come for American industry to be toilet-trained'.

FURTHER READING

A comprehensive outline was given in:

Douglas, I. (1983) *The Urban Environment*. Arnold, London.

A thought-provoking read is still provided by:

McHarg, I.L. (1969) *Design with Nature*. Natural History Press, New York.
McHarg, I.L. (1992) *Design with Nature*. Wiley, Chichester.

TOPIC

1 Ian McHarg stated (1996: 91) that geomorphology should have been more involved in urban planning and design with nature. His book was published in 1969 – why did geomorphology ignore the opportunity for so long?

PART V

MANAGEMENT OF THE LAND SURFACE – FUTURE PROSPECTS

12

FUTURE LAND SURFACES – MANAGEMENT OF CHANGE

Many books on geomorphology devote their final chapter to applications or management; it is very beguiling to think of saving the land surface of the Earth but what can be managed and what are the management issues that interest the geomorphologist? Some pointers for urban areas were given at the end of the previous chapter, but to answer these questions we can focus on past, present and future. *Past*, because it draws our attention to what was perceived to be required and what has already been done; *present*, for the knowledge of what is now being done and how; and *future* so that we might learn from past and present imperatives and mistakes in order to ask what are we, and should we be, trying to achieve?

12.1 THE PAST – CONTAINING AND CONSTRAINING MANAGEMENT

Management, involving control of the land surface, has been undertaken almost as long as humans have occupied the Earth. Over several thousand years control has been necessary for reasons of resources, hazards or risks, and more recently for preservation or conservation. The need for control is well illustrated by management of rivers on the Earth's surface: six broadly chronological but overlapping phases (see Table 12.1) indicate how control has progressed, although not all phases may apply in any one geographical region. Control and diversion of river flows initially occurred when agricultural communities included irrigation systems, exemplified by hydraulic civilizations. Dam construction and land drainage also occurred prior to 2000 years ago (Phase 1, Table 12.1). A second phase after the end of the Roman Empire (in the fifth century AD) lasted for a thousand years when deforestation was extensive, but local and small-scale river use devolved upon agriculture, fishing, drainage, water mills and navigation; irrigation for agriculture had long been deployed and flood farming widely used in North America and North Africa for example. A third phase, beginning with the industrial revolution, saw watermills, for example along rivers in the United States and in Europe at the end of the eighteenth century, transport of industrial materials along rivers, the innovation of hydroelectricity generation, and then thermal generating stations utilizing river water for cooling; technological developments enabled the establishment of water supply systems, involving impoundment and direct extraction of river water.

Table 12.1 CHRONOLOGICAL PHASES OF RIVER USE (DOWNS AND GREGORY, 2004) PROVIDE A CONTEXT FOR PHASES OF RIVER CHANNEL MANAGEMENT

Chronological phase	Characteristic developments in	Management methods involved
1 Hydraulic civilizations	River flow regulation Irrigation Land reclamation	Dam construction River diversions Ditch building Land drainage
2 Pre-industrial revolution	Flow regulation Drainage schemes Fish weirs Watermills Navigation Timber transport	Land drainage In channel structures River diversions Canal construction Dredging Local channelization
3 Industrial revolution	Industrial mills Cooling water Power generation Irrigation Water supply	Dam construction Canal building River diversions Channelization
4 Late nineteenth to early–mid twentieth century	River flow regulation Conjunctive and multiple use river projects Flood defence	Large dam construction Channelization River diversions Structural revetment River basin planning
5 Second part of twentieth century	River flow regulation Integrated use river projects Flood control Conservation management Re-management of rivers	Large dam construction River basin planning Channelization Structural and bioengineered revetments River diversions Mitigation, enhancement and restoration techniques

Table 12.1   *(Continued)*

Chronological phase	Characteristic developments in	Management methods involved
6 Late twentieth and early twenty-first century	Conservation management Re-management of rivers Sustainable use river projects	Integrated river basin planning Re-regulation of flow Mitigation, enhancement, and restoration techniques Hybrid and bioengineered revetments

A fourth period (see Table 12.1) in the late nineteenth and the first half of the twentieth century benefited from new technology, perhaps being characterized by the imperative 'technology can fix it' (Leopold, 1977). This applied to more integrated water supply systems, catchment-based management, waste water disposal, more effective flood defence, big dams as well as small ones, and large multipurpose schemes such as the Tennessee Valley Authority (TVA) scheme of river basin management. In that case, in order to achieve the two original primary purposes of flood control and navigation development, the TVA developed the Tennessee River and its tributaries into one of the most controlled river systems in the world, with hydro-electric power generation facilities at dams, soil erosion control, reforestation, improvement in agricultural land use, and increased and diversified industrial development throughout the basin. In a fifth phase, in the second half of the twentieth century, flood control projects became more extensive together with clearing of river corridors often requiring extensive channelization; dam building increased, with the 1960s having the greatest number of dams built in any decade of the twentieth century; and rivers were extensively regulated and systems fragmented into sections by dams, at a time when the further progress of integrated river basin management included land and water management together under a unified administration. Concerns over the impact of channelization schemes accompanied increased **environmental awareness** and the growth of conservation movements was reinforced by legislation enacted from the 1970s, so that channel management alternatives were sought, including techniques for mitigating the impact of management projects, culminating in river channel restoration. The major phase of dam building finished at the end of the 1980s. The very late twentieth century saw the beginning of a sixth phase with alternative strategies invented for managing the legacy of past uses of the river, for dealing with contemporary hazards and for considering sustainability, especially in relation to the impacts of global change. Although some dam building continues, the great phase of dam building has been succeeded by retirement of dams or at least by changes in their operating rules, with recognition that 75,000 dams in the USA now fragment the natural flow characteristics and ecology of once integrated river systems (Graf, 2001). Schemes for water resources development and for flood control and hazard avoidance now require integral conservation measures, are often constrained or challenged by previous management operations, and public participation in decision making is a regular occurrence.

Table 12.2   HAZARDS FOR WHICH MANAGEMENT HAS BEEN UNDERTAKEN

Earth hazard

Floods and flash floods
Soil erosion
Landslides
Avalanches
Subsidence
Coastal erosion
Glacier hazards
Endogenetic: earthquakes, volcanic eruptions
Climatic hazards: drought, hurricanes, tornadoes, lightning and severe thunderstorms, hailstorms, snow storms, frost hazards
Wildfires
Dam disasters
Desertification
Tsunamis
Soil heave and collapse

These six phases (see Table 12.1) are echoed by management related to other aspects of the land surface, with many intended to avoid, or reduce the impact of hazards. Of 21 natural hazards (Alexander, 1999), some have been the reason for major management modifications of the land surface, although others may have prompted changes in particular areas (see Table 12.2). Floods and flash floods have required protection of river channels by channelization (see, for example, Figure 11.6) and of coastal zones by coastal defences, in addition to implementation of catchment management methods designed to reduce the severity of flood impact. Management to combat soil erosion has been extensively used in those vulnerable parts of the land surface that are subject to farming for livestock, for agricultural crops or that are urbanized. Drainage schemes and structures have been installed to mitigate the effects of landslides, avalanches and subsidence, and protection of the coast has required structures designed to reduce or manipulate the processes of coastal erosion. Coastal management was required because in 2000 it was estimated that 75% of the world's population live in a 60 km wide strip along the coastal zone (Viles and Spencer, 1995).

Management of the land surface related to resources and hazards is long-established, but environmental awareness and conservation are more recent. Concerted moves to preserve and then to conserve areas of the land surface were made in the nineteenth century, but it was in the mid and late twentieth century that conservation really took hold. Growing environmental awareness was derived at least in part from Rachel Carson's (1962) book *Silent Spring*; subsequent contributions were the creation of Green Parties in many countries, many significant books (for example, Nash, 1967), and the growth of environmental ethics recognizing new concepts such as 'deep ecology', contending that all species have an intrinsic right to exist in the natural environment (Naess, 1973), in contrast to 'shallow ecology' where nature is valued only from an anthropogenic viewpoint (Lemons, 1987, 1999). Legislation enacted from the 1970s

in different countries began to reflect heightened environmental awareness
when projects were being considered. Sustainability, defined by the United
Nations in 1987, was reinforced by awareness of the limits of the world's
resources (*The Limits to Growth*), defined in *Our Common Future* (UNWCED,
1987). These were aided by *Caring for the Earth: A Strategy for Sustainable
Living* (IUCN et al., 1991) which developed from, and built upon, the World
Conservation Strategy published in 1980, including its third Principle as
'Conserve the Earth's vitality and diversity'. In addition growing awareness of
global change stimulated international conferences such as the United Nations
Conference on Environment and Development (UNCED) in Rio de Janeiro in
June 1992.

National responses included legislation directed towards conservation management of landscapes, including the National Environmental Policy Act
(NEPA) of 1969 in the USA which encouraged '... productive and enjoyable harmony between man and his environment; to promote efforts which will prevent
or eliminate damage to the environment and biosphere ...'. Environmental
Impact Statements (EIS), arising from Section 102 (c) of NEPA, had profound
significance for development proposals. The purpose of **Environmental Impact
Assessment** (EIA) analysis is to estimate the effect on land, water, biota and
atmosphere that a particular development is likely to have and then to indicate
possible mitigations. From the USA, EIA was extended to other countries including Canada, New Zealand, China and the European Union. Specific approaches
to environmental management, including those necessary to control environmental hazards, increasingly required working in harmony with the environment, employing softer rather than hard engineering approaches. Thus in river
channel management, wherever possible, techniques including channelization
were replaced by softer techniques collectively described as 'working with the
river rather than against it'.

This illustrates how a context was provided for managing the land surface up
to the end of the twentieth century but how had geomorphologists been
involved? First by extending research to encompass more applied aspects, second by focusing on problems amenable to geomorphological solutions, and
third by identifying approaches which could complement investigations by
other scientists. Thus environmental geomorphology was defined (Coates,
1971) as '... the practical use of geomorphology for the solution of problems
where man wishes to transform landforms or to use and change surficial
processes', and geomorphic engineering combined 'the talents of the geomorphology and engineering disciplines. ... The geomorphic engineer is interested
in maintaining ... the maximum integrity and balance of the total land-water
ecosystem as it relates to landforms, surface materials and processes' (Coates,
1976: 6). Environmental geomorphology and geomorphic engineering enabled
significant progress but potential contributions of geomorphology to management and planning were outlined in collections of applications (e.g. Hooke,
1988) or in books such as *Geomorphology in Environmental Management*
(Cooke and Doornkamp, 1990). The need to interface with engineering was

reflected in *A Handbook of Engineering Geomorphology* (Fookes and Vaughn, 1986), later developed to *Geomorphology for Engineers* (Fookes et al., 2005). The latter volume introduces engineering geomorphology for practitioners and academics in branches of engineering as well as for physical geographers, geologists and hydrologists, contending that 'Engineering geomorphology complements engineering geology in providing a spatial context for explaining the nature and distribution of particular ground-related problems and resources ... concerned with evaluating landform changes for society and the environment' (Fookes and Lee, 2005: 26). Other books deal with specific parts of the land surface such as the coast (Viles and Spencer, 1995) and beach problems (Bird, 1996) in particular. However, what is viewed as geomorphology in one country may be partly or completely embraced by other disciplines elsewhere, so that *Environmental Geology* (e.g. Keller, 1996), covers material associated with the work of geologists in some countries, but traditionally covered by other disciplines including geomorphology elsewhere.

12.2 THE PRESENT – MANAGEMENT AND PLANNING APPROACHES

Planning is taken to involve an element of design in control and management so that the present can build upon the achievements of the past, learning from any mistakes that are perceived to have been made. Although attitudes to planning vary considerably from one country to another, four broad categories of applied research (Gregory, 2000) indicate the types of applications currently undertaken to which geomorphologists can contribute (see Table 12.3).

12.2.1 THE AUDIT STAGE

The audit stage (see Table 12.3), involving description, depiction and auditing of the environment as relevant to planning, is particularly concerned with how the land surface, its landforms and processes, can be described in ways directly appropriate to planning problems. This requires not only selecting relevant ways of characterizing landforms and processes but also of communicating information in such a way that it can readily be understood by planners, with GIS as an extremely appropriate tool for assembling and providing necessary information. Mapping is a key technique for auditing the environment, but morphological or geomorphological mapping (p. 25) may not be the most appropriate for planning investigations. Landscape ecology depicts physical environment in ways which indicate how the environment may be utilized for agriculture, forestry, residential or other purposes, differing from land evaluation which relates to a specific purpose. Such methods can support a regional survey but techniques are also necessary to describe the general context of a specific project site, or the

Table 12.3 EXAMPLES OF ASPECTS OF PLANNING TO WHICH
GEOMORPHOLOGICAL CONTRIBUTIONS CAN BE MADE

Type of planning activity	Examples
Audit stage: Description, depiction and auditing of physical environment	Erosion potential of coast Scenic character of landscape Slope erosion classification Land classification Landscape ecology Classification of land according to potential for soil erosion Growing season defined according to specified criteria
Environmental impacts: Investigation and analysis of environmental impacts	Extent of coastal floods Effects of dams on flows and channels downstream Drought impact in particular areas Effects of crop practices on soils Accumulation of metal contaminants in soils Impact on wetlands Impact of pressures on national parks, nature reserves
Evaluation of environment and environmental processes: Evaluation of environment to show how certain characteristics are appropriate for a particular form of action	Characterization of avalanche slopes in relation to transport and to settlement location Evaluation of soil capability for agricultural and other land uses Land use–vegetation systems in relation to environmental management
Prediction, design and policy-related issues: Prediction and design concerned with future and policy-related uses	Specific highway location and design according to slope stability Vegetation and textiles to control soil erosion Buffer strips along rivers

nature of sites for engineering construction of roads, buildings, or dams. In such cases it is often necessary to audit geomorphological processes and particularly those that might occasion natural hazards. The drainage network can be delineated into segments according to their character and sensitivity so that a study of Fountain Hills, Arizona, demonstrated the distribution of six types of channel ranging from near-natural, through adjusting, to different styles of channelization (Chin and Gregory, 2005), showing how stream channel landform can be characterized according to degree of change. Audit of spatial distribution is important in the way that river channelization 1930–1980 for England and Wales (Brookes et al., 1983) provided a benchmark for investigations. Process data can be provided in terms of mean values or frequency of occurrence, but in the case of river flows it is frequently necessary to collate and present hydrological data, in a way pertinent to planning situations, by ascertaining flood frequencies of a particular recurrence interval related to a selected design flood. For nature conservation it was necessary to audit a great range of possible locations prior to identifying those to be included in the UK Geological Conservation Review volumes (e.g. Gregory, 1997).

12.2.2 ENVIRONMENTAL IMPACTS

Environmental impacts (see Table 12.3) have been investigated as an integral part of studies of relationships between earth processes and landforms. Techniques involved have been developed for Environmental Impact Assessment (EIA) which is the process of determining and evaluating the positive and negative effects that a proposed action could have on the environment before a decision is taken to proceed or not. This originated in NEPA in the USA in 1969 (see above) and subsequently Environmental Impact Statements (EIS) have been required in many countries prior to planning proposals being approved. The International Association for Impact Assessment (IAIA) defines an environmental impact assessment as 'the process of identifying, predicting, evaluating and mitigating the biophysical, social, and other relevant effects of development proposals prior to major decisions being taken and commitments made' and many manuals are available (e.g. Petts, 1999). However as not all of the effects that a proposed plan may have on the land surface or on earth surface processes are always considered, there is an opportunity for the geomorphologist to assess potential impacts. This can require reviewing the purpose and need for the proposed action, identifying the environment that could be affected, evaluating the alternatives available, and analysing the impacts of the possible alternatives. Examples of the geomorphological impact of particular schemes are included in Table 12.3.

12.2.3 EVALUATION OF ENVIRONMENT AND ENVIRONMENTAL PROCESSES

Evaluation of environment and environmental processes (see Table 12.3) establishes how certain characteristics of environment are appropriate for particular forms of utilization. Land evaluation is the estimation of the potential of land for specific kinds of use, including productive uses such as arable and other types of farming, forestry, uses for water catchment areas, recreation, tourism and wildlife conservation. Land evaluation, founded on an audit or inventory of the components of the particular landscape, has to take account of the requirements of legislation in the particular country and also of any judgements or preferences which may reflect the local culture, before proceeding to complete an evaluation for a particular purpose. Three scales of landscape research for planning purposes range from evaluation maps classifying terrain for a particular purpose, prediction maps indicating the modifications likely to arise, to recommendation maps showing the measures which could be used to change the environment (Isachenko, 1973). Landscape evaluation can be achieved by developing information from systematic data sources so that

Table 12.4 EXAMPLES OF ASPECTS OF ENVIRONMENT THAT HAVE BEEN THE SUBJECT FOR LANDSCAPE EVALUATION

Environmental characteristic	Examples of purposes for which evaluation undertaken
Geology/rock type	Suitability for residential development and for use of septic systems
Soil type	Necessary drainage treatment for arable land use Land use capability classification
Relief	Slope categories in relation to agricultural implements
Vegetation type	Map primary productivity, possibly based on evapotranspiration and soil capability classes
Climate	For agriculture accumulated day degrees in excess of a threshold value; potential water deficit; accumulated frost as degree-days below freezing
Integrated evaluations	For cultivation of vines; based upon soil type, possible sunshine, danger of late frost, general aptitude of land Scenery as a resource

national soil surveys can be utilized to produce land capability surveys, by proceeding from maps of soil bodies, to soil quality maps and soil limitation maps, and subsequently to land classification in terms of soil crop response, present use, use capabilities, and recommended use. Land capability surveys were used by Ian McHarg (1969, 1992; see Box 11.1) to indicate the value of specific areas for particular kinds of land use in the course of evaluation of land suitability for urban extension and development. The range of land evaluation applications (see Table 12.4) has included the value of scenery as a resource, and all landscape evaluation procedures are greatly facilitated by utilising GIS (p. 26).

12.2.4 PREDICTION, DESIGN AND POLICY–RELATED ISSUES

Prediction, design and policy (see Table 12.3) are all interrelated because prediction involves forecasting how the land surface and its processes may develop and design is a mental plan, both requiring policies for their implementation. Planning and management essentially embrace, both implicitly and explicitly, the intention to design landscapes making up the land surface, but how far should the geomorphologist be interested in, and extend research to the question of landscape design? Involvement could help address the paradigm lock (see Figure 12.1) which describes the possible gulf between scientists who do not grasp what managers require, and managers and stakeholders who do not appreciate the scientific alternatives available (Bonell and Askew, 2000; Endreny, 2001; Gregory, 2004a). In order to be involved in design procedures, geomorphologists do not require an eclectic super-scientist approach, but rather to participate in a multidisciplinary team as one member who has the advantage of knowledge of the evolution of land surface systems and landscapes. Such teams can contribute to the application of

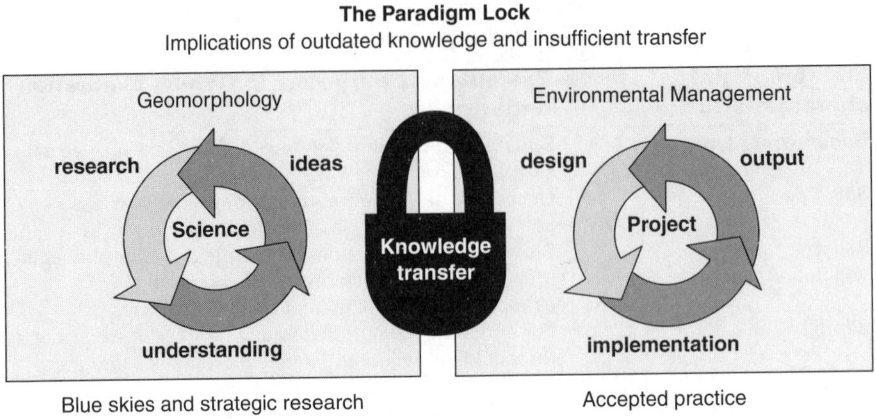

The Paradigm Lock
Implications of outdated knowledge and insufficient transfer

Figure 12.1 Paradigm lock – an expression of the gulf which can exist between researchers and managers (developed from Bonell and Askew, 2000; Endreny, 2001)

hydrology, geomorphology and biology to the effective management of rivers (Calow and Petts, 1992) and to management of other land surface systems. This accords with the suggestion (McHarg, 1996) that inputs from hydrology, geomorphology and other environmental sciences were needed to remedy the absence of geomorphology in design. The professionalization of geomorphology requires geomorphologists to develop and refine a design science, codifying a body of information, tools and skills for licensing or certification of programmes (Rhoads and Thorn, 1996). Geomorphological designs are more vital as awareness of the need for alternatives to hard engineering has encouraged a softer management approach, including restoration of some landscapes to a more 'natural' character, encouraging a focus on alternative strategies for management.

The potential contribution of a geomorphologist is emphasized by a 'design with nature' approach; the origins of which can be traced to Europe in the mid-nineteenth century (Petts et al., 2000), although efforts were very sporadic until the mid 1960s when awareness of the need for an alternative to hard engineering encouraged a general movement towards a softer management approach which included restoring the landscape to a more 'natural' character. The disciplines of landscape architecture and ecology were the first to become involved, succeeded by ecology; Ian McHarg (1969), in his book *Design with Nature* (see p. 272, Chapter 11), contended that geomorphology was the integrative device for physical processes and ecology was the culminating integrator for the biophysical. In ecology and ecological engineering, ecological design principles have been enumerated (Mitsch and Jorgensen, 2004). Geomorphological design should be founded on sustainable landforms and environments appropriate for a particular area. In addition to design solutions for specific areas or problems, there are opportunities for design in landscape restoration, for example to restore the land surface after surface mining when it is desirable to restore the landforms to a character appropriate to the particular location. In addition there are cases where earlier hard engineering, of rivers for example, has been restored to a former or original condition. A series of different definitions has emerged (e.g. Downs and Gregory, 2004: 240; Gregory, 2000: 265) and the critical distinction is between whether it is a matter of making the restored condition look more natural or whether it is appropriate for the particular location (see Chapter 11, p. 273). To ensure the latter option a geomorphological input to the design can be necessary and appropriate, and examples of contributions made to design of the land surface together with design predictions are collated in Table 12.5.

Table 12.5 EXAMPLES OF CONTRIBUTIONS TO DESIGN OF LAND SURFACE

Specific examples are fluvial; a stimulating discussion of the subject is in Newson (1995). Elements of design for river channel landscapes are itemized in Gregory (2006) and a comprehensive review integrating geomorphological tools in ecological and management studies is in Kondolf et al. (2003).

Subject	Specific approach	Reference
Identify unique area to preserve from development	When the Federal Power Commission studied applications for a permit to construct one or more additional dams for electric power in the Hell's Canyon area of the Snake River, Idaho, it was necessary to consider how the attributes of the landscape could be ranked so that some, possibly the most unique, could be preserved from development. A uniqueness method was developed (Leopold, 1969) to identify which of a number of sites was the most unique and therefore to suggest which of several alternatives could be developed without losing some particular qualities	Leopold (1969). This method was developed for the evaluation of riverscape (Leopold and Marchand, 1968) and a related matrix method can be extended to environmental impact (Leopold et al., 1971)
Devise plan to implement to resolve a particular problem	Tilmore Brook, Hampshire UK to manage flooding and erosion and to produce aesthetically effective channel (see Figure 12.3) involved geomorphologist leading the design	Brookes et al., 2005
Propose most expedient restoration for reaches of river channel	Analyse recovery condition in a temporal context	Fryirs and Brierley, 2000
Rehabilitate salmonid spawning habitat	Use catchment survey methods in Deer Creek, California to propose a scheme which reconnects the flood plain and channel and improves habitat conditions	Kondolf et al., 2003: 644–7

Ways in which these applications of geomorphology (see Table 12.3) are demonstrated include:

1   Most obviously by employment of geomorphologists in situations where, using their geomorphological training and skills, they contribute directly towards decision making. Training is achieved during higher degree course programmes but subsequent practical applications and experience is recognized for geomorphologists by chartered status now available (see, for example, www.rgs. org/pdf/CGeogApplication).

2   Through contributions by individual geomorphologists as consultants to bring the advantage of a holistic view of the land surface, a perspective not always achieved by other disciplines. Such a holistic approach is epitomized by the Bahrain surface materials resources survey (Brunsden et al., 1979, 1980), undertaken between 1974 and 1976 at the request of the Ministry of Development and Engineering Services, Government of Bahrain and involving a team comprising 10 geologists, 7 geomorphologists, 2 pedologists, 2 surveyors and a cartographer. The survey produced a series of maps at a scale of 1:10,000 and an extensive report so that the final volume was then 'probably the most intensive and comprehensive view of the surface materials of any state within the arid lands of the world' (Brunsden et al., 1980). Many benefits accrued from this survey and from others undertaken by members of the same team (e.g. Brunsden, 1999), including knowledge of the consequences of environmental processes that otherwise would not have been appreciated. In drylands this is well exemplified by the salinity of groundwater and by salt weathering, shown (Cooke et al., 1982) to be a complex hazard which reflects relations between local environmental conditions, the types of salt present, the nature of susceptible materials, and the design and nature of the structures built in hazardous areas. There are now many examples of the contributions by geomorphologists to the solution of specific problems; the projects undertaken by one individual were listed by Coates (1990).

3   By geomorphologists working in specific environmental organizations, often associated with institutions of Higher Education, which have encouraged the formation of advisory units or groupings of scientists specifically founded to undertake consultancy research and to survive on 'soft money'. This advisory unit model was well established for engineering disciplines but embraced geomorphology as awareness of environmental problems increased. In the UK, units such as the Flood Hazard Research Centre at the University of Middlesex or the Geodata Institute at the University of Southampton are examples. Such units are usually staffed by scientists from a number of disciplines, including geologists, biologists, civil engineers, hydrologists, remote sensers and possibly planners. In addition, geomorphologists can be associated with independent consulting firms or, as in the USA for example, to have an academic contract which provides a salary for less than 12 months of the year because it is assumed that the balance will be made up from income from research grants or consultancy. Graf (1988) showed how geomorphologists in dryland western United States deal with legal issues surrounding the management of rivers: he cited cases where the stability of a river-defined boundary was an issue and in

western Wyoming whether or not locational changes would reasonably be expected in the course of normal river processes or whether recent changes were unusual; and along the Agua Fria River near Phoenix, Arizona where explanation was required from a geomorphologist of river channel change and its implications for the use of engineering models.

4 Contributions made by individuals, for example in a role influencing policy-making, as exemplified by W.L. Graf (see Box 12.1) as Chair of the Committee on Watershed Management in the USA, which was charged to:

- Review the range of scientific and institutional problems related to water-sheds; especially water quality, water quantity and ecosystem integrity.
- Evaluate selected examples of watershed management in a search for the common elements of successful management.
- Recommend ways for local, state, regional and federal water managers to integrate ecological, social and economic dimensions of watershed management.

This work led to the publication of *New Strategies for America's Watersheds* (National Research Council, 1999) including suggestions to guide the reauthorization of the Clean Water Act and 15 conclusions 'to steer the nation toward improved strategies for watershed management'.

5 Through publication of research papers and books. An increasing number of research papers include planning or management considerations exemplifying the types of contribution suggested in Table 12.3. In fluvial geomorphology Gregory and colleagues (2008: table 1) listed many recent examples of appli-cations of fluvial geomorphology where recommendations for management are explicitly given, and Chin and Gregory (2009: table 3) cited 47 research investigations which made explicit management recommendations pertinent to urban river channels. There are also an increasing number of reports which, although having restricted circulation, exemplify contributions by geo-morphologists to planning problems. Specific books show the contribution that geomorphology can make in particular areas such as river channel man-agement (Downs and Gregory, 2004), urban geomorphology in drylands (Cooke et al., 1982), and coastal geomorphology (Viles and Spencer, 1995).

Whatever type of application is adopted, geomorphologists need to communi-cate their conclusions in order to overcome the paradigm lock (see Figure 12.1). Whereas blue skies and strategic research are integral parts of geomorphology research, accepted practice is affected by results from applied research. In a review of hydrology, geomorphology and public policy as employed in the management of river resources in the USA, Graf (1992) showed how endeavours are often poorly connected to each other. The ultimate challenge is the need to raise aware-ness of the function of geomorphological processes in landscape and environmen-tal management in the minds of policy makers and of the general public (Higgitt and Lee, 2001). Various strategies have been employed which can be listed under the headings of applicable research output, applied research output and educa-tional outreach (Gregory et al., 2008: table 3). Methods used to communicate applied geomorphology results include guidelines, online toolkits, diagrammatic representations, checklists and protocols (Gregory et al., 2008). A protocol can provide a basic set of rules for managers to consider. It is generally agreed that continuous hydrological records are not of sufficient length or monitored at enough locations to provide enough data to analyse past hydrological systems in relation to possible future changes. Collaborative research by six international research groups enabled a checklist (Gregory, 2004b) and guidelines for river channel management (Gregory, 2003) to be developed to a protocol providing recommendations about the use of past hydrological events related to understanding global change. A protocol developed for managers to take account of adjustments to urban river channels was devised (Chin and Gregory, 2009).

The above categories demonstrate many of the potential applications available for the geomorphologist: opportunities to provide research, understanding and interpretations which complement those of other scientists and are necessary for the holistic view which is increasingly part of management and planning agendas. Rapid urban expansion can give environmental consequences which require management (see Figure 12.2). Two reflections on management and planning approaches are that first there are many current examples of applications of applied research illustrated by fluvial geomorphology (Gregory et al., 2008) and urbanizing channels (Chin and Gregory, 2009) require that geomorphologists should be aware of engineering techniques employed such as culverts, restoration techniques and design of schemes (see Figure 12.3). Applications are often multidisciplinary, reflecting international collaboration, as illustrated by the use of palm-mat geotextiles to control gullying in the São Luís area of Brazil (see Figure 12.4).

Second, a number of recent reports have reviewed opportunities for geomorphology. The US National Research Council, at the request of the US National Science Foundation, is conducting a study to assess (1) the state of the art of the multidisciplinary field of earth surface processes, (2) the fundamental research questions for the field in the future, and (3) the challenges and opportunities related to answering these questions and advancing the field. This enquiry elicited views from a wide range of sources (e.g. Eide, 2008), some of which have been published and include interesting suggestions such as can we develop 'Earthcasts' analogous to weather forecasts, of both gradual changes and extreme landscape-changing events (Murray et al., 2009)? The draft report is available NRC (2009).

12.3 THE FUTURE – SUSTAINABLE SOLUTIONS AND FUTURE SCENARIOS

How should present applications of geomorphology and study of the surface and its landforms evolve in the future? Some things can be learnt from the past (see section 12.3.1), although as Norman MacCaig wryly said in *A World of Difference* (1983: 42) 'Experience teaches us that it doesn't'. In addition some more recent considerations arise (see section 12.3.2).

Figure 12.2   Rapid urban expansion in a tropical environment

Figure 12.3   Tilmore Brook, Petersfield, UK – a restoration scheme devised to reduce flooding and erosion during construction (see Table 12.5; Brookes et al., 2005)

Figure 12.4   Sacavém, São Luís city, Brazil: (above) the gully system in 2006, where erosion has been monitored since 2000; (below) gully reclamation work in progress in São Luís, February 2008 with geotextile mats constructed from the leaves of local Buriti (*Mauritia flexuosa*) palms used for stabilizing gully head and slopes

Source: Both photographs were taken by Fernando Bezerrra, a PhD student, under supervision of Professor A. Guerra´s, a geomorphologist at the Federal University of Rio de Janeiro who is co-ordinator of the Brazilian group, one of 10 countries contributing to the multidisciplinary BORASSUS Project (http://www.borassus-project.net/), directed by Professor M.A. Fullen, University of Wolverhampton, UK.

12.3.1 WHAT HAVE WE LEARNT?

Geomorphological research knowledge on the land surface should be considered in management and planning. As indicated in Chapter 1, although there may have been a tendency to avoid focusing on the land surface explicitly, this aspect of environment is demonstrably very significant. This has been underlined by Graf (1992) when he contended that effective science and well-informed public policy are the avenues to successful management of environmental resources. In the management of the river resources of the western United States, geomorphology, hydrology and public policy have been poorly connected to each other (Graf, 1992) so that, as geomorphic and hydrologic understanding of river behaviour improved throughout the past century, the complexity of the systems has become more apparent. Graf (1992: 17) concluded that: 'It is now apparent that public policy needs and can use the explanations provided by science. The remaining question is whether or not geomorphologists and hydrologists can address the useful and socially significant issues with convincing answers'.

In some cases a geomorphological contribution gives opportunity to correct erroneous perceptions. Thus Schumm (1994) drew attention to three types of misperceptions of fluvial hazards which were:

- of stability – the idea that any change is not natural
- of instability – that change will not cease
- of excessive response – that change is always major.

He concluded that such misperceptions can lead to litigation and may be the reason for unnecessary engineering works.

Geomorphological contributions are significant for several reasons, including:

1   The *importance of place* – geomorphological contributions can audit and evaluate place which is pertinent because many environmental problems are highly context-dependent (Trudgill and Richards, 1997). Consequently if policy making is to be based upon environmental sciences, it needs to reflect the nature and characteristics of those sciences, based on a dialogue between generalization and specific contexts rather than simply on generalization (Trudgill and Richards, 1997: 11). It is necessary to be aware of several scientific approaches to a problem even if a single one is adopted; to propose solutions where these are required but with the appropriate precautionary caveats; and to have an awareness of the implementation of recommendations and the implications that these may have in relation to policy.

2   *The historical imperative* is necessary to evaluate the evolving dialogue between people and environment in the context of environmental change in the human domain in the recent past (e.g. Cooke, 1992), so that the consequences of human actions can be judged, and if necessary blame apportioned, management responses improved and future needs assessed. Numerous geomorphological research investigations have shown how analysis of past

circumstances can provide important inputs to modern management, for example palaeoflood data extending the instrumental records (p. 150).

3   *Enhanced modelling* methods available can now be used to implement geomorphological understanding; Hooke (1999) showed how major developments in the contribution of geomorphology to engineering and environmental management had occurred in the United Kingdom, particularly in the coastal and fluvial spheres, reflecting understanding of interconnectedness in geomorphic systems and the long-term variability of processes and landforms and achieving a 'work with nature' objective. She concluded that engineering geomorphology, in a phase involving much case-study work, was providing geomorphological information, and implementing management in accordance with the principles advocated, but is poised to progress to a further phase of development involving modelling and predicting responses in ways that adequately deal with complexity, positive feedback, non-linearity and holism.

4   A *holistic point of view* is increasingly valued, and geomorphological investigations have demonstrated its importance. Downs and Gregory (2004) argued that river channel management could progress beyond design with nature to a more holistic river channel management with nature. This was suggested to require understanding of the past and the present; incorporating future conditions; coping with uncertainties; rationalizing risk to support decision making; management with stakeholders; and management as a reflection of institutional structure. Such an approach potentially applies throughout geomorphology.

Identifying the geomorphological contribution more precisely requires:

- emphasis on a *multidisciplinary approach* – now essential for the management and planning of the environment. Geomorphology contributes as one of the environmental and earth sciences, and multidisciplinary collaboration is beneficial – for example with hydrology.
- raising awareness of the importance of *uncertainty, showing how it is valuable to specify the limits of risk in relation to proposals and solutions.* Sear et al. (2007) show that river restoration projects in gravel-bed rivers are becoming increasingly sophisticated and complex, as river managers and scientists attempt to deliver the goals of catchment-scale ecosystem restoration. Sources of uncertainty have been surveyed in relation to river restoration (Darby and Sear, 2008) including the benefits of considering long-term response in relation to the sustainability of restored rivers (Gregory and Downs, 2008).
- realizing that *environmental policy now develops in ways more consistent with the prevailing scientific knowledge*, so that the paradigm lock (see Figure 12.1) becomes less problematic because it is clearer how recognized uncertainties and the distinctiveness of the spatially distributed and temporally variable conditions are associated with environmental problems (e.g. Owens et al., 1997: 4). This has been expressed as *adaptive science* by Graf (2003), whereby we can identify significant questions, seek to answer them and then, in consultation with managers, redefine the questions. Adaptive management in decision making enables a system to adjust through time (Clark, 2002).

- Any geomorphological contribution has to be made with knowledge of the *significance of culture*, for the area concerned. This was addressed in relation to changes of river channels (Gregory, 2006) but ways in which this applies to the land surface as a whole are suggested in Table 12.6. Instead of thinking about ecosystems, and therefore the land surface, as physical objects, they can be visualized in terms of attributes with value for people as natural assets or 'natural capital' – an approach employed to combine the scientific and cultural traditions of landscape ecology in managing landscapes, providing an understanding of how the physical and biological processes associated with landscapes can have value in an economic and cultural context (Haines Young, 2000).

Table 12.6   CULTURAL ISSUES TO CONSIDER WHEN DEVELOPING MANAGEMENT AND PLANNING OF THE LAND SURFACE (DEVELOPED FROM GREGORY, 2006: TABLE 7; EXAMPLES RELATE TO RIVERS AND RIVER CHANNELS)

Cultural strand	Illustrative implications/questions
Knowledge of land surface and its perception	Assumptions about rivers and their mechanics Attitude to rivers: water, sediment, biota, channel, floodplain, riverscape Are rivers revered? Associated with religion, mythology, customs, beliefs? Language of rivers How are rivers portrayed in literature, visual arts, the media? Presentation in terms of threats, risks?
Valuation of the land surface	Past and present valuation of rivers, commercially and aesthetically Economic value, expenditure on maintenance and management Pollution control Risk tolerant and allowing for continuing change
Attitude to the management of the land surface	Non-interference or structural solutions or non-structural options Upstream and downstream effects and spatial context considered? Understanding that rivers have a history Individual decision makers or public involvement and shared vision? Embrace post-project appraisal and adaptive management
Expression of aesthetics in relation to environment	Leave undisturbed Interpretation of 'design with nature' Restore to 'garden' character or to perceived natural condition
Involvement of ethical considerations in unwritten codes to guide action	Human ecology viewpoint? Shallow or deep ecology perspective? Conservation and sustainability pre-eminent?
Legislation for rivers	Codes of practice and laws; institutions with responsibility for rivers Local, regional, national or international controls upon decision-making or some amalgam of these? How political systems react to, and reflect the above issues

12.3.2 IMPLICATIONS OF GLOBAL CHANGE AND A HIGH CO$_2$ WORLD

Global warming gets more than 35 million hits on a search engine, giving an indication of the interest that it generates. It is now generally agreed to be occurring, although debate continues about the rapidity of change, and it may be preferable to refer to a high CO$_2$ world. Many international discussions, such as one including more than 20 Nobel Laureates (26–28 May 2009) that issued the St James's Palace Memorandum, urge that action is necessary. However in many internet sites and calls for action the land surface is not explicitly prominent. Thus the US Environmental Protection Agency site (http://www.epa.gov/climatechange/) identifies 13 categories (Health, Agriculture and Food Supply, Forests, Ecosystems and Biodiversity, Coastal Zones and Sea Level Rise, Water Resources, Energy Production and Use, Public Lands and Recreation, US Regions, Polar Regions, International, Extreme Events, Adaptation) under Health and Environmental Effects, but there is no specific mention of the Earth's land surface, although Coastal Zones and Sea Level Rise, and Polar Regions are directly relevant. Is this a reflection of the perception of the land surface referred to in Chapter 1 (p. 3)?

Inevitably major consequences of a high CO$_2$ world need to be analysed, and the significance for natural disasters (e.g. Houghton, 1997: table 1.1) is an obvious subject. However, as we are now entering a new era of anthropogenic influence (Goudie, 2006), we need to focus upon ways in which geomorphological contributions can be made and are necessary in relation to future scenarios (e.g. Slaymaker et al., 2009).

12.4 HOW FIRM IS TERRA FIRMA?

Terra firma, mentioned at the beginning of Chapter 1, means literally firm ground or firm earth. How firm is the Earth in the light of changes now known to be under way? A similar theme was addressed by Hare (1980) who concluded that the Earth was resilient, just as James Lovelock was optimistic about Gaia, although now less so (see Box 1.1) as he argues that although the planet will look after itself, it is humans who need to be saved – and soon. Some hope offered could arise from geoengineering, by reflecting insolation with giant sunshades or artificial clouds, or by fertilizing the oceans to grow more algae and remove more CO$_2$ from the atmosphere (Lovelock, 2009), but geomorphologists should contemplate changes to the land surface and sustainable measures that could minimize the effects. There is no doubt that past change is recorded in all landscapes of the Earth's surface, on some more than others, that human impacts will continue to be responsible for enormous changes, but with the breadth of holistic understanding achieved can we contribute to managing the land surface in a more sustainable way?

BOX 12.1

PROFESSOR W.L. GRAF

Professor Will Graf, who is University Foundation Distinguished Professor at the University of South Carolina, has contributed in virtually all of the ways listed (see section 12.2) as methods of applying geomorphology; his research contributions have not only advanced geomorphological science but also progressed the practical applications of the subject. He obtained his PhD at the University of Wisconsin in 1974 and has held appointments at the University of Iowa and Arizona State University. Amongst the awards he has received have been the John Wesley Powell Award, US Geological Survey, 2005 (awarded in recognition of contributions to the US Geological Survey in the geographical sciences); the Founder's Medal of the Royal Geographical Society, 2001 (in recognition of research on rivers and contributions to the use of environmental science in public policy); the David Linton Research Award, British Geomorphological Research Group, 2000 (in recognition of geomorphology and environmental change research); and the Kirk Bryan Award, Geological Society of America, 1999 (given in recognition of his contributions to the science of geomorphology). He was President of the Association of American Geographers in 1999 and became a National Associate of the National Academy of Sciences in 2003. He describes his research specialties as including fluvial geomorphology and policy for public land and water, with emphasis on river channel change, human impacts on river processes and morphology, contaminant transport and storage in river sediments, and the downstream impacts of large dams. Much of his work has focused on dryland rivers, though in recent years his contribution has been national in scale.

Professor Graf is an outstanding academic geomorphologist, reflected in his publications: more than 140 papers, articles, book chapters and reports and 7 books authored or edited; he is internationally renowned for his contributions on dryland areas. His other contributions have demonstrated how geomorphology can and should be applied: these have arisen from multidisciplinary national committees that he has chaired so that he is the primary author of *New Strategies for America's Watersheds; Hydrology, Ecology, and Fishes of the Klamath River Basin; Progress Toward Restoring the Everglades;* and *Endangered and Threatened Species of the Platte River.* He has been member or chair of committees that have produced an additional 8 books. He is principal author of *Dam Removal: Science and Decision Making*, and is preparing *Dam the*

(Continued)

(Continued)

Consequences: An Environmental History of Dams and American Rivers. He has served as consultant and expert witness in 30 legal cases related to environmental issues and management, as a science/policy adviser on more than 35 committees for federal, state and local agencies and organizations. President Clinton appointed him to the Presidential Commission on American Heritage Rivers to advise the White House on river management and he serves on the Environmental Advisory Board of the Chief of the US Army Corps of Engineers and chairs the National Academies of Science committee on geographical sciences.

Such involvement gives the flavour of his important contributions – not only in relation to national policy on rivers, dams and watersheds but also in indicating how geomorphologists can contribute. Many innovative ideas, such as the rate law, have been taken up by others and have been extremely influential in the development of understanding of the Earth's land surface.

FURTHER READING

General coverage is provided by:

Cooke, R.U. and Doornkamp, J.C. (1990) *Geomorphology in Environmental Management.* Clarendon Press, Oxford.

An approach to global environmental change is offered by:

Slaymaker, O., Spencer, T. and Embleton Hamann, C. (eds) (2009) *Geomorphology and Global Environmental Change.* CUP, Cambridge.

A thought-provoking read is:

Graf, W.L. (2001) Damage control: restoring the integrity of America's rivers. *Annals of the Association of American Geographers* 91: 1–27.

TOPICS

1 Examples in Tables 12.1, 12.5 and 12.6 relate to rivers and river channels (reflecting the inclination of the author!). Can you complement these tables with other geomorphological contributions?
2 What opportunities for geomorphological research are provided by the implications of a high CO_2 world?

GLOSSARY OF KEY ADVANCES

Many definitions have been included throughout the preceding chapters, often placed in tables to avoid fragmenting the text. However studies of the land surface of the Earth over the last 100 years have reflected a number of fundamental changes in how we approach, investigate and value the land surface – all reflected in the way that geomorphology has developed. Some of these changes are reflected in foundation milestones (see Table 2.1), in debates (see Table 2.5), or in conceptual thinking (see Table 5.4), but the definitions provided here reflecting major changes and also linking mentions in the text, could therefore be thought of as forming a conclusion – to which you may wish to add further items.

Adaptive cycle (Figure 5.3) acknowledges that systems may exist in multiple steady states, flipping from one state to another as thresholds are transgressed (Table 5.4). A related idea is adaptive management which proceeds in the face of uncertainty, does not assume that the system being managed is understood, and adapts management in the light of experience.

Anthropogeomorphology is the study of humans as geomorphological agents, so that investigation of the land surface included human impacts in the noosphere and anthroposphere (see Tables 3.1, 8.3 and 8.6).

Biomes are major ecological communities characterized by a dominant type of vegetation, and their association with major climates was basic for climatic and climatogenetic geomorphology (p. 159) and, recently, 18 anthropogenic biomes have been identified as a basis for integrating human and ecological systems p. 67, Chapter 3).

Catastrophism is the notion that land surface processes are affected by sudden and violent events; it complements uniformitarianism and gradualism (e.g. see Chapter 2, p. 22).

Closure applies to the limits of disciplines. The land surface is studied by several disciplines, each one cannot study all aspects, and closure refers to the extent of a single discipline.

Culture is associated with a particular form or stage of civilisation, now realised to affect how the land surface is perceived, managed and developed (see p. 296, Table 12.6).

Databases are an integrated collection of logically-related records or files which consolidate information in a consistent way, can easily be accessed, managed and updated, and can provide material for many applications.

Dating methods for time estimation are necessary to establish stages of land surface development and the age of landforms. They became more precise with the advent of radiocarbon dating establishing the age of organic materials up to c.40,000 years old based on the decay of the radioactive isotope ^{14}C within organic materials. Recent advances (see Tables 2.1 and 2.3) have greatly advanced knowledge of rates of land surface development including oxygen isotope stages and cosmogenic methods (see Table 5.3).

Design with nature is managing the land surface to imitate nature as much as possible rather than using hard engineering. It is sometimes referred to as soft engineering and can be achieved by restoration or rehabilitation of environment (see Chapter 11).

Digital elevation model (DEM) models land surface height whereas a digital terrain model (DTM) models shape of the land surface; both are used in Geographic Information Systems.

Ecological footprint is the amount of land required to sustain cities and expresses their environmental impact (see http://www.gdrc.org/uem/footprints: Chapter 11).

Electronic distance measurement (EDM) is a highly accurate method of measuring distance between two points based on transit time of electromagnetic waves between an emitting instrument, a reflector and back again; it displaced traditional distance measuring survey instruments.

Environmental awareness, now often thought of as awareness of environmental issues, was important as the environmental movement developed in the 1960s with greater public knowledge of the land surface encouraging public participation in decision-making.

Environmental change occurs naturally in response to forcing factors operating over a range of time scales, embracing human impacts and global climate change. Appreciation of the range of changes advanced understanding of the land surface and has embraced concepts of resilience (pp. 134), vulnerability, and panarchy (p. 135).

Environmental hazards are natural processes that can damage the land surface or property and take human lives; management is required to manage and mitigate hazards (see Tables 3.7 and 12.2).

Environmental impact assessment (EIA) is the process of determining and evaluating the positive and negative effects of a proposed action on the environment before a decision is taken to proceed or not. Environmental Impact Statements (EIS) are required in many countries prior to planning proposals being approved so that implications of impacts need to be estimated.

Equilibrium, the state of balance created by a variety of forces so that the state remains unchanged over time unless the controlling forces change, was developed for different types of equilibrium and succeeded by recognition of non-equilibrium situations, realizing that characteristic (steady-state) equilibrium, zonal, and mature forms may be complemented by systems that have multiple potential characteristic or equilibrium forms, and others having no particular normative state at all (see Table 5.4).

Factor of safety, the ratio of resistance to force, means that when <1 failure can occur; it is used most frequently in studying slope stability (Chapter 4, p. 71).

Geographical Information Systems (GIS), the collection, analysis, storage and display of data spatially referenced to the surface of the Earth, became possible with digital computers; they enable discovery of relationships and patterns between characteristics and processes, for example in land systems.

Global Positioning Systems (GPS), a navigation system using a constellation of 24 orbiting satellites, available since 1994, enables determination of a specific location anywhere on the surface of the Earth and so speeds up survey and measurement.

Global warming, the rise in near-surface atmospheric temperature due to human activities, especially increased amounts of greenhouse gases, has consequences for the land surface as a result of the greenhouse effect, part of a high CO_2 world.

Historical contingency recognizes that the state of a system or environment is partially dependent on one or more process states or upon events in the past, thus including inheritance which relates to features inherited from previous conditions, so that management of the land surface has to be designed for the particular area.

Holistic view acknowledges the need to understand how a land surface system operates as well as to know the sum of its component parts. Thus any problem should not be considered exclusively by a single discipline but in the wider context, by looking at relationships as a whole.

International System of Uunits (SI) is the most widely used metric system of measurement since 1960.

Land system, an area with a recurring pattern of topography, soils and vegetation, was used for resource surveys but is now also employed for the grouping of landforms in type areas (for example, Tables 8.1 and 8.8).

Modelling expresses theories, processes, events or systems in concepts or mathematical terms; models are any abstraction or simplification of reality, enabling progress in studies of the land surface since the quantitative revolution of the 1960s (see Table 5.9).

Natural environment is the environment pre human activity, unlike the built environment which consists of the areas and components strongly influenced by humans; it is the physical character of a place including physical environment which can apply to cities.

Non-linear methods, unlike linear methods which assume direct relationships between cause and effect, do not have easily derived solutions, have responses which can be chaotic and do not settle down to a fixed equilibrium condition or value. Realization that relatively small events can trigger large and rapid changes encouraged use of non-linear systems; chaos theory and catastrophe theory can account for sudden shifts of a system from one state to another as a result of the system being moved across a threshold condition (see Table 5.4).

Noosphere, or anthroposphere, is the realm of human consciousness in nature or the 'thinking' layer arising from the transformation of the biosphere under the influence of human activity. Importance of direct and indirect impacts of human activity realized to be significant for the land surface and its management (see Table 3.1).

Palimpsest, inheritance where a suite of landforms are 'written' on the landscape beneath, so that new landforms combine with remnants of a previous surface (see p. 165, Chapter 7; p. 187, Chapter 8).

Paradigms, practices within which science operates, recognition of which facilitated debates progressing understanding of the land surface (see Table 2.5).

Plate tectonics involves movement of rigid plates of the lithosphere causing seismic and tectonic activity along their boundaries.

Process domains are combinations of land surface processes often associated with particular zones.

Quaternary chronology is the sequence of stages of the Pleistocene and Holocene (about the last 2 million years) now identified universally from isotopic studies of deep sea sediments and glacier cores, succeeding chronologies that were difficult to correlate from one continent to another.

Remote sensing, often thought of as the collection of data from orbital satellites, is the use of electromagnetic sensors to record images of the environment which, since the 1960s, complement ground survey information. Developments continue to occur such as LiDAR (light detection and ranging) which allows accurate measurements of topography, can measure changes of glaciers or of land uplift and, like other techniques, can provide data not accessible by traditional survey methods.

Resilience is the capacity of a system to absorb disturbance and to reorganize while undergoing change but still retaining essentially the same function, structure, identity and feedbacks.

Restoration involves restoring part of the land surface to a former or original condition and can be employed as an alternative to hard engineering, with various terms used to signify different ways of making an area appear more 'natural'.

Sensitivity is the likelihood of response to slight changes, expressed as the ratio between the magnitude of adjustment and the magnitude of change in the stimulus causing the adjustment (see Figure 5.2).

Systems are a set of components, and relationships between them, functioning to act as a whole, providing an approach (since 1962) that focuses on the total land surface rather than on selected aspects. Self-organizing systems increase in complexity, their dynamics being a function of positive and negative feedbacks, thus precluding linear explanations and meaning that uncertainty and limited predictability prevail (see Table 3.3).

Thresholds are stages or tipping points at which essential characteristics change dramatically, they are boundary conditions separating two distinct phases or equilibrium conditions. Although not easy to isolate they potentially indicate where and how much change will occur and may give clues about when and why.

Uniformitarianism is the present as the key to the past, so that land surface processes have operated for much of Earth's history (see Table 2.1); it includes actualism (effects of present processes) and gradualism (surface changes require long periods of time).

PLATE 1

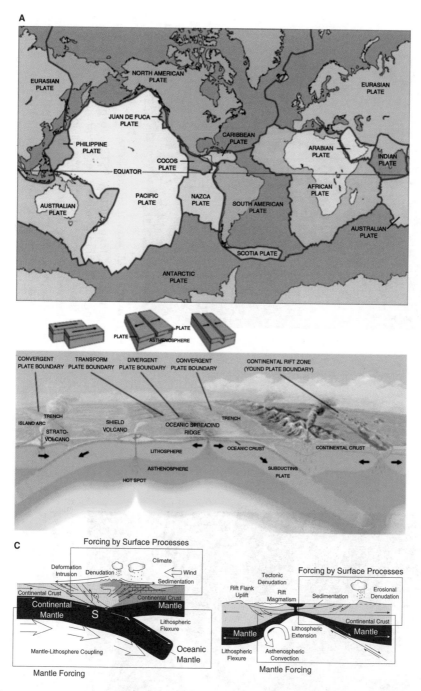

Figure 3.2 Tectonic plates in (A); main types of plate boundaries in (B); and processes driving landscape evolution in (C) showing (left) convergent settings and (right) extensional (rifting) settings

Sources: (A) http://pubs.usgs.gov/publications/text/slabs.html, last updated 25 November 2008; (B) http://pubs.usgs.gov/publications/text/Vigil.html, last updated 5 May 1999: Contact: jmwatson@usgs.gov; (C) from Bishop, 2007: 335 and Beaumont et al., 2000, artwork by USGS.

PLATE 2

Figure 4.7 Mississippi River Delta

Image taken 24 May 2001 by ASTER: turbid waters spill out into the Gulf of Mexico where their suspended sediment is deposited to form the Mississippi River Delta. Like the webbing on a duck's foot, marshes and mudflats prevail between the shipping channels that have been cut into the delta. This image can be found on ASTER Path 21 Row 40, centre: 29.45 N, -89.28 W. From USGS Eros Image Gallery Earth as Art. With permission from USGS.

Figure 4.4 Mam Tor, Derbyshire – the road on Edale shales subject to frequent mass movement was eventually closed after repeated sliding

Credit: Stockphoto 2181642.

PLATE 3

Figure 5.1   The geologic time spiral

Source: (from USGS designed by Joseph Graham, William Newman and John Stacy – see http://pubs.usgs.gov/gip/geotime/age.html)

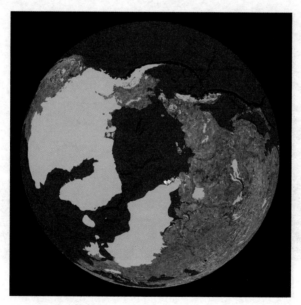

Figure 5.5   Last glacial maximum in the northern hemisphere (compiled by Jürgen Ehlers from Ehlers and Gibbard, 2004).

Evidence from both the land and the ocean floors now demonstrates that the major continental glaciations, outside the polar regions, rather than occurring throughout the 2.6 million years of the Quaternary were markedly restricted to the last 1 million years to 800,000 years or less (Ehlers and Gibbard, 2007).

PLATE 4

Figure 7.3 Patterned ground (top) – polygonal ground (ice wedge polygons), Yukon territory, Canada; (below) – sorted nets, Northwest Territories, Canada

Sources: http://sis.agr.gc.ca/cansis/taxa/landscape/ground/yukon.html; Agriculture and Agri-Food Canada at www.agr.gc.ca.

Reprinted with permission.

PLATE 5

Figure 7.4   Pingos in the Tuktoyaktuk Peninsula 1988

Above is the setting of the Tuktoyaktuk pingo with another large pingo, Split Hill, in the background. Below shows the geomorphic setting with the exposed ice core with seasonal growth bands. Photographs by Professor Hugh French, with permission (see French, 2007).

PLATE 6

December 30, 1990

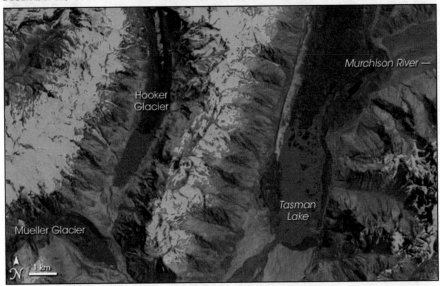

December 6, 2007

Figure 7.5 Tasman glacier, Southern Alps, New Zealand, 1990 and 2007

In the false-colour images, pale blue indicates ice and snow, electric blue indicates water, green indicates vegetation, and brown indicates rock or bare ground.

Sources: NASA 30 December 1990 (Landsat 4 TM; 28.5 m resolution; 3.6 MB JPEG); 6 December 2007 (ASTER; 15 m resolution; 4.2 MB JPEG). NASA image created by Jesse Allen, using Landsat data provided by the University of Maryland's Global Land Cover Facility and ASTER data provided courtesy of NASA/GSFC/METI/ERSDAC/JAROS, and US/Japan ASTER Science Team.

PLATE 7

Figure 7.6 Huascaran ice avalanches, 1962 and 1970

On 10 January 1962, an ice avalanche occurred with an estimated starting volume of 10 million m³; it travelled 16 km and destroyed the city of Ranrahirca, where 4000 people died and 6 villages were destroyed. It began as a displacement of up to 3 million m³ of ice, but as it progressed it descended vertically 4000 m, was transformed into a mudflow of 13 million m³ of debris and ice with a velocity of over 100 km.hour⁻¹. On 31 May 1970, the most catastrophic rock-ice avalanche known in history was triggered by a strong earthquake with a magnitude of 7.7. The avalanche originated from a partially overhanging cliff at 5400 to 6500 m above sea level, where the fractured granite rock of the peak was covered by a 30 m thick glacier. The avalanche, which had an estimated volume of 50 to 100 million m³, travelled 16 km to Rio Santa, proceeded 4000 m vertically, overrode a hill in the downstream area and completely destroyed the city of Yungay, with 18,000 lives lost.

Sources: Map: Hugo Ahlenius, UNEP/GRID-Arendal, see http://maps.grida.no/go/graphic/ice-avalanches-of-the-nevados-huascar-n-in-peru: *Photo:* Courtesy of Servicio Aerofotografico Nacional de Peru, 13 June 1970.

PLATE 8

Figure 9.4 Alluvial fan between the Kunlun and Altun mountain ranges, southern Taklimakan Desert, XinJiang Province, China

The active part of the fan on the left appears blue. The image was acquired 2 May 2002, covers an area of 56.6 × 61.3 km, and is centred near 37.4°N, 84.3°E. *Sources:* NASA/GSFC/METI/ERSDAC/JAROS and US/Japan ASTER Science Team. see http://asterweb.jpl.nasa.gov/gallery-detail.asp?name=fan.

Figure 11.2 Palm Jumeirah, human-made island on the Dubai coast

Palm Jumeirah's construction from 2000–09 was observed by the Advanced Spaceborne Thermal Emission and Reflection Radiometer (ASTER) on NASA's Terra satellite and this image (February 2009) shows vegetation on most of the palm fronds, and numerous buildings on the tree trunk. The construction process for the Palm Islands involves dredging sand from the bottom of the Persian Gulf and then spraying that sand over the appropriate areas to create the desired shapes. (NASA images courtesy of the NASA/GSFC/MITI/ERSDAC/JAROS and US/Japan ASTER Science Team).

REFERENCES

Adams, J., Maslin, M. and Thomas, E. (1999) Sudden climate transitions during the Quaternary. *Progress in Physical Geography* 23: 1–36.

Alexander, D.E. (1982) Leonardo da Vinci and fluvial geomorphology. *American Journal of Science* 282: 735–755.

Alexander, D. (1989) Urban landslides. *Progress in Physical Geography* 13: 157–191.

Alexander, D.E. (1999) Natural hazards. In D.E. Alexander and R.W. Fairbridge (eds), *Encyclopedia of Environmental Science*. Kluwer Academic Publishers, Dordrecht, Boston, London, pp. 421–425.

Alexander, J.F. (1979) A global systems ecology model. In R.A. Fazzolare and C.B. Smith (eds), *Changing Energy Use Futures*. Pergamon, New York and Oxford, pp. 443–456.

Allen, J.R.L. (1970) *Physical Processes of Sedimentation*. Allen and Unwin, London.

Arctic Climate Impact Assessment (ACIA) (2004) *Impacts of a Warming Arctic* (a 140-page synthesis report of the Arctic Climate Impact Assessment).

Ashman, M.R. and Puri, G. (2002) *Essential Soil Science*. Blackwell Publishing, Malden USA.

Bagnold, R.A. (1941) *The Physics of Blown Sand and Desert Dunes*. Methuen, London (2nd edn 1954).

Bagnold, R.A. (1960) Sediment discharge and stream power: A preliminary announcement. *US Geological Survey Circular* 421.

Bagnold, R.A. (1990) *Sand, Wind and War: Memoirs of a Desert Explorer*. Tucson, University of Arizona Press .

Baker, V.R. (1978a) Palaeohydraulics and hydrodynamics of scabland floods. In V.R. Baker and D. Nummedal (eds), *The Channeled Scabland*. NASA, Washington, pp. 59–79.

Baker, V.R. (1978b) Large-scale erosional and depositional features of the channeled scabland. In V.R. Baker and D. Nummedal (eds), *The Channeled Scabland*. NASA, Washington, pp. 81–115.

Baker, V.R. (1981) *Catastrophic Flooding: The Origin of the Channeled Scabland*. Dowden, Hutchinson & Ross, Stroudsburg, PA.

Baker, V.R. (1986) http://geoinfo.amu.edu.pl/wpk/geos/geo_1/GEO_CHAPTER_1.HTML *Geomorphology from Space* is an out of print 1986 NASA publication edited by Nicholas M. Short, Sr. and Robert W. Blair.

Baker, V.R. (1996) Hypotheses and geomorphological reasoning. In B.W. Rhoads and C.E. Thorn (eds), *The Scientific Nature of Geomorphology*. Wiley, Chichester, pp. 57–85.

Baker, V.R. (1998) Palaeohydrology and the hydrological sciences. In G. Benito, V.R. Baker and K.J. Gregory (eds), *Palaeohydrology and Environmental Change*. Wiley, Chichester, pp. 1–10.

Baker, V.R. (2008a) Paleoflood hydrology: origin, progress, prospects. *Geomorphology* 101: 1–13.

Baker, V.R. (2008b) Planetary landscape systems: a limitless frontier. *Earth Surface Processes and Landforms* 33: 1341–1353.

Baker, V.R. and Bunker, R.C. (1985) Cataclysmic late Pleistocene flooding from Glacial Lake Missoula: a review. *Quaternary Science Reviews* 4: 1–41.

Baker, V.R. and Costa, J.E. (1987) Flood power. In L. Mayer and D. Nash (eds), *Catastrophic Flooding*. Allen and Unwin, Boston, pp. 1–21.

Baker, V.R. and Twidale, C.R. (1991) The reenchantment of geomorphology. *Geomorphology* 4: 73–100.

Baker, V.R., Bowler, J.M., Enzel, Y. and Lancaster, N. (1995) Late Quaternary Palaeohydrology of arid and semi-arid regions. In K.J. Gregory, L. Starkel and V.R. Baker (eds), *Global, Continental Palaeohydrology*. Wiley, Chichester, pp. 203–231.

Ballantyne, C.K. (2002a) Paraglacial geomorphology. *Quaternary Science Reviews* 21: 1935–2017.

Ballantyne, C.K. (2002b) A general model of paraglacial landscape response. *The Holocene* 12: 371–376.

Ballantyne, C.K. (2003) Paraglacial land systems. In D.J.A. Evans (ed.), *Glacial Landsystems*. Hodder Arnold, London, pp. 432–461.

Barnes, C.P. (1954) The geographic study of soils. In P.E. James and C.F. Jones (eds), *American Geography Inventory and Prospect*. Syracuse University Press: Association of American Geographers, pp. 382–395.

Barry, R.G. (1997) Palaeoclimatology, climate system processes and the geomorphic record. In D.R. Stoddart (ed.), *Process and Form in Geomorphology*. Routledge, London and New York, pp. 187–214.

Bates, R.L. and Jackson, J.A. (1980) *Glossary of Geology*. American Geological Institute, Falls Church Virginia.

Beaumont, C., Fullsack, P. and Hamilton, J. (2000) Erosional control of active compressional orogens. In K. McClay (ed.), *Thrust Tectonics*. Chapman Hall, London, pp. 1–18.

Beauvais, A., Parisot, J.C. and Savin, C. (2007) Ultramafic rock weathering and slope erosion processes in a South West Pacific tropical environment. *Geomorphology* 83: 1–13.

Beckinsale, R.P. (1997) Richard J. Chorley: A reformer with a cause. In D.R. Stoddart (ed.), *Process and Form in Geomorphology*. Routledge, London and New York, pp. 3–12.

Beckinsale, R.P. and Chorley, R.J. (1991) *The History of the Study of Landforms or The Development of Geomorphology Vol. 3: Historical and Regional Geomorphology 1890–1950*. Routledge, London.

Beer, T. (1983) *Environmental Oceanography*. Pergamon, Oxford.

Benito, G., Baker, V.R. and Gregory, K.J. (eds) (1998) *Palaeohydrology and Environmental Change*. Wiley, Chichester.

Benito, G., Thorndycraft, V.R., Rico, M., Sánchez-Moya, Y. and Sopeña, A. (2008) Palaeoflood and floodplain records from Spain: Evidence for long-term climate variability and environmental changes. *Geomorphology* 101: 68–77.

Benn, D.I. and Evans, D.J.A. (1998) *Glaciers and Glaciation*. Arnold, London.

Bennett, M.R. and Glasser, N.F. (1996) *Glacial Geology. Ice Sheets and Landforms*. Wiley, Chichester.

BGS (2009) Rock classification scheme at www.bgs.ac.uk/bgsrcs/download.html, accessed 17 April 2009.

Bierman, P.R., Reuter, J.M., Pavich, M., Gellis, A.C., Caffee, M.W. and Larsen, J. (2005) Using cosmogenic nuclides to contrast rates of erosion and sediment yield in a semi-arid, arroyo-dominated landscape, Rio Puerco Basin, New Mexico. *Earth Surface Processes and Landforms* 30: 935–953.

Bird, E.C.F. (1996) *Beach Management*. Wiley, Chichester.

Bird, E.C.F. (2000) *Coastal Geomorphology: An Introduction*. Wiley, Chichester.

Birkeland, P.W. (1984) *Pedology, Weathering and Geomorphological Research*. Oxford University Press, New York.

Bishop, P. (2007) Long-term landscape evolution: linking tectonics and surface processes. *Earth Surface Processes and Landforms* 32: 329–365.

Biswas, A. (1970) *History of Hydrology*. Amsterdam, North Holland Publishing Company.

Blackwelder, E. (1931) Desert plains. *Journal of Geology* 39: 133–140.

Bolling, D.M. (1994) *How to Save a River. A Handbook for Citizen Action*. Island Press, Washington, DC.

Bonell, M. and Askew, A. (2000) Report of 2nd International Conference on Climate and Water, Espoo, Finland.

Boulton, G.S., Smith, G.D., Jones, A.S. and Newsome, J. (1985) Glacial geology and glaciology of the last mid-latitude ice sheets. *Journal of the Geological Society* 142: 447–474.

Bowler, S. (2002) *Restless Earth*. DK Publishers, New York.

Brabec, E., Schulte, S. and Richards, P.L. (2002) Impervious surfaces and water quality: a review of current literature and its implications for watershed planning. *Journal of Planning Literature* 16: 499–514.

Bravard, J.P. and Gilvear, D.J. (1996) Hydrological and geomorphological structure of hydrosystems. In G.E. Petts and C. Amoros (eds), *Fluvial Hydrosystems*. Chapman and Hall, London, pp. 98–116.

Bretz, J.H. (1923) The channelled scabland of the Columbia plateau. *Journal of Geology* 3(6): 17–49.

Bridge, J.S. (2003) *Rivers and Floodplains*. Blackwell Publishing, Oxford.

Brookes, A. (1988) *Channelized Rivers: Perspectives for Environmental Management*. Wiley, Chichester.

Brookes, A., Chalmers, A. and Vivash, R. (2005) Solving an urban river erosion problem on the Tilmore Brook, Hampshire (UK). *Journal of the Chartered Institution of Water and Environmental Management* 19(3): 199–206.

Brookes, A., Gregory, K.J. and Dawson, F.H. (1983) An assessment of river channelization in England and Wales. *The Science of the Total Environment* 27: 97–111.

Brooks, S.M. (2003) Modelling in physical geography. In A. Rogers and H.A. Viles (eds), *The Student's Companion to Geography*. Blackwell Publishing, Oxford, pp. 161–166.

Brown, A.G. (1987) Long term sediment storage in the Severn and Wye catchments. In K.J. Gregory, J. Lewin and J.B. Thornes (eds), *Palaeohydrology in Practice*. Wiley, Chichester, pp. 307–332.

Brown, A.G. (1991) Hydrogeomorphological changes in the Severn Basin during the last 15,000 years: orders of change in a maritime catchment. In L. Starkel, K.J. Gregory and J.B. Thornes (eds), *Temperate Palaeohydrology*. Wiley, Chichester, pp. 147–170.

Brown, A.G. (1997) *Alluvial Geoarchaeology: Floodplain Archaeology and Environmental Change*. Cambridge University Press, Cambridge.

Brown, A.G. (2003) Time, space and causality in floodplain palaeoecology. In A.J. Howard, M.G. Macklin and D.G. Passmore (eds), *Alluvial Archaeology in Europe*. Rotterdam, Swets & Zeitlinger, pp. 15–26.

Brown, A.G. and Quine, T. (1999) Fluvial processes and environmental change: an overview. In A.G. Brown and T. Quine (eds), *Fluvial Processes and Environmental Change*. Wiley, Chichester, pp. 1–27.

Brown, A.G., Carey, C., Erkens, G., Fuchs, M., Hoffmann, T., Macaire, J.J., Moldenhauer, K.M. and Walling, D.E. (2009) From sedimentary records to sediment budgets: multiple approaches to catchment sediment flux. *Geomorphology* 108: 35–47.

Brown, H.I. (1996) The methodological roles of theory in science. In B.W Rhoads and C.E. Thorn (eds), *The Scientific Nature of Geomorphology*. Wiley, Chichester, pp. 3–20.

Brown, J., Ferrians, O.J. Jr., Heginbottom, J.A. and Melnikov, E.S. (1998) (revised February 2001) *Circum-Arctic map of permafrost and ground-ice conditions*. National Snow and Ice Data Center/World Data Center for Glaciology Digital Media, Boulder, CO.

Brunsden, D. (1999) Geomorphology in environmental management: an appreciation. *East Midland Geographer* 22: 63–77.

Brunsden, D. and Thornes, J.B. (1979) Landscape sensitivity and change. *Transactions Institute of British Geographers* NS4, 463–484.

Brunsden, D., Doornkamp, J.C. and Jones, D.K.C. (1979) The Bahrain surface materials resources survey and its application to regional planning. *Geographical Journal* 145: 1–35.

Brunsden, D., Doornkamp, J.C. and Jones, D.K.C. (eds) (1980) *Geology, Geomorphology and Pedology of Bahrain*. Geo Books, Norwich.

Budel, J. (1957) Die 'Doppelten Einebnungsflachen, in den feuchten Tropen. *Zeitschrift fur Geomorphologie* 1: 201–228.

Budel, J. (1963) Klima-genetische Geomorphologie. *Geographische Rundschau* 15: 269–285.

Budel, J. (1969) Das System der klima-genetischen Geomorphologie. *Erdkunde* 23: 165–182.

Budel, J. (1977) *Klima-Geomorphologie*. Berlin/Stuttgart: Borntraeger.

Budel, J. (1982) *Climatic Geomorphology*. Trans. L. Fischer and D. Busche. Princeton University Press, Princeton.

Budyko, M.I. (1958) *The Heat Balance of the Earth's Surface*. Trans. N. Steepanova from original dated 1956. Washington, US Weather Bureau.

Budyko, M.I. (1974) *Climate and Life*. Academic Press, New York.

Budyko, M.I. (1986) *The Evolution of the Biosphere*. D. Reidel Publishing Company, Dordrecht.

Bullard, J.E. and Nash, D.J. (2000) Valley-marginal sand dunes in the south-west Kalahari: their nature, classification and possible origins. *Journal of Arid Environments* 45: 369–383.

Bunge, W. (1973) The Geography. *Professional Geographer* 25: 331–337.

Burbank, D.W., Leland, J., Fielding, E., Anderson, R.S., Brozovic, N. and Reid, M.R. (2003) De-coupling of erosion and precipitation in the Himalayas. *Nature* 379: 505–510.

Burt, T.P., Chorley, R.J., Brunsden, D., Cox, N.J. and Goudie, A.S. (2008) *The History of the Study of Landforms or the Development of Geomorphology volume 4: Quaternary and Recent Processes and Forms (1890–1965) and the Mid-Century Revolutions*. Geological Society, London.

Butler, B.E. (1959) *Periodic Phenomena in Landscapes as a Basis for Soil Studies*. Soil Publication 14, CSIRO, Australia.

Butzer, K.W. (1976) *Geomorphology from the Earth*. Harper Row, New York, London.

Calow, P. and Petts, G.E. (eds) (1992) *The Rivers Handbook*, vol. 1. Blackwell Science, Oxford.

Campbell, R.W. (ed.) (2001) *Mount St. Helens, Washington: 1973, 1983, 1988, 1992, 1999. Earthshots: Satellite Images of Environmental Change*. U.S. Geological Survey. http://earthshots.usgs.gov. This article was released 14 February 1997 and last revised 14 August 2001.

Carey, P. (2001) *The History of the Kelly Gang*. Faber and Faber, London.

Carson, R. (1962) *Silent Spring*. Houghton Mifflin, Boston.

Carter, R.W.G. (1988) *Coastal Environments: An Introduction to the Physical, Ecological and Cultural Systems of Coastlines.* Academic Press, London.

CATENA (2005) Gully Erosion: A Global Issue, Volume 63, Issues 2–3, 31 October 2005.

Changming, L. (2000) A remarkable event of human impacts on the ecosystems: the Yellow River drained dry. Paper read to 29th International Geographical Congress, 17th August, Seoul, Korea.

Charlesworth, J. K. (1957) *The Quaternary Era.* 2 volumes. Arnold, London.

Charman, J. and Lee, M. (2005) Mountain environments. In P.G. Fookes, E.M. Lee and G. Milligan (eds), *Geomorphology for Engineers.* Whittles Publishing Dunbeath, Scotland, pp. 501–534.

Chepil, W.S. (1945) Dynamics of wind erosion: II Initiation of soil movement. *Soil Science* 60: 397–411.

Chin, A. (2006) Urban transformation of river landscapes in a global context. *Geomorphology* 79: 460–487.

Chin, A. and Gregory, K.J. (2005) Managing urban river channel adjustments. *Geomorphology* 69: 28–45.

Chin, A. and Gregory, K.J. (2009) From research to application: management implications from studies of urban river channel adjustment. *Geography Compass* 3/1: 297–328.

Chocat, B., Ashley, R., Marsalek, J., Matos, M.R., Rauch, W., Schilling, W. and Urbonas, B. (2007) Towards the sutainable management of urban storm water. *Indoor and Built Environment* 16: 273–285.

Chorley, R.J. (1962) Geomorphology and general systems theory. *US Geological Survey Professional Paper* 500-B: 1–10.

Chorley, R.J. (ed.) (1969) *Water, Earth and Man.* Methuen, London.

Chorley, R.J. (1978) Bases for theory in geomorphology. In C. Embleton, D. Brunsden and D.K.C. Jones (eds), *Geomorphology. Present Problems and Future Prospects.* Oxford, Oxford University Press.

Chorley, R.J., Dunn, A.J. and Beckinsale, R.P. (1964) *The History of the Study of Landforms, Vol. I, Geomorphology before Davis.* Methuen, London.

Chorley, R.J. and Kennedy, B.A. (1971) *Physical Geography: A Systems Approach.* Prentice Hall, London.

Chorley, R.J., Beckinsale, R.P and Dunn, A.J. (1973) *The History of the Study of Landforms Vol. II The Life and Work of William Morris Davis.* Methuen, London.

Chorley, R.J., Dunn, A.J. and Beckinsale, R.P. (1964) *The History of the Study of Landforms, Vol. I, Geomorphology before Davis.* Methuen, London.

Chorley, R.J., Schumm, S.A. and Sugden, D.E. (1984) *Geomorphology.* Methuen, London and New York.

Christian, C.S. and Stewart, G.A. (1953) *Survey of the Katherine-Darwin region 1946. CSIRO Land Research Series* 1, Melbourne.

Church, M. (2005) Continental drift. *Earth Surface Processes and Landforms* 30: 129–130.

Church, M. and Ryder, J.M. (1972) Paraglacial sedimentation, a consideration of fluvial processes conditioned by glaciations. *Geological Society of America Bulletin* 83: 3059–3072.

Church, M. and Slaymaker, O. (1989) Disequilibrium of Holocene sediment yield in glaciated British Columbia. *Nature* 337: 452–454.

CIA World Factbook (2003) At https://www.cia.gov/library/publications/the-world-factbook/ accessed 5 March 2009.

CIRIA (Construction Industry Research and Information Association) (2001) *Sustainable Urban Drainage Systems. Best Practice Manual for England, Scotland, Wales and Northern Ireland.* CIRIA C523 Westminster, London.

Clark, M.J. (2002) Dealing with uncertainty:adaptive approaches to sustainable management. *Aquatic Conservation: Marine and Freshwater Ecosystems* 12: 347–363.

Clayton, J.A. and Knox, J.C. (2008) Catastrophic flooding from Glacial Lake Wisconsin. *Geomorphology* 93: 384–397.

Clements, T. (1981) Leonardo da Vinci as a geologist. In F.H.T. Rhodes (ed.), *Language of the Earth*. Pergamon Press, New York, pp. 310–314.

Clifford, N.J. (2007) River restoration: paradigms, paradoxes and the urban dimension. *Water Science and Technology: Water Supply* 7: 57–68.

Clifford, N.J. and Richards, K.S. (2005) Earth system science: an oxymoron? *Earth Surface Processes and Landforms* 30: 379–383.

Coates, D.R. (ed.) (1971) *Environmental Geomorphology*. State University of New York Publications in Geomorphology, Binghamton.

Coates, D.R. (1976) *Geomorphic Engineering. Geomorphology and Engineering.* Dowden, Hutchinson and Ross, Stroudsburg, pp. 3–21.

Coates, D.R. (1990) Perspectives on environmental geomorphology. *Zeitschrift fur Geomorphologie* Supplement 79: 83–117.

Cockburn, H.A.P. and Summerfield, M.A. (2004) Geomorphological applications of cosmogenic isotope analysis. *Progress in Physical Geography* 28: 1–42.

Colgan, P.M., Mickelson, D.M. and Cutler, P.M. (2003) Ice-marginal terrestrial landsystems: southern Laurentide ice sheet margin. In D.J.A. Evans (ed.), *Glacial Landsystems*. Hodder Arnold, London, pp. 111–142.

Commoner, B. (1972) *The Closing Circle: Confronting the Environmental Crisis*. Cape, London.

Conacher, A.J. (1988) The geomorphic significance of process measurements in an ancient landscape. In A.M. Harvey and M. Sala (eds), *Geomorphic Processes in Environments with Strong Seasonal Contrasts. Vol. 2 Geomorphic Systems*. Catena Supplement 13: 147–164.

Conacher, A.J. and Dalrymple, J.B. (1977) The nine–unit landsurface model: an approach to pedogeomorphic research. *Geoderma* 18: 1–154.

Cooke, R.U. (1992) Common ground, shared inheritance: research imperatives for environmental geography. *Transactions Institute of British Geographers* NS 17: 131–151.

Cooke, R.U. and Doornkamp, J.C. (1974) *Geomorphology in Environmental Management*. Oxford University Press, Oxford.

Cooke, R.U. and Doornkamp, J.C. (1990) *Geomorphology in Environmental Management. A New Introduction*. Clarendon Press, Oxford, 2nd edn.

Cooke, R.U., Brunsden, D., Doornkamp, J.C. and Jones, D.K.C. (1982) *Urban Geomorphology in Drylands*. Oxford University Press, Oxford.

Cooke, R.U., Warren, A. and Goudie, A.S. (1993) *Desert Geomorphology*. UCL Press, London.

Costa, J.E. (1987) A comparison of the largest rainfall-runoff floods in the United States with those of the People's Republic of China and the world. *Journal of Hydrology* 96: 101–115.

Coulthard, T.J., Kirkby, M.J. and Macklin, M.G. (1999) Modelling the impacts of Holocene environmental changes in an upland river catchment, using a cellular automation approach. In A.G. Brown and T.A. Quine (eds), *Fluvial Processes and Environmental Changes*. Wiley, Chichester, pp. 31–46.

Cowell, P.J. and Thom, B.G. (1997) Morphodynamics of coastal evolution. In R.W.G. Carter and C.D. Woodroffe (eds), *Coastal Evolution, Late Quaternary Shoreline Morphodynamics*. Cambridge University Press, Cambridge, pp. 33–86.

Crozier, M.J. (1986) *Landslides: Causes, Consequences and Environment*. Croom Helm, London.

Czudek, T. and Demek, K.J. (1970) Thermokarst in Siberia and its influence on the development of lowland relief. *Quaternary Research* 1: 103–120.

Dackcombe, R.V. and Gardiner, V. (1983) *Geomorphological Field Manual*. George Allen and Unwin, London.

Dalyrmple, J.B., Conacher, A.J. and Blong, R.J. (1969) A nine–unit hypothetical land-surface model. *Zeitschrift fur Geomorphologie* 12: 60–76.

Darby, S. and Sear, D. (eds) (2008) *River Restoration. Managing the Uncertainty in Restoring Physical Habitat*. Wiley, Chichester.

Davies, J.L. (1980) *Geographical Variations in Coastal Development*. Longman, London.

Davis, W.M. (1884) Gorges and waterfalls. *American Journal of Science* 28: 123–132.

Davis, W.M. (1900) The physical geography of the lands. *Popular Science Monthly* 57: 157–170.

de Wit, M. and Stankiewicz, J. (2006) Changes in surface water supply across Africa with predicted climate change. *Science* 311: 1917–1921.

Dearing, J.A. (2008) Landscape change and resilience theory: a palaeoenvironmental assessment from Yunnan, SW China. *The Holocene* 18: 117–127.

Derbyshire, E., Gregory, K.J. and Hails (1979) *Geomorphological Processes*. Dawson, Folkestone.

Dietrich, W.E. and Perron, J.T. (2006) The search for a topographic signature of life. *Nature* 439: 411–418.

Douglas, I. (1983) *The Urban Environment*. Arnold, London.

Douglas, I. (2005a) Hot wetlands. In P.G. Fookes, E.M. Lee and G. Milligan (eds), *Geomorphology for Engineers*. Whittles Publishing, Dunbeath, Caithness, pp. 473–500.

Douglas, I. (2005b) Urban geomorphology. In P.G. Fookes, E.M. Lee and G. Milligan (eds), *Geomorphology for Engineers*. Whittles Publishing, Dunbeath, Caithness, pp. 757–779.

Douglas, I. and Lawson, N. (2000) The human dimensions of geomorphological work in Britain. *Journal of Industrial Ecology* 4: 9–33.

Douglas, I. and Spencer, T. (eds) (1985) *Environmental Change and Tropical Geomorphology*. George Allen and Unwin, London.

Downs, P.W. and Gregory, K.J. (1995) Approaches to river channel sensitivity. *Professional Geographer* 47: 168–175.

Downs, P.W. and Gregory, K.J. (2004) *River Channel Management. Towards Sustainable Catchment Hydrosystems*. Arnold, London.

Drew, F. (1873) Alluvial and lacustrine deposits and glacial records of the upper Indus basin. Pt 1: Alluvial deposits. *Geological Society of London Quarterly Journal* 29: 441–471.

Driver, T.S. and Chapman, G.P. (eds) (1996) *Time-scales and Environmental Change*. Routledge, London and New York.

Duchaufour, P. (1977) *Pédologie: 1. Pédogenèse et Classification*. Masson, Paris.

Dury, G.H. (1965) Theoretical implications of underfit streams. *US Geological Survey Professional Paper* 452C.

Ehlers, J. and Gibbard, P. (2004) *Quaternary Glaciations – Extent and Chronology. Part I: Europe. Part II: North America. Part III: South America, Asia, Africa, Australasia, Antarctica*. Elsevier, Amsterdam.

Ehlers, J. and Gibbard, P.L. (2004) *Quaternary Glaciations: Extent and Chronology 2: Part II: North America*. Elsevier, Amsterdam.

Ehlers, J. and Gibbard, P.L. (2007) The extent and chronology of Cenozoic Global Glaciation. *Quaternary International* 164–165: 6–20.

Eide, E. (2008) Input sought on earth surface processes. *EOS, Transactions American Geophysical Union* 89: 10.1029/2008EO230003.

Ellis, E.C. and Ramankutty, N. (2008) Putting people in the map: anthropogenic biomes of the world. *Frontiers in Ecology and the Environment* 6: 439–447.

Elsenbeer, H. (2001) Hydrologic flowpaths in tropical rainforest soilscapes – a review. *Hydrological Processes* 15: 1751–1759.

Emanuel, W.R., Shugart, H.H. and Stevenson, M.P. (1985) Climatic change and the broad-scale distribution of terrestrial ecosystem complexes. *Climatic Change* 7: 29–43.

Embleton, C. and King, C.A.M. (1968) *Glacial and Periglacial Geomorphology.* Arnold, London.

Embleton–Hamann, C. (2004) Proglacial landforms. In A.S. Goudie (ed.), *Encyclopedia of Geomorphology Vol. 2*, Routledge, London, pp. 810–813.

Endreny, T.A. (2001) A global impact for hydro-socio-ecological watershed research. *Water Resources Impact* 3: 20–25.

Environmental Research Funders Forum (ERFF) (2003) *Analysis of Environmental Science in the UK*. At http://www.erff.org.uk, accessed 21 January 2010.

Evans, N.C. and King, J.P. (1997) *The Natural Terrain Landslide Study: Debris Avalanche Susceptibility.* GEO Technical note TN 1/98.

Evans, D.J.A. (ed.) (2003) *Glacial Landsystems.* Arnold, London.

Eyles, N., Dearman, W.R. and Douglas, T.D. (1983) The distribution of glacial land systems in Britain and North America. In N. Eyles (ed.), *Glacial Geology.* Pergamon, Oxford, pp. 213–228.

Ezcurra, E. (ed.) (2006) *Global Deserts Outlook.* United Nations.

Fairbridge, R.W. (1968) Mountain and hilly terrain, mountain systems: mountain types. In R.W. Fairbridge (ed.), *Encyclopedia of Geomorphology.* Reinhold, New York, pp. 745–761.

Fairbridge, R.W. (1999) Von Richthofen, Ferdinand, Baron (Freiherr) (1833–1905). In D.E. Alexander (ed.), *Encyclopedia of Environmental Science.* Kluwer Academic Publishers, Dordrecht, pp. 662–663.

FAO Millenium assessment 2005. At http://www.millenniumassessment.org/en/Index.aspx, accessed 25 June 2009.

Favis-Mortlock, D.T. (2007) The soil erosion website. www.soilerosion.net/.

Fenneman, N.M. (1931) *Physiography of the Western United States.* McGraw-Hill, New York.

Fenneman, N.M. (1938) *Physiography of the Eastern United States.* McGraw-Hill, New York.

Ferrigno, J.G., Cook, A.J., Mathie, A.M., Williams, R.S., Swithinbank, C., Foley, K.M., Fox, A.J., Thomson, J.W. and Sievers, J. (2008) Coastal change and glaciological map of the LKarsen ice shelf area, Antarctica 1940–2005. *USGS Geologic Investigations Series* Map I-2600-B.

Findlay, S.J. and Taylor, M.P. (2006) Why rehabilitate river systems? *Area* 38: 312–325.

Finnegan, D.A. (2004) The work of ice: glacial theory and scientific culture in early Victorian Edinburgh. *British Journal for the History of Science* 37: 29–52.

Firestone, R.B., West, A., Kennett, J.P., Becker, L., Bunch, T.E., Revay, Z S., Schultz, P.H., Belgya, T., Kennett, D.J., Erlandson, J.M., Dickenson, O.J., Goodyear, A.C., Harris, R.S., Howard, G.A., Kloosterman, J.B., Lechler, P., Mayewski, P.A., Montgomery, J., Poreda, R., Darrah, T., Que Hee, S.S., Smith, A.R., Stich, A., Topping, W., Wittke, J.H. and Wolbach, W.S. (2007) Evidence for an extraterrestrial impact 12,900 years ago that contributed to the megafaunal extinctions and the Younger Dryas cooling. *Proceedings National Academy of Sciences* 104: 16016–16021.

Fischer, L. and Busche, D. (eds) (1982) *Climatic Geomorphology*. Princeton University Press, Princeton.

Folk, R.L., Roberts, H.H. and Moore, C.H. (1973) Black phytokarst from Hell, Cayman Islands, British West Indies. *Bulletin Geological Society of America* 84: 2351–2360.

Fookes, P.G. (1976) Road geotechnics in hot deserts. *Highway Engineer* 23: 11–29.

Fookes, P.G. (1997) *Tropical Residual Soils*. Geological Society, London.

Fookes, P. and Lee, M. (2005) Introduction to engineering geomorphology. In P.G. Fookes, E.M. Lee and G. Milligan (eds), *Geomorphology for Engineers*. Whittles Publishing, Dunbeath, Caithness, pp. 1–28.

Fookes, P.G. and Vaughn, P.R. (eds) (1986) *A Handbook of Engineering Geomorphology*. Surrey University Press, London.

Fookes, P.G., Lee, E.M. and Milligan, G. (eds) (2005) *Geomorphology for Engineers*. Whittles Publishing, Dunbeath, Caithness.

Ford, D.C. and Williams, P.W. (1989) *Karst Geomorphology and Hydrology*. Unwin Hyman, London.

Francis, P. (1993) *Volcanoes: A Planetary Perspective*. Oxford University Press, Oxford.

Francis, R.A., Petts, G.E. and Gurnell, A.M. (2008) Wood as a driver of past landscape change along river corridors. *Earth Surface Processes and Landforms* 33: 1622–1626.

Freerain Rainwater Management Solutions (2006) At http://www.freerain.co.uk, accessed 18 December 2006.

French, H.M. (1987) Permafrost and ground ice. In K.J. Gregory and D.E. Walling (eds), *Human Activity and Environmental Processes*. Wiley, Chichester, pp. 237–269.

French, H.M. (1996) *The Periglacial Environment*. Addison Wesley Longman, London, 2nd edn.

French, H.M. and Williams, P. (2007) *The Periglacial Environment*. Wiley, Chichester, 3rd edn.

Fryirs, K. and Brierley, G. (2000) A geomorphic approach to the identification of river recovery potential. *Physical Geography* 21: 244–277.

Gabrovšek, F. (2009) On concepts and methods for the estimation of dissolutional denudation rates in karst areas. *Geomorphology* 106: 9–14.

Garland, G.G. (1999) Soil erosion. In D.E. Alexander and R.W. Fairbridge (eds), *Encyclopedia of Environmental Science*. Kluwer Academic Publishers, Dordrecht, pp. 570–573.

Garvin, C.D., Hanks, T.C., Finkel, R.C. and Heimsath, A.J. (2005) Episodic incision of the Colorado River in Glen Canyon, Utah. *Earth Surface Processes and Landforms* 30: 973–984.

Gerrard, J. (1990) *Mountain Environments: An Examination of the Physical Geography of Mountains*. Cambridge, MA: MIT Press.

Gibbard, P. and van Kolfschoten, Th. (2005) The Pleistocene and Holocene Epochs. In F. Gradstein, J. Ogg and A. Smith (eds), *A Geologic Time Scale 2004*. Cambridge University Press, Cambridge, pp. 441–452.

Gilbert, G.K. (1875) Report on the Geology of portions of Nevada, Utah, California and Arizona (1871–72). US Geological and Geographical Survey, 3: 17–187.

Gilbert, G.K. (1914) The transportation of debris by running water. *US Geological Survey Professional Paper* 86: 263.

Glantz, M.H. (1995) In Central Asia, a sea dies: a sea also rises. *Climate-Related Impacts International Network Newsletter* 10(2): 1.

Glasser, N.F. and Bennett, M.R. (2004) Glacial erosional landforms: origins and significance for palaeoglaciology. *Progress in Physical Geography* 28: 43–75.

Goswami, B.N., Venugopal, V., Sengupta, D., Madhusoodanan, M.S. and Xavier, P.K. (2006) Increasing trend of extreme rain events over India in a warming environment. *Science* 314: 1442–1445.

Goudie, A.S. (1986) *The Human Impact: Man's Role in Environmental Change*. Blackwell, Oxford, 2nd edn.

Goudie, A.S. (1988) The geomorphological role of termites and earthworms in the tropics. In H.A. Viles (ed.), *Biogeomorphology*. Blackwell, Oxford, pp. 166–192.

Goudie, A.S. (1992) *Environmental Change*. Clarendon Press, Oxford.

Goudie, A.S. (1995) *The Changing Earth: Rates of Geomorphological Processes*. Blackwell, Oxford.

Goudie, A.S. (1999) Weathering. In D.E. Alexander and R.W. Fairbridge (eds), *Encyclopedia of Environmental Science*. Kluwer Academic Publishers, Dordrecht, Boston and London, pp. 693–695.

Goudie, A.S. (2003) Long-term environmental change: Quaternary climate oscillations and their impacts on the environment. In A. Rogers and H.A. Viles (eds), *The Student's Companion to Geography*. Blackwell Publishing, Oxford, pp. 13–17.

Goudie, A.S. (2006) Global warming and geomorphology. *Geomorphology* 79: 384–394.

Goudie, A.S. (2008) The history and nature of wind erosion in deserts. *Annual Review of Earth and Planetary Sciences* 36: 97–119.

Goudie, A.S. and Viles, H. (1997) *The Earth Transformed*. Blackwell, Oxford.

Grabau, W.E., Walker, H.J. and Lee, M. (2005) Estuaries and deltas. In P.G. Fookes, E.M. Lee and G. Milligan (eds), *Geomorphology for Engineers*. Whittles Publishing, Dunbeath, pp. 535–565.

Gradstein, F.M., Ogg, J.G. and Smith, A.G. (eds) (2005) *A Geologic Time Scale*. Cambridge University Press, Cambridge.

Graf, W.L. (1977) The rate law in fluvial geomorphology. *American Journal of Science* 277: 178–191.

Graf, W.L. (ed.) (1987) *Geomorphic Systems of North America*. Boulder, CO, Geological Society of America, Centennial Special Volume 2.

Graf, W.L. (1988) *Fluvial Processes in Dryland Rivers*. Springer-Verlag, Berlin, Heidelberg, New York, London, Paris and Tokyo.

Graf, W.L. (1992) Science, public policy and western American rivers. *Transactions Institute of British Geographers* NS 17: 5–19.

Graf, W.L. (2001) Damage control: restoring the physical integrity of America's rivers. *Annals of the Association of American Geographers* 91: 1–27.

Graf, W.L. (ed.) (2003) Dam Removal Research: Status and Prospects. The H. John Heinz III Center for Science, Economics and the Environment, Washington, DC.

Graf, W.L. (2006) Downstream hydrologic and geomorphic effects of large dams on American rivers. *Geomorphology* 79: 336–360.

Gregory, K.J. (1977) Channel and network metamorphosis in northern New South Wales. In K.J. Gregory (ed.), *River Channel Changes*. Wiley, Chichester, pp. 389–410.

Gregory, K.J. (1979) Drainage basin processes. In E. Derbyshire, K.J. Gregory and J.R. Hails (eds), *Geomorphological Processes*. Dawson, Folkestone.

Gregory, K.J. (1985) *The Nature of Physical Geography*. Arnold, London.

Gregory, K.J. (1987) The power of nature – energetics in physical geography. In K.J. Gregory (ed.), *Energetics of Physical Environment: Energetic Approaches to Physical Geography*. Wiley, Chichester, pp. 1–31.

Gregory, K.J. (1995) Human activity in palaeohydrology. In K.J. Gregory, L. Starkel and V.R. Baker (eds), *Global Continental Palaeohydrology*. Wiley, Chichester, pp. 151–172.

Gregory, K.J. (ed.) (1997) *Fluvial Geomorphology of Great Britain*. Geological Conservation Review Series, Joint Nature Conservation Committee. Chapman and Hall, London.

Gregory, K.J. (2000) *The Changing Nature of Physical Geography*. Arnold, London.

Gregory, K.J. (2002) Urban channel adjustments in a management context. *Environmental Management* 29: 620–633.

Gregory, K.J. (2003) Palaeohydrology, environmental change and river channel management. In K.J. Gregory and G. Benito (eds), *Palaeohydrology: Understanding Global Change*. Wiley, Chichester, pp. 357–378.

Gregory, K.J. (2004a) Human activity transforming and designing river landscapes: a review perspective. *Geographica Polonica* 77: 5–20.

Gregory, K.J. (2004b) Palaeohydrology and river channel management. *Journal of the Geological Society of India* 64: 383–394.

Gregory, K.J. (2005a) Temperate environments. In P.G. Fookes, E.M. Lee and G. Milligan (eds), *Geomorphology for Engineers*. Whittles Publishing, Dunbeath, Caithness, pp. 400–418.

Gregory, K.J. (ed.) (2005b) Editor's Introduction. In *Physical Geography*. Sage, London, 4 volumes, pp. xix–lviii.

Gregory, K.J. (2006) The human role in changing river channels. *Geomorphology* 79: 172–191.

Gregory, K.J. (2009) Place: The management of sustainable physical environments. In N. Clifford, S.L. Holloway, S.P. Rice and G. Valentine (eds), *Key Concepts in Geography*. Sage, London, 2nd edn, pp. 173–198.

Gregory, K.J. and Benito, G. (eds) (2003) *Palaeohydrology: Understanding Global Change*. Wiley, Chichester.

Gregory, K.J. and Downs, P.W. (2008) The sustainability of restored rivers: catchment scale perspectives on long term response. In S. Darby and D. Sear (eds), *River Restoration: Managing the Uncertainty in Restoring Physical Habitat*. Wiley, Chichester, pp. 253–286.

Gregory, K.J. and Walling, D.E. (1973) *Drainage Basin Form and Process*. Arnold, London.

Gregory, K.J., Benito, G. and Downs, P.W. (2008) Applying fluvial geomorphology to river channel management: background for progress towards a palaeohydrology protocol. *Geomorphology* 98: 153–172.

Gregory, J.M., Huybrechts, P. and Raper, S.C.B. (2004) Climatology: threatened loss of the Greenland ice-sheet. *Nature* 428: 616.

Gregory, K.J., Macklin, M.G. and Walling, D.E. (2006) Past hydrological events related to understanding global change. Catena 66: 1–187.

Gregory, K.J., Simmons, I.G., Brazel, A.J., Day, J.W., Keller, E.A., Sylvester, A.G. and Yanez-Arancibia, Y. (2009) *Environmental Sciences: A Student's Companion*. Sage, London.

Gregory, K.J., Starkel, L. and Baker, V.R. (eds) (1995) *Global Continental Palaeohydrology*. Wiley, Chichester.

Guerra, A., Marcal, M., Polivanov, H., Sathler, R., Mendonça, J., Guerra, T., Bezerra, F., Furtado, M., Lima, N., Souza, U., Feitosa, A., Davies, K., Fullen, M.A. and Booth, C.A. (2009) Environment management and health risks of soil erosion gullies in São Luís (Brazil) and their potential remediation using palm-leaf geotextiles. *Transactions of the Wessex Institute*, 28 May 2009, Paper DOI: 10.2495/EHR050461, see http://library. witpress.com/pages/PaperInfo.asp?PaperID=15704.

Gunderson, L.H. and Holling, C.S. (2001) *Panarchy: Understanding Transformations in Systems of Humans and Nature*. Island Press, New York.

Gunn, J. (1986) Solute processes and karst landforms. In S.T. Trudgill (ed.), *Solute Processes*.Wiley, Chichester, pp. 363–437.

Gupta, A. (1993) The changing geomorphology of the humid tropics. *Geomorphology* 7: 165–186.

Gupta, A. and Ahmad, R. (1999) Geomorphology and the urban tropics: building an interface between research and usage. *Geomorphology* 31: 133–149.

Gurnell, A.M. (1998) The hydrogeomorphological effects of beaver dam-building activity. *Progress in Physical Geography* 22: 167–189.

Gurney, S.D. (1998) Aspects of the genesis and geomorphology of pingos: perennial permafrost mounds. *Progress in Physical Geography* 22: 307–324.

Haines-Young, R. (2000) Sustainable development and sustainable landscapes: defining a new paradigm for landscape ecology. *Fennia* 178: 7–14.

Haines-Young, R. and Petch J.R. (1986) *Physical Geography: Its Nature and Methods*. Harper Row, London.

Hanson, G.N. (2007) Did an extra-terrestrial impact over the Laurentian Ice Sheet cause the extinction of North America's mega-fauna and the Clovis culture? At http://www.geo.sunysb.edu/openight/fall07.html#hanson, accessed 29 June 2009.

Harbor, J. (1999) Engineering geomorphology at the cutting edge of land disturbance: erosion and sediment control. *Geomorphology* 31: 247–263.

Hare, F.K. (1980) The planetary environment: fragile or sturdy. *Geographical Journal* 146: 379–395.

Harris, S.A. (1988) The alpine periglacial zone. In M.J. Clark (ed.), *Advances in Periglacial Geomorphology*. Wiley, Chichester, pp. 369–413.

Harvey, A.M. (2002) Effective timescales of coupling within fluvial systems. *Geomorphology* 44: 175–201.

Hawking, S. (1988) *A Brief History of Time: From the Big Bang to Black Holes*. Bantam Books, New York.

Hayakawa, Y.S. and Matsukura, Y. (2010) Factors influencing the recession rate of Niagara Falls since the 19th century. *Geomorphology* 110: 212–216.

Heimsath, A.M. and Ehlers, T.A. (2005) Quantifying rates and timescales of geomorphic processes. *Earth Surface Processes and Landforms* 30: 917–921.

Heinrich, H. (1988) Origin and consequences of cyclic ice rafting in the Northeast Atlantic Ocean during the past. *Quaternary Research* 29: 142–152.

Herrington Consulting (2006) Retrieved 18 December 2006 from http://www.herrington consulting.co.uk/SUDS.htm

Heywood, I., Cornelius, S. and Carver, S.C. (1998) *An Introduction to Geographical Information Systems*. Pearson Education, Harlow, 3rd edn.

Higgitt, D.L. and Lee, E.M. (eds) (2001) *Geomorphological Processes and Landscape Change: Britain in the last 1000 years*. Blackwell, Oxford.

Holdridge, L.R. (1947) Determination of world plant formations from simple climatic data. *Science* 105: 367–368.

Holdridge, L.R. (1967) *Life Zone Ecology*. Tropical Science Center, San Jose, Costa Rica.

Holling, C.S. (2001) Understanding the complexity of economic and social systems. *Ecosystems* 4: 390–405.

Holmes, A. (1944) *Principles of Physical Geology*. Nelson, London and New York.

Holzner, L. and Weaver, G.D. (1965) Geographic evaluation of climatic and climatogenetic geomorphology. *Annals Association of American Geographers* 55: 592–602.

Hooke, J.M. (ed.) (1988) *Geomorphology in Environmental Planning*. Wiley, Chichester.

Hooke, J.M. (1999) Decades of change: contributions of geomorphology to fluvial and coastal engineering and management. *Geomorphology* 31: 373–389.

Hooke, J.M. and Kain, R.J.P. (1982) *Historical Change in the Physical Environment: A Guide to Sources and Techniques*. Butterworth, London.

Horrocks, N.K. (1954) *Physical Geography and Climatology*. Longman, London.

Houghton, J. (1997) *Global Warming: The Complete Briefing*. Cambridge University Press, Cambridge, 2nd edn.

Howard, J.A. and Mitchell, C.W. (1985) *Phytogeomorphology*. Wiley, New York.

Huggett, R.J. (1975) Soil landscape systems: a model of soil genesis. *Geoderma* 13: 1–22.

Huggett, R.J. (1980) *Systems Analysis in Geography*. Clarendon Press, Oxford.

Huggett, R.J. (1985) *Earth Surface Systems*. Springer Verlag, Berlin, Heidelberg and New York.

Huggett, R.J. (1995) *Geoecology: An Evolutionary Approach*. Routledge, London.

Huggett, R.J. (1997) *Environmental Change. The Evolving Ecosphere*. Routledge, London.

Hupp, C.R. (1999) Relations among riparian vegetation, channel incision processes and forms, and large woody debris. In S.E. Darby and A. Simon (eds), *Incised River Channels*. Wiley, Chichester, pp. 219–245.

Huybrechts, P. (2009) Global change: West-side story of Antarctic ice. *Nature* 458: 295–296.

Isachenko, A.G. (1973) On the method of applied landscape research. *Soviet Geography* 14: 229–243.

Iturrizaga, L. (2008) Paraglacial landform assemblages in the Hindukush and Karakoram Mountains. *Geomorphology* 95: 1–102.

IUCN, UNEP and WWF (1991) *Caring for the Earth: A Strategy for Sustainable Living*. International Union for the Conservation of Nature, Earthscan, London.

Jamieson, S.S.R. and Sugden, D.E. (2008) Landscape evolution of Antarctica. In A.K. Cooper, P.J. Barrett, H. Stagg, B. Storey, E. Stump, W. Wise and the 10th ISAES editorial team (eds), *Antarctica: A Keystone in a Changing World*. Proceedings of the 10th International Symposium on Antarctic Earth Sciences. Washington, DC: The National Academies Press, pp. 39–54.

Johnsson, M. (2000) Chemical weathering and soils. In W.G. Ernst (ed.), *Earth Systems: Processes and Issues*. CUP, Cambridge, pp. 119–132 .

Kale, V.S. (2002) Fluvial geomorphology of Indian rivers: an overview. *Progress in Physical Geography* 26: 400–433.

Kale, V.S. (2003) Geomorphic effects of monsoon floods on Indian rivers. *Natural Hazards* 28: 65–84.

Kale, V.S., Gupta, A. and Singhvi, A.K. (2003) Late Pleistocene-Holocene palaeo-hydrology of Monsoon Asia. *Journal Geological Society of India* 64: 403–417.

Kalicki, T. and Sanko, A.F. (1998) Palaeohydrological changes in the upper Dneper valley, Belarus, during the last 20,000 years. In G. Benito, V.R. Baker and K.J. Gregory (eds), *Palaeohydrology and Environmental Change*. Wiley, Chichester, pp. 125–135.

Kearey, P. and Vine, F.J. (1990) *Global Tectonics*. Blackwell Scientific, Oxford.

Keller, E.A. (1996) *Environmental Geology*. Prentice Hall, Upper Saddle River, NJ, 7th edn.

Keller, E.A. and Pinter, N. (1996) *Active Tectonics. Earthquakes, Uplift and Landscape*. Prentice Hall, Upper Saddle River, NJ.

Kennett, D.J., Kennett, J.P., West, A., Mercer, C., Que Hee, S.S., Bement, L., Bunch, T.E., Sellers, M. and Wolbach, W.S. (2009) Nanodiamonds in the Younger Dryas boundary sediment layer. *Science* 323: 94.

Kerr, R.A. (2009) Did the mammoth slayer leave a diamond calling card? *Science* 323: 26.

Kilburn, C.R.J. and Petley, D.N. (2003) Forecasting giant, catastrophic slope collapse: lessons from the Vajont, northern Italy. *Geomorphology*, 54: 21–32.

Kimball, D. (1948) Denudation chronology. The dynamics of river action. *Occasional Paper No. 8, University of London, Institute of Archaeology*.

King, L.C. (1953) Canons of landscape evolution. *Bulletin Geological Society of America* C4: 721–752.

King, L.C. (1962) *The Morphology of the Earth*. Oliver & Boyd, Edinburgh.

Kleman, J. and Borgstrom, I. (1996) Reconstruction of palaeo-ice sheets: the use of geomorphological data. *Earth Surface Processes and Landforms* 21: 893–909.

Knox, J.C. (1995) Fluvial systems since 20,000 years BC. In K.J. Gregory, L. Starkel and V.R. Baker (eds), *Global Continental Palaeohydrology*. Wiley, Chichester, pp. 87–108.

Knox, J.C. (1999) Long-term episodic changes in magnitudes and frequency of floods in the Upper Mississippi River Valley. In A.G. Brown and T. Quine (eds), *Fluvial Processes and Environmental Change*. Chichester, Wiley, pp. 255–282.

Kondolf, G.M., Piegay, H. and Sear, D. (2003) Integrating geomorphological tools in ecological and management studies. In G.M. Kondolf and H. Piegay (eds), *Tools in Geomorphology*. Wiley, Chichester, pp. 633–660.

Kong, P., Na, C., Fink, D., Ding, L. and Huang, F. (2007) Erosion in northwest Tibet from in-situ-produced cosmogenic ^{10}Be and ^{26}Al in bedrock. *Earth Surface Processes and Landforms* 32: 116–125.

Kooi, H. and Beaumont, C. (1996) Large-scale geomorphology: classical concepts reconciled and integrated with contemporary ideas via a surface processes model. *Journal of Geophysical Research* 101: 3361–3386.

Koppen, W. (1931) *Grundriss der Klimakunde*. Walter de Gruyter, Berlin.

Kowal, J.M. and Kassam, A.H. (1976) Energy load and instantaneous intensity of rainstorms at Samaru, Northern Nigeria. *Tropical Agriculture* (Trinidad) 53: 185–197.

Kuhn, T.S. (1970) *The Structure of Scientific Revolutions*. University of Chicago Press, Chicago.

Lancaster, N. (2008) Desert dune dynamics and development: insights from luminescence dating. *Boreas* 37: 559–573.

Lane, S. (2003) Environmental modeling. In A. Rogers and H.A. Viles (eds), *The Student's Companion to Geography*. Blackwell Publishing, Oxford, pp. 64–76.

Larsen, L. and Harlan, S.L. (2006) Desert dreamscapes: residential landscape preference and behaviour. *Landscape and Urban Planning* 78: 85–100.

Larsen, M.C. (1997) Tropical geomorphology and geomorphic work: A study of geomorphic processes and sediment and water budgets in montane humid-tropical forested and developed watersheds, Puerto Rico. Unpublished PhD thesis, University of Colorado Geography Department, accessed online 20 December 2008.

Lee, M. and Fookes, P. (2005) Hot drylands. In P.G. Fookes, E.M. Lee and G. Milligan (eds), *Geomorphology for Engineers*. Whittles Publishing, CRC Press, Dunbeath, pp. 419–453.

Lemons, J. (ed.) (1987) Special focus on environmental ethics. *Environment Professional* 9.

Lemons, J. (1999) Environmental ethics. In D.E. Alexander and R.W. Fairbridge (eds), *Encyclopedia of Environmental Science*. Kluwer Academic, Dordrecht, pp. 204–206.

Leopold, L.B. (1969) Landscape esthetics. *Natural History* Oct: 37–44.

Leopold, L.B. (1973) River channel change with time: an example. *Bulletin Geological Society of America* 84: 1845–1860.

Leopold, L.B. (1974) *Water – A Primer*. W.H. Freeman and Co., San Francisco.

Leopold, L.B. (1977) A reverence for rivers. *Geology* 5: 429–430.

Leopold, L.B. (1994) *A View of the River*. Harvard University Press, Cambridge, MA.

Leopold, L.B. and Maddock, T., Jr. (1953) The hydraulic geometry of stream channels and some physiographic implications. *US Geological Survey Professional Paper* 252, Washington, DC.

Leopold, L.B. and Marchand, M.O. (1968) On the quantitative inventory of riverscape. *Water Resources Research* 4: 709–717.

Leopold, L.B. and Miller, J.P. (1954) Postglacial chronology for alluvial valleys in Wyoming. *US Geological Survey Water Supply Paper* 1261: 61–85.

Leopold, L.B., Clarke, F.E., Hanshaw, B.B. and Balsley, J.R. (1971) A procedure for evaluating environmental impact. *US Geological Survey Circular* 645.

Leopold, L.B., Wolman, M.G. and Miller, J.P. (1964) *Fluvial Processes in Geomorphology*. Freeman, San Francisco.

Leopoldo, P.R., Franken, W.K. and Villa Nova, N.A. (1995) Real evapotranspiration and transpiration through a tropical rain forest in central Amazonia as estimated by the water balance method. *Forest Ecology and Management* 73: 185–195.

Lewin, J. (1980) Available and appropriate time scales in geomorphology. In R.A. Cullingford, D.A. Davidson and J. Lewin (eds), *Timescales in Geomorphology*. Wiley, Chichester, pp. 3–10.

Lewin, J. (1987) Stable and unstable environments – the example of the Temperate zone. In M.J. Clark, K.J. Gregory and A.M. Gurnell (eds), *Horizons in Physical Geography*. Macmillan, Basingstoke, pp. 200–212.

Li, Y.H. (1976) Denudation of Taiwan Island since the Pliocene epoch. *Geology* 4: 105–107.

Lillesand, T.M., Kiefer, R.W. and Chipman, J.W. (2004) *Remote Sensing and Image Interpretation*. Wiley, Hoboken NJ, 5th edn.

Linnaeus, C. (1735) *Systema Naturae*.Theodor Haak, Leiden.

Linton, D.L. (1951) The delimitation of morphological regions. In L.D. Stamp and S.W. Wooldridge (eds), *London Essays in Geography*. LSE, London, pp. 199–218.

Lioubimtseva, E. (2004) Climate change in arid environments: revisiting the past to understand the future. *Progress in Physical Geography* 28: 502–530.

Lovelock, J. (1979) *Gaia: A New Look at Life on Earth*. Oxford University Press, Oxford.

Lovelock, J. (1988) *The Ages of Gaia*. Oxford University Press, Oxford.

Lovelock, J. (2000) *Homage to GAIA – The Life of an Independent Scientist*. Oxford University Press, Oxford.

Lovelock, J. (2006) *The Revenge of Gaia*. Allen and Lane, London.

Lovelock, J. (2009) *The Vanishing Face of Gaia: A Final Warning*. Basic Books, New York.

Lozinski, M.W. (1909) On the mechanical weathering of sandstones in temperate climates (in German). *Acad. sci. Cracovie Bull. Internat., cl. sci. math et naturelles* 1: 1–25.

Lyell, C. (1830) *The Principles of Geology*. John Murray, London.

Lynch, K. (1960) *The Image of the City*. Harvard University Press, Cambridge, MA.

Mabbutt, J.A. (1977) *Desert Landforms*. Australian National University Press, Canberra.

MacCaig, N. (1983) *A World of Difference*. Chatto & Windus, London.

Macdonald, G.A. (1972) *Volcanoes*. Prentice Hall, Englewood Cliffs, NJ.

Mackay, A., Battarbee, R.W., Birks, H.J.B. and Oldfield, F. (eds) (2003) *Global Change in the Holocene*. Arnold, London.

Macklin, M.G. and Lewin, J. (2008) Alluvial responses to the changing Earth system. *Earth Surface Processes and Landforms* 33: 1374–1395.

Macklin, M.G., Fuller, I.C. and Lewin, J. (2002) Correlation of fluvial sequences in the Mediterranean basin over the last 200ka and their relationship to climate change. *Quaternary Science Reviews* 21, 1633–1641.

Macmillan, B. (1989) Modelling through: an afterword to Remodelling Geography. In B. Macmillan (ed.), *Remodelling Geography*: 291–313.

Magee, J.W., Miller, G.H., Spooner, N.A. and Questiaux, D. (2004) Continuous 150 k.y. monsoon record from Lake Eyre, Australia: insolation-forcing implications and unexpected Holocene failure. *Geology* 32: 885–888.

Magilligan, F.J., Gomez B., Mertes L.A.K., Smith, L.C., Smith N.D., Finnegan D. and Garvin J.B. (2002) Geomorphic effectiveness, sandur development, and the pattern of landscape response during jökulhlaups: Skeiðarársandur, southeastern Iceland. *Geomorphology* 44: 95–113.

Major, J.J. (2004) Posteruption suspended sediment transport at Mount St. Helens: Decadal-scale relationships with landscape adjustments and river discharges. *Journal of Geophysical Research* 109: F01002, doi:10.1029/2002JF000010.

Mangerud, J., Ehlers, J. and Gibbard, P. (2004) *Quaternary Glaciations: Extent and Chronology 1: Part I Europe*. Elsevier, Amsterdam.

Manning, A.D., Fischer, J., Felton, A., Newell, B., Steffen, W. and Lindenmayer, D.B. (2009) Landscape fluidity – a unifying perspective for understanding and adapting to global change. *Journal of Biogeography* 36: 193–199.

Maroulis, J.C., Nanson, G.C., Price, D.M. and Pietsch, T. (2007) Aeolian-fluvial interaction and climate change: source-bordering dune development over the past ~100 ka on Cooper Creek, central Australia. *Quaternary Science Reviews* 26: 386–404.

Marren, P.M. (2005) Magnitude and frequency in proglacial rivers: a geomorphological and sedimentological perspective. *Earth-Science Reviews* 70: 203–251.

Marsh, G.P. (1864) *Man and Nature or Physical Geography as Modified by Human Action*. Charles Scribner, New York.

Martin, B. (1998) See http://www.uow.edu.au/arts/sts/bmartin/pubs/98il/index.html, accessed 27 July 2009.

Maslin, M., Seidov, D. and Lowe, J. (2001) Synthesis of the nature and causes of rapid climate transitions during the Quaternary. *Geophysical Monograph* 126: 9–52 (see http://www.essc.psu.edu/~dseidov/pdf_copies/maslin_seidov_levi_agu_book_2001.pdf).

Massachusetts Government (2006) Retrieved 18 December 2006 from http://www.mass.gv/envir/smart_growth_toolkit/pages/glossary.html

Mathews, J.A. (ed.) (2001) *The Encyclopaedic Dictionary of Environmental Change*. Arnold, London.

May, R.M. (1977) Thresholds and breakpoints in ecosystems with a multiplicity of stable states. *Nature* 269: 471–477.

McFarlane, M.J. and Whitlow, R. (1991) Key factors affecting the initiation and progress of gullying in dambos in parts of Zimbabwe and Malawi. *Land Degradation & Development* 2: 215–235.

McGee, W.J. (1888) The classification of geographic forms by genesis. *National Geographic Magazine* 1: 27–36.

McGee, W.J. (1897) Sheetflood erosion. *Geological Society of America Bulletin* 8: 87–112.

McGuire, B., Mason, I. and Kilburn, C. (2002) *Natural Hazards and Environmental Change*. Arnold, London.

McHarg, I.L. (1969) *Design with Nature*. Natural History Press, New York.

McHarg, I.L. (1992) *Design with Nature*. Wiley, Chichester.

McHarg, I.L. (1996) *A Quest for Life: An Autobiography*. Wiley, New York.

McHarg, I.L. and Steiner, F.R. (eds) (1998) *To Heal the Earth: Selected Writings of Ian L. McHarg*. Island Press, Washington, DC.

McKee, E.D. (ed.) (1979) A study of global sand seas. *US Geological Survey Professional Paper* 1052.

McKibben, B. (1989) *The End of Nature*. Random House, New York.

McTainsh, G. and Strong, C. (2007) The role of aeolian dust in ecosystems. *Geomorphology* 89: 39–54.

Mead, W.R. (1953) The language of place. *Geographical Studies* 1: 63–68.

Meigs, P. (1953) The world distribution of arid and semi-arid homoclimates. *Reviews of Research on Arid Zone Hydrology*, UNESCO, Paris pp. 203–209.

Mercier, D. and Atienne, S. (2008) Paraglacial geomorphology: processes and paraglacial context. *Geomorphology* 95: 1–2.

Mickelson, D.M. (1987) Central lowlands. In W.L. Graf (ed.), *Geomorphic Systems of North America*. Geological Society of America, Boulder, CO, pp. 111–118.

Mitchell, C.W. (1991) *Terrain Evaluation. An Introductory Handbook to the History, Principles and Methods of Practical Terrain Assessment*. Longman Scientific and Technical, Harlow.

Mitsch, W.J. and Jorgensen, S.E. (2004) *Ecological Engineering and Ecosystem Restoration*. Wiley, New Jersey.

Montgomery, D.R. (1999) Process domains and the River Continuum. *Journal of the American Water Resources Association* 35: 397–410.

Moon, B.P. (1984) Refinement of a technique for determining rock mass strength for geomorphological purposes. *Earth Surface Processes and Landforms* 9: 189–193.

Mörner, N.-A. (2007) *Terra Nova* 3: 408–413.

Morretti, S. and Rodolfi, G. (2000) A typical 'calanchi' landscape on the Eastern Apennine margin (Atri, Central Italy): geomorphological features and evolution. *Catena* 40: 217–228.

Mosley, M.P. and Zimpfer, G.L. (1978) Hardware models in geomorphology. *Progress in Physical Geography* 2: 438–461.

Murray, B., Lazarus, E., Ashton, A., Baas, A., Coco, G., Coulthard, T., Fondstad, M., Haff, P., McNamara, D., Paola, C., Pelletier, J. and Rheinhardt, L. (2009) Geomorphology, complexity, and the emerging science of the Earth's surface. *Geomorphology* 103: 496–505.

Murton, J.B. (2008) Global warming and thermokarst. In R. Margesin (ed.), *Permafrost Soils*. Springer, Amsterdam, pp. 185–204.

Nace, R.L. (1969) World water inventory and control. In R.J. Chorley (ed.), *Water, Earth, and Man*. Methuen, London, pp. 31–42.

Naess, A. (1973) The shallow and the deep, long range ecology movement: a summary. *Inquiry* 16: 95–100.

Nanson, G.C., Price, D.M. and Short, S.A. (1992) Wetting and drying of Australia over the past 300 ka. *Geology* 20: 791–794.

Nasar, J.L. (1988) *Environmental Aesthetics*. CUP, New York.

Nash, R.F. (1967) *Wilderness and the American Mind*. Yale University Press, New Haven, CT.

National Earthquake Information Center (NEIC) at http://earthquake.usgs.gov/regional/neic/.

National Research Council (1999) *New Strategies for America's Watersheds*. National Academy Press, Washington, DC.

National Research Council (NRC) (2009) *Landscapes on the Edge: New Horizons for Research in Earth Surface Processes*. National Research Council, Washington DC.

Nelson, F.E. (2003) (Un)frozen in Time. *Science* 299: 1673–1675.

Newson, M.D. (1995) Fluvial geomorphology and environmental design. In A.M. Gurnell and G.E. Petts (eds), *Changing River Channels*. Wiley, Chichester, pp. 413–432.

Ng, K.Y. (2006) Landslide locations and drainage network development: a case study of Hong Kong. *Geomorphology* 76: 229–239.

Nicholas, A.P., Ashworth, P.J., Kirkby, M.J., Macklin, M.G. and Murray, T. (1995) Sediment slugs: large scale fluctuations in fluvial sediment transport rates and storage volumes. *Progress in Physical Geography* 19: 500–519.

Nishiizumi, K., Caffee, M.W., Finkel, R.C., Brimhall, G. and Mote, T. (2005) Remnants of a fossil alluvial fan landscape of Miocene age in the Atacama Desert of northern Chile using cosmogenic nuclide exposure age dating. *Earth and Planetary Science Letters* 237: 499–507.

Northcote, K.H. (1978) Soils and land use. In *Atlas of Australian Resources*, Division of National Mapping, Canberra.

Oberlander, T.M. (1997) Slope and pediment systems. In D.S.G. Thomas (ed.), *Arid Zone Geomorphology: Process, Form and Change in Drylands.* Wiley, Chichester, pp. 135–163.

Odum, H.T. and Odum, E.C. (1976) *Energy Basis for Man and Nature.* Wiley, New York.

Oerlemans, J. (2001) *Glaciers and Climate Change.* Balkema, Rotterdam.

Ogg, J.G., Ogg, G.M. and Gradstein, F.M. (2008) *The Concise Geologic Time Scale.* Cambridge University Press, Cambridge.

Oguchi, T. and Wasklewicz, T. (2010) Geographic information systems in geomorphology. In K.J. Gregory (ed.), *Handbook of Geomorphology.* Sage, London.

Oldfield, F. (1987) The future of the past – a perspective on palaeoenvironmental study. In M.J. Clark, K.J. Gregory and A.M. Gurnell (eds), *Horizons in Physical Geography.* Macmillan, Basingstoke, pp. 10–26.

Oldfield, F. (2005) *Environmental Change. Key Issues and Alternative Perspectives.* Cambridge University Press, Cambridge.

Ollier, C.D. (1979) Evolutionary geomorphology of Australia and Papua-New Guinea. *Transactions Institute of British Geographers* NS4: 516–539.

Ollier, C.D. (1981) *Tectonics and Landforms.* Oliver & Boyd, Edinburgh.

Ollier, C.D. (1995) Classics in physical geography revisited: L.C. King 1953. Canons of landscape evolution. *Progress in Physical Geography* 19: 371–377.

Ollier, C. and Pain, C. (1996) *Regolith, Soils and Landforms.* Wiley, Chichester.

Olson, J., Watts, J.A. and Allison, L.J. (1985) *Major World Ecosystem Complexes Ranked by Carbon in Live Vegetation: A Database.* Report NDP–017. Oak Ridge National Laboratory, Oak Ridge, TN. (Identified 46 ecosystems of the world)

Orban, C.E. (2006) *Terra Firma.* Florida Academic Press, Gainesville, FL.

Osterkamp, W.R. and Friedman, J.M. (1997) Research considerations for biogeomorphology. Proceedings of the US Geological Survey (USGS) Sediment Workshop 'Expanding Sediment Research Capabilities in Today's USGS'.

Osterkamp, W.R. and Hedman, E.R. (1982) Perennial streamflow characteristics related to channel geometry characteristics in Missouri River Basin. *US Geological Survey Professional Paper 1242*, Washington, DC.

Osterkamp, W.R. and Hupp, C.R. (1996) The evolution of geomorphology, ecology and other composite sciences. In B.L. Rhoads and C.E. Thorn (eds), *The Scientific Nature of Geomorphology.* Wiley, Chichester, pp. 415–441.

Owen, L.A. and Derbyshire, E. (2005) Glacial environments. In P.G. Fookes, E.M. Lee and G. Milligan (eds), *Geomorphology for Engineers.* Whittles Publishing, Dunbeath, Caithness, pp. 345–377.

Owens, S., Richards, K. and Spencer, T. (1997) Managing the earth's surface: science and policy. *Transactions Institute of British Geographers* NS22: 3–5.

PAGES at http://www.pages.unibe.ch/.

Paillard, D. (2001) Glacial cycles: toward a new paradigm. *Reviews of Geophysics* 39: 325–346.

Peltier, L.C. (1950) The geographic cycle in periglacial regions as it is related to climatic geomorphology. *Annals Association of American Geographers* 40: 214–236.

Peltier, L.C. (1975) The concept of climatic geomorphology. In W.N. Melhorn and R.C. Flemal, *Theories of Landform Development.* State University of New York, Binghamton, pp. 87–102.

Penck, A. and Bruckner, E. (1901–9) *DieAlpen im Eiszeitalter.* Tauchnitz, Leipzig.

Peschel, O. (1870) *Neue Probleme der Vergleichende Erdkunde als Versuch einer Morphologie der Erdoberflache.* Duncker and Humblot, Leipzig.

Pethick, J. (1984) *An Introduction to Coastal Geomorphology.* Arnold, London.

Petts, G.E., Sparks, R. and Campbell, I. (2000) River restoration in developed economies. In P.J. Boon, B.R. Davies and G.E. Petts (eds), *Global Perspectives on River Conservation: Science, Policy and Practice*. Wiley, Chichester, pp. 493–508.

Petts, J. (ed.) (1999) *Handbook of Environmental Impact Assessment*, Vols 1 & 2. Blackwell, Oxford.

Phillips, J.D. (1995) Biogeomorphology and landscape evolution: the problem of scale. *Geomorphology* 13: 387–405.

Phillips, J.D. (1997) A short history of a flat place: three centuries of geomorphic change in the Creatan, National Forest. *Annals Association of American Geographers* 87: 197–216.

Phillips, J.D. (1999) *Earth Surface Systems. Complexity, Order, and Scale*. Blackwell, Oxford.

Phillips, J.D. (2001) Human impacts on the environment: unpredictability and primacy of place. *Physical Geography* 27: 1–23.

Phillips, J.D. (2006a) Evolutionary geomorphology: thresholds and nonlinearity in landform response to environmental change. *Hydrology and Earth System Sciences Discussions* 3: 365–394.

Phillips, J.D. (2006b) Deterministic chaos and historical geomorphology: a review and look forward. *Geomorphology* 76: 109–21.

Phillips, J.D. (2006c) The perfect landscape. *Geomorphology* 84: 159–169.

Phillips, J.D. (2009) Changes, perturbations, and responses in geomorphic systems. *Progress in Physical Geography* 33: 1–14.

Pimentel, D. (2006) Soil erosion: a food and environmental threat. *Journal of the Environment, Development and Sustainability* 8: 119–137.

Plit, F. (2006) Desertification: on significance of the term and research of the phenomenon. *Miscellanea Geographica* 12: 147–153.

Poesen, J.W.A. and Hooke, J.M. (1997) Erosion, flooding and channel management in Mediterranean environments of southern Europe. *Progress in Physical Geography* 21: 157–179.

Powell, J.W. (1875) *Exploration of the Colorado River of the West (1869–72)*. Government Printing Office, Washington.

Powell, R.L. and Roberts, D. (2008) Characterizing variability of the urban physical environment for a suite of cities in Rondônia, Brazil. *Earth Interactions* 12: 1–32.

Prentice, I.C., Cramer, W., Harrison, S.P., Leemans, R., Monserud, R.A. and Solomon, A.M. (1992) A global biome model based on plant physiology and dominance, soil properties and climate. *Journal of Biogeography* 19: 117–134.

QRA at http://www.qra.org.uk/.

Ramage, C.S. (1971) *Monsoon Meteorology*. Academic Press, New York

Reading, A.J., Thompson, R.D. and Millington, A.C. (1995) *Humid Tropical Environments*. Blackwell, Oxford.

Reinfelds, I. and Nanson, G.C. (2001) 'Torrents of Terror': the August 1998 storm and the magnitude, frequency and impact of major floods in the Illawarra region of New South Wales. *Australian Geographical Studies* 39: 335–352.

Reynolds, J.F., Stafford Smith, D.M., Lambin, E.F., Turner II, B.L., Mortimore, M., Batterbury, S.P.J., Downing, T.E., Walker, B. (2007) Ecology: global desertification: building a science for dryland development. *Science* 316: 847–851.

Rhoads, B.L. and Thorn, C.E. (1996) Towards a philosophy of geomorphology. In B.L. Rhoads and C.E. Thorn (eds), *The Scientific Nature of Geomorphology*. Wiley, Chichester, pp. 115–143.

Rhodes, F.H.T. and Stone, R.O. (eds) (1981) *Language of the Earth*. Pergamon Press, New York.

Rhodes, F.H.T., Stone, R.O. and Malamud, B.D. (eds) (2008) *Language of the Earth*. Blackwell, Oxford.

Richards, K.S. and Clifford, N. (2008) Science, systems and geomorphologies: why LESS may be more. *Earth Surface Processes and Landforms* 33: 1323–1340.

Riley, A.L. (1998) *Restoring Streams in Cities*. Island Press, Washington, DC.

Riley, S., Luscombe, G. and Williams, A. (1986) Urban stormwater design: some lessons from the 8 November 1984 Sydney storm. *Australian Geographer* 17: 40–50.

Rinaldo, A., Dietrich, W.E., Rigon, R., Vogel, G.K. and Rodriguez-Iturbe, I. (2002) Geomorphological signatures of varying climate. *Nature* 374: 632–635.

Rivas, V., Cendrero, A., Hurtado, M., Cabral, M., Gimenez, J., Forte, L., del Rio, L. and Becker, A. (2006) Geomorphic consequences of urban development and mining activities; an analysis of study areas in Spain and Argentina. *Geomorphology* 73 (3–4): 185–206.

Roach, W.J., Heffernan, J.B., Grimm, N.B., Arrowsmith, J.R., Eisinger, C. and Rychener, T. (2008) Unintended consequences of urbanization for aquatic ecosystems: a case study from the Arizona desert. *BioScience* 58: 715–727.

Roberts, N. (1998) *The Holocene: An Environmental History*. Blackwell Publishing, Oxford.

Rodionov, S.N. (1990) A climatological analysis of the unusual recent rise in the level of the Caspian Sea. *Soviet Geography* 31: 265–275.

Rogers, A. and Viles, H.A. (eds) (2003) *The Student's Companion to Geography*. Blackwell Publishing, Oxford.

Rotnicki, K. (1991) Retrodiction of palaeodischarges of meandering and sinuous alluvial rivers and its palaeoclimatic implications. In L. Starkel, K.J. Gregory and J.B. Thornes (eds), *Temperate Palaeohydrology*. Wiley Chichester, pp. 431–471.

Rudoy, A. (1998) Mountain ice-dammed lakes of southern Siberia and their influence on the development and regime of the intracontinental runoff systems of North Asia in the late Pleistocene. In G. Benito, V.R. Baker and K.J. Gregory (eds), *Palaeohydrology and Environmental Change*. Wiley, Chichester, pp. 215–234.

Russell, A.J., Roberts, M.J., Fay, H., Marren, P.M., Cassidy, N.J., Tweed, F.S. and Harris, T. (2006) Icelandic jokulhlaup impacts: implications for ice-sheet hydrology, sediment transfer and geomorphology. *Geomorphology* 75: 33–64.

Rutimeyer, L. (1869) *Ueber Thal- und See-bildung*. Schweighauser, Basle.

Rutter, N.W. (1987) Glacial processes in the central Canadian Rocky Mountains. In W.L. Graf (ed.), *Geomorphic Systems of North America*. Geological Society of America, Boulder, CO, pp. 228–238.

Ryder, J.M. (1971) The stratigraphy and morphology of paraglacial alluvial fans in south central British Columbia. *Canadian Journal of Earth Sciences* 8: 279–298.

Sarnthein, M. (1978) Sand deserts during glacial maximum and climatic optimum. *Nature* 272: 43–46.

Scheffer, M., Carpenter, S., Foley, J.A., Folke, C. and Walker, B. (2001) Catastrophic shifts in ecosystems. *Nature* 413: 591–596.

Schellekensa, J., Scatenab, F.N., Bruijnzeela, L.A. and Wickela, A.J. (1999) Modelling rainfall interception by a lowland tropical rain forest in northeastern Puerto Rico. *Journal of Hydrology* 225: 168–184.

Schick, A.P. (1977) A tentative sediment budget for an extremely arid watershed in the southern Negev. In D.O. Doehring (ed.), *Geomorphology in Arid Regions*. Allen and Unwin, London, pp. 139–163.

Schulte, L., Julia, R., Burjachs, F. and Hilgers, A. (2008) Middle Pleistocene to Holocene geochronology of the River Aguas terrace sequence (Iberian peninsula): fluvial response to Mediterranean environmental change. *Geomorphology* 98: 13–33.

Schumm, S.A. (1979) Geomorphic thresholds: the concept and its applications. *Transactions Institute of British Geographers* NS4: 485–515.

Schumm, S.A. (1994) Erroneous perception of fluvial hazards. *Geomorphology* 10: 129–138.

Schumm, S.A. and Lichty, R.W. (1965) Time, space and causality in geomorphology. *American Journal of Science* 263: 110–119.

Sear, D.A., Wheaton, J.M. and Darby, S.R. (2007) Uncertain restoration of gravel-bed rivers and the role of geomorphology. *Developments in Earth Surface Processes* 11: 739–760.

Selby, M.J. (1980) Rock mass strength classification for geomorphic purposes: with tests from Antarctica and New Zealand. *Zeitschrift fur Geomorphologie* 24: 31–51.

Selby, M.J. (1985) *Earth's Changing Surface. An Introduction to Geomorphology.* Clarendon Press, Oxford.

Seppala, M. (2004) *Wind as a Geomorphic Agent in Cold Climates.* Cambridge University Press, Cambridge.

Shackleton, N.J. and Opdyke, N.D. (1973) Oxygen isotope and paleomagnetic stratigraphy of equatorial Pacific Core V28-238: oxygen isotope temperatures and ice volume on a 105 year and 106 year scale. *Quaternary Research* 3: 39–55.

Shakesby, R.A. and Whitlow, J.B. (1991) Perspectives on prehistoric and recent gullying in Central Zimbabwe. *GeoJournal* 23: 49–58.

Shaw, P.A. and Thomas, D.S.G. (1997) Pans, playas and salt lakes. In D.S.G. Thomas (ed.), *Arid Zone Geomorphology,* Wiley, Chichester, 2nd edn, pp. 293–317.

Sherlock, R.L. (1922) *Man as a Geological Agent.* Witherby, London.

Sherman, D.J. (1996) Fashion in geomorphology. In B.L. Rhoads and C.E. Thorn (eds), *The Scientific Nature of Geomorphology.* Wiley, Chichester, pp. 87–114.

Siegert, M.S. (2001) *Ice Sheets and Late Quaternary Environmental Change.* Wiley, Chichester.

Slaymaker, H.O. (1997) A pluralist problem-focused geomorphology. In D.R. Stoddart (ed.), *Process and Form in Geomorphology.* Routledge, London, pp. 328–339.

Slaymaker, O. (2007) Criteria to discriminate between proglacial and paraglacial environments. *Landform Analysis* 5: 72–74.

Slaymaker, O. and Kelly, R.E.J. (2007) *The Cryosphere and Global Environmental Change.* Blackwell Publishing, Malden, MA.

Slaymaker, H.O. and Spencer, T. (1998) *Physical Geography and Global Environmental Change.* Longman, Harlow.

Slaymaker, O., Spencer, T. and Embleton Hamann, C. (eds) (2009) *Geomorphology and Global Environmental Change.* CUP, Cambridge.

Smil, V. (1991) *General Energetics. Energy in the Biosphere and Civilization.* Wiley, New York.

Smith, D.E., Cullingford, R.A. and Firth, C.A. (2000) Patterns of isostatic land uplift during the Holocene: evidence from mainland Scotland. *The Holocene* 10: 489–501.

Smith, G.I. and Street Perrott, F.A. (1983) Pluvial lakes of the western United States. In S.C. Porter (ed.), *Late Quaternary Environments of the United States.* Longman, Harlow, pp. 190–212.

So, C.L. (1971) Mass movements associated with the rainstorms of June 1966 in Hong Kong. *Transactions Institute of British Geographers* 53: 55–65.

Soja, R. and Starkel, L. (2007) Extreme rainfalls in Eastern Himalaya and southern slope of Meghalaya Plateau and their geomorphologic impacts. *Geomorphology* 84: 170–180.

Soper, K. (1995) *What is Nature?: Culture, Politics and the Non-Human.* Blackwell, Oxford.

Sowers, S., Noller, J.S. and Lettis, W.R. (2000) Methods for dating Quaternary sediments. In J.S. Noller, S. Sowers and W.R. Lettis (eds), *Quaternary Geochronology: Methods and Applications.* American Geophysical Union, Washington DC.

Spencer, T. and Douglas, I. (1985) The significance of environmental change: diversity, disturbance and tropical ecosystems. In I. Douglas and T. Spencer (eds), *Environmental Change and Tropical Geomorphology.* George Allen and Unwin, London, pp. 13–38.

Sperber, I. (1990) *Fashions in Science: Opinion Leaders and Collective Behaviour in the Social Sciences*. University Minnesota Press, Minneapolis.

Stallins, J.A. (2006) Linking geomorphology and ecology. *Geomorphology* 77: 207–216.

Stanley, E.H. and Boulton, A.J. (2000) River size as a factor in conservation. In P.J. Boon, B.R. Davies and G.E. Petts (eds), *Global Perspectives on River Conservation: Science, Policy and Practice*. Wiley, Chichester, pp. 403–414.

Starkel, L. (1983) The reflection of hydrological change in the fluvial environment of the temperate zone during the last 15,000 years. In K.J. Gregory (ed.), *Background to Palaeohydrology.* Wiley, Chichester, pp. 213–235.

Starkel, L. (1991) The Vistula river valley: a case study for central Europe. In L. Starkel, K.J. Gregory and J.B. Thornes (eds), *Temperate Palaeohydrology*. Wiley, Chichester, pp. 171–188.

Starkel, L., Gregory, K.J. and Thornes, J.B. (eds) (1991) *Temperate Palaeohydrology*. Wiley, Chichester.

Stoddart, D.R. (1968) Climatic geomorphology: review and assessment. *Progress in Geography* 1: 160–222.

Stoddart, D.R. (1997) Richard J Chorley and modern geomorphology. In D.R. Stoddart (ed.), *Process and Form in Geomorphology*. Routledge, London, pp. 383–399.

Stokes, S. and Walling, D.E. (2003) Radiogenic and isotopic methods for the direct dating of fluvial sediments. In G.M. Kondolf and H. Piegay (eds), *Tools in Fluvial Geomorphology*. Wiley, Chichester, pp. 233–267.

Stokes, S., Thomas, D.S.G. and Washington, R. (1997) Multiple episodes of aridity in southern Africa since the last interglacial period. *Nature* 388: 154–158.

Strahler, A.N. (1956) The nature of induced erosion and aggradation. In W.L. Thomas (ed.), *Man's Role in Changing the Face of the Earth*. University of Chicago Press, Chicago, pp. 621–638.

Strakhov, N.M. (1967) *Principles of Lithogenesis, Vol.1*. Consultants Bureau, New York.

Strudley, M.W. and Murray, A.B. (2007) Sensitivity analysis of pediment development through numerical simulation and selected geospatial query. *Geomorphology* 88: 329–351.

Sugden, D.E. and John, B.S. (1976) *Glaciers and Landscape: A Geomorphological Approach*. Arnold, London.

Summerfield, M.A. (1981) Macroscale geomorphology. *Area* 13: 3–8.

Summerfield, M.A. (1991) *Global Geomorphology*. Longman, Harlow.

Summerfield, M.A. (ed.) (2000) *Geomorphology and Global Tectonics*. Wiley, Chichester.

Summerfield, M.A. (2005a) The changing landscape of geomorphology. *Earth Surface Processes and Landforms* 30: 779–781.

Summerfield, M.A. (2005b) A tale of two scales, or the two geomorphologies. *Transactions of the Institute of British Geographers* 30: 402–415.

Summerfield, M.A. and Hulton, N.J. (1994) Natural controls of fluvial denudation rates in major world drainage basins. *Journal of Geophysics Research* 99(B7): 13,871–13,883.

Surrell, A. (1841) *Etude sur les torrents des Hautes-Alpes*. Carilin-Goewy, Paris.

Tang, Z., Engel, B.A., Lim, K.J., Pijanowski, B.C. and Harbor, J. (2005) Minimizing the impact of urbanization on long term runoff. *Journal of the American Water Resources Association* 41: 1347–1359.

Tansley, A.G. (1935) The use and abuse of vegetational concepts and terms. *Ecology* 16: 284–307.

Tansley, A.G. (1946) *Introduction to Plant Ecology*. Allen and Unwin, London.

Taylor, K.C., Lamorey, G.W., Doyle, G.A., Alley, R.B., Grootes, P.M., Mayewski, P.A., White, J.W.C. and Barlow, L.K. (1993) The 'flickering switch' of late Pleistocene climate change. *Nature* 361: 432–436.

Teller, J.T. (1995) The impact of large ice sheets on continental palaeohydrology. In K.J. Gregory, L. Starkel and V.R. Baker (eds), *Global Continental Palaeohydrology*. Wiley, Chichester, pp. 109–129.

Teller, J.T. (2003) Subaquatic landsystems: large proglacial lakes. In D.J.A. Evans (ed.), *Glacial Landsystems*. Hodder Arnold, London, pp. 348–371.

Thomas, D.S.G. (ed.) (1997) *Arid Zone Geomorphology: Process, Form and Change in Drylands*. Wiley, Chichester.

Thomas, D.S.G. and Wiggs, G.F.S. (2008) Aeolian system responses to global change: challenges of scale, process and temporal integration. *Earth Surface Processes and Landforms* 33: 1396–1418.

Thomas, D.S.G., Knight, M. and Wiggs, G.F.S. (2005) Remobilization of southern African desert dune systems by twenty first century global warming. *Nature* 435: 1218–1221.

Thomas, M.F. (1965) Some aspects of the geomorphology of domes and tors in Nigeria. *Zeitschrift fur Geomorphologie* 9: 63–81.

Thomas, M.F. (1974) *Tropical Geomorphology*. Macmillan, Basingstoke.

Thomas, M.F. (1980) Preface. In H. Hagedorn and M. Thomas (eds), *Perspectives in Geomorphology. Zeitschrift fur Geomorphologie Supplementband* 36: V–VI.

Thomas, M.F. (1994) *Geomorphology of the Tropics*. Wiley, Chichester.

Thomas, M.F. (2004) Landscape sensitivity to rapid environmental change – a Quaternary perspective with examples from Tropical Areas. *Catena* 55: 107–124.

Thomas, M.F. (2005) Savanna. In P.G. Fookes, E.M. Lee and G. Milligan (eds), *Geomorphology for Engineers*. Whittles Publishing, Dunbeath, Caithness, pp. 454–472.

Thomas, M.F. (2006) Lessons from the tropics for a global geomorphology. *Singapore Journal of Tropical Geography* 27: 111–127.

Thomas, M.F. (2008a) Geomorphology in the tropics 1895–1965. In T.P. Burt (ed.), *The History of the Study of Landforms or Development of Geomorphology Volume 4: Quaternary and recent processes and forms (1890–1965) and the mid-century revolutions*. The Geological Society, London, pp. 677–725.

Thomas, M.F. (2008b) Understanding the impacts of late Quaternary climate change in tropical and sub-tropical regions. *Geomorphology* 101: 146–158.

Thomas, M.F. and Thorp, M.B. (1995) Geomorphic response to rapid climatic and hydrologic change during the late Pleistocene and early Holocene in the humid and sub-humid tropics. *Quaternary Science Reviews* 14: 193–207.

Thorndycraft, V.R., Benito, G. and Gregory, K.J. (2008) Fluvial geomorphology: a perspective on current status and methods. *Geomorphology* 98: 2–12.

Thornes, J.B. (1979) Processes and interrelationships, rates and changes. In C. Embleton and J. Thornes (eds), *Process in Geomorphology*. Arnold, London, pp. 378–387.

Thornes, J.B. (2003) Time: change and stability in environmental systems. In S.L. Holloway, S.P. Rice and G. Valentine (eds), *Key Concepts in Geography*. Sage, London, pp. 131–150.

Thornes, J.B. (2008) Time: change and stability in environmental systems. In N.J. Clifford, S.L. Holloway, S.P. Rice and G. Valentine (eds), *Key Concepts in Geography*. Sage, London, pp. 119–139.

Thornes, J.E. (1999) *John Constable's Skies*. University of Birmingham Press, Birmingham.

Thornthwaite, C.W. (1948) An approach towards a rational classification of climate. *Geographical Review* 38: 55–94.

Thwaites, R.N. (2006) Development of Soil Geomorphology as a Sub-discipline of Soil Science. 18th World Congress of Soil Science, July 9–15, Philadelphia, PA, USA.

Tikkanen, M. and Juha, O. (2002) Late Weichselian and Holocene shore displacement history of the Baltic Sea in Finland. *Fennia* 180(1–2): S141–149.

Tinkler, K.J. (1985) *A Short History of Geomorphology*. Croom Helm, London and Sydney.

Tinkler, K.J. (1993) *Field Guide Niagara Peninsula and Niagara Gorge*. McMaster University, Hamilton, Ontario.

Tomasella, J., Hodnett, M.G., Cuartas, L.A., Nobre, A.D., Waterloo, M.J. and Oliveira, S.M. (2007) The water balance of an Amazonian micro-catchment: the effect of interannual variability of rainfall on hydrological behaviour. *Hydrological Processes* 22: 2133–2147.

Tooley, M.J. and Smith, D.E. (2005) Relative sea-level change and evidence for the Holocene Storegga Slide tsunami from a high-energy coastal environment: Cocklemill Burn, Fife, Scotland, UK. *Quaternary International* 133–134: 107–119.

Tooth, S. (2007) Arid geomorphology: investigating past, present and future changes. *Progress in Physical Geography* 33: 319–335.

Tooth, S. (2008) Arid geomorphology: recent progress from an Earth System Science perspective. *Progress in Physical Geography* 32: 81–101.

Tooth, S. (2009) Invisible geomorphology. *Earth Surface Processes and Landforms* 34: 752–754.

Tricart, J. (1957) Application du concept de zonalite a la geomorphologie. *Tijdschrift van het Koninklijk Nederlandsch Aarddrijiskundig Geomootschap*: 422–434.

Tricart, J. (1965) *The Landforms of the Humid Tropics, Forests and Savannas* (trans C.J. KieweitdeJonge, 1972). Longman, London.

Tricart, J. and Cailleux, A. (1965) *Introduction à la géomorphologie climatique*. Sedes, Paris.

Tricart, J. and Cailleux, A. (1972) *Introduction to Climatic Geomorphology* (trans C.J.K. De Jonge). Longman, London.

Trimble, S. (1983) A sediment budget for Coon Creek basin in the driftless area, Wisconsin 1853–1977. *American Journal of Science*, 283: 454–474.

Trimble, S.W. (1999) Decreased rates of alluvial sediment storage in the Coon Creek basin, Wisconsin, 1975–1993. *Science* 285: 123–124.

Trimble, S.W. (2008) The use of historical data and artifacts in geomorphology. *Progress in Physical Geography* 32: 3–29.

Trudgill, S.T. and Richards, K.S. (1997) Environmental science and policy: generalizations and context sensitivity. *Transactions Institute of British Geographers* NS 22: 5–12.

Twidale, C.R. (2002) The two-stage concept of landform and landscape development involving etching: origin, development and implications of an idea. *Earth-Science Reviews* 57: 37–74.

Twidale, C.R. and Lageat, Y. (1994) Climatic geomorphology: a critique. *Progress in Physical Geography* 18: 319–334.

Udvardy, M.D.F. (1981) The riddle of dispersal: dispersal theories and how they affect vicariance biogeography. In G. Nelson and D.E. Rosen (eds), *Vicariance Biogeography: A Critique*. Columbia University Press, New York, pp. 6–29.

UNEP (1992) *World Atlas of Desertification*. Arnold, Sevenoaks.

UNEP (2006) The Global Runoff Data Center. See http://www.earthobservations.org/documents/sbas/wa/74_Global%20Runoff%20Data%20Centre.pdf.

UNESCO (1979) *Map of the World Distribution of Arid Regions*. MAP Technical note 4. UNESCO, New York.

United Nations World Commission on Environment and Development (UNWCED) (1987) *Our Common Future*. Oxford University Press, Oxford.

United Nations (1987) *The Limits to Growth*. Report of the World Commission on Environment and Development. General Assembly Resolution 42/187, 11 December.

Urban, M. and Rhoads, B. (2003) Conceptions of nature: implications for an integrated geography. In S. Trudgill and A. Roy (eds), *Contemporary Meanings in Physical Geography*. Hodder, London, pp. 211–231.

US Department of Agriculture, Natural Resources Conservation Service (2006) *Keys to Soil Taxonomy*. USDA, Washington, DC.

US Environmental Protection Agency (1973) *Methods for Identifying and Evaluating the Nature and Extent of Nonpoint Sources of Pollutants*. US Department of Agriculture, Washington, DC.

USGS (2009) http://volcanoes.usgs.gov/activity/ accessed 13 August 2009.

Van der Wateren, F.M. (2003) Ice-marginal terrestrial landsystems: southeren Scandinavian ice sheet margin. In D.J.A. Evans (ed.), *Glacial Landsystems*. Hodder Arnold, London, pp. 166–203.

Van Dyne, G.M. (1980) Reflections and projections. In A.I. Breymeyer and G.M. Van Dyne (eds), *Grasslands, Systems Analysis and Man IBP* Vol. 19. Cambridge University Press, Cambridge.

Varnes, D.J. (1978) Landslide types and processes. In *Landslides: Analysis and Control, Special Report 176*.Transportation Research Board, Washington, pp. 11–33.

Verstappen, H.Th. (1983) *Applied Geomorphology: Geomorphological Surveys for Environmental Development*. Elsevier, Amsterdam.

Viles, H. (ed.) (1988) *Biogeomorphology*. Wiley, Chichester.

Viles, H. and Spencer, T. (1995) *Coastal Problems: Geomorphology, Ecology and Society at the Coast*. Arnold, London.

Viles, H.A., Naylor, L.A., Carter, N.E.A. and Chaput, D. (2008) Biogeomorphological disturbance regimes: progress in linking ecological and geomorphological systems. *Earth Surface Processes and Landforms* 33: 1419–1435.

von der Heyden, C.J. (2004) The hydrology and hydrogeology of dambos: a review. *Progress in Physical Geography* 28: 544–564.

Von Richthofen, B.F. (1886) *Fuhrer fur Forschungsreisende, Anleitung Beobachtungen uber Gegenstande der Physischen Geographie und Geologie*. Robert Oppenheim, Berlin.

Wakasa, S., Matsuzaki, H., Tanaka, Y. and Matsukura, Y. (2006) Estimation of episodic exfoliation rates of rock sheets on a granite dome in Korea from cosmogenic nuclide analysis. *Earth Surface Processes and Landforms* 31: 1246–1256.

Walker, B., Holling, C., Carpenter, S. and Kinzig, A. (2004) Resilience, adaptability, and transformability. *Ecology and Society* 9(2): 5 http://www.ecologyandsociety.org/vol9/iss2/art5/.

Walling, D.E. (1977) Rainfall, runoff and erosion of the land: a global view. In K.J. Gregory (ed.), *Energetics of Physical Environment*. Wiley, Chichester, pp. 89–117.

Walling, D.E. (1981) Yellow River that never runs clear. *Geographical Magazine* 53: 568–575.

Walling, D.E. (1987) Hydrological processes. In K.J. Gregory and D.E. Walling (eds), *Human Activity and Environmental Processes*.Wiley, Chichester, pp. 53–85.

Walling, D.E. (2006) Human impact on land-ocean sediment transfer by the world's rivers. *Geomorphology* 79(3–4): 192–216.

Walling, D.E. (2008) The changing sediment loads of the world's rivers. *Sediment Dynamics in Changing Environments* (Proceedings of a symposium held in Christchurch, New Zealand, December). IAHS Publ. 325: 323–338.

Walling, D.E. and He, Q. (1999) Changing rates of overbank sedimentation on the floodplains of British rivers during the past 100 years. In A.G. Brown and T. Quine (eds), *Fluvial Processes and Environmental Change*. Chichester, Wiley, pp. 207–222.

Waltham, T. (2005) Subsidence. In P.G. Fookes, E.M. Lee and G. Milligan (eds), *Geomorphology for Engineers*. Whittles Publishing, Dunbeath, Scotland, pp. 318–342.

Wang, X., Chen, F., Hasi, E. and Li, J. (2008) Desertification in China: an assessment. *Earth-Science Reviews* 88: 188–206.

Wang, X., Dong, Z., Liu, L. and Qu, J. (2004) Sand sea activity and interactions with climatic parameters in the Taklimakan Sand Sea, China. *Journal of Arid Environments* 57: 225–238.

Wanner, H., Beer, J., Butikofer, J., Crowley, T.J., Cubasch, U., Fluckiger, J., Goosse, H. and Widmann, M. (2008) Mid- to Late Holocene climate change: an overview. *Quaternary Science Reviews* 27: 1791–1828.

Ward, J.V. and Stanford, J.A. (1983) The serial discontinuity concept of lotic ecosystems. In T.D. Fontaine and S.M. Bartell (eds), *Dynamics of Lotic Ecosystems*. Ann Arbor Science, Ann Arbor, MI, pp. 29–41.

Warner, R.F. (2000) The role of stormwater management in Sydney's urban rivers. In S. Brizga and B. Finlayson (eds), *River Management: The Australian Experience*. Wiley, Chichester, pp. 173–196.

Warren, A. (1979) Aeolian processes. In C. Embleton and J. Thornes (eds), *Process in Geomorphology*. Arnold, London, pp. 325–351.

Watts, D.A. (1971) *Principles of Biogeography: An Introduction to the Functional Mechanisms of Ecosystems*. McGraw Hill, London.

Wells, S.G., Bullard, T.F., Smith, L.N. and Gardner, T.W. (1983) Chronology, rates and magnitudes of Late Quaternary landscape changes in the southeast Colorado Plateau. In S.G. Wells, D.W. Love and T.W. Gardner (eds), *Chaco Canyon Country*. American Field Geomorphology Group, Albuquerque, pp. 177–186.

Whalley, W.B., Marshall, J.R. and Smith, B.J. (1982) The origin of desert loess: some experimental observations. *Nature* 300: 433–435.

White, G.F. (1789) *Natural History of Selborne*. Walter Scott, London.

Wiggs, G.F.S. (2001) Desert dune processes and dynamics. *Progress in Physical Geography* 25: 53–79.

Wild, A. (1993) *Soils and the Environment*. Cambridge University Press, Cambridge.

Wilde, S.A., Valley, J.W., Peck, W.H. and Graham, C.M. (2001) Evidence from detrital zircons for the existence of continental crust and oceans on Earth 4.4 Gyr ago. *Nature* 409: 175–178.

Williams, M., Dunkerley, D., De Decker, P., Kershaw, P. and Chappell, J. (1998) *Quaternary Environments*. Arnold, London, 2nd edn.

Williams, R.S. and Ferrigno, J.G. (eds) (1999) *Satellite Image Atlas of Glaciers of the World*. Chapter A: Introduction. US Geological Survey Professional Paper 1386–A.

Wilson, I.G. (1973) Ergs. *Sedimentary Geology* 10: 77–106.

Wiltshire, H.G. (1980) Human causes of accelerated wind erosion in California's deserts. In D.R. Coates and J.D. Vitek (eds), *Thresholds in Geomorphology*. George Allen and Unwin, London, Boston, Sydney, pp. 415–433.

Wirthmann, A. (1987) *Geomorphology of the Tropics*, Trans. D. Barsch, 2000. Springer, Heidelberg.

Wischmeier, W.H. and Smith, D.D. (1965) *Predicting Rainfall Erosion from Cropland East of the Rocky Mountains*. Agriculture Handbook No. 282. US Department of Agriculture, Washington, DC.

Wohl, E.E. (2004) *Disconnected Rivers: Linking Rivers to Landscapes*. Yale University Press, New Haven, CT.

Wohl, E. and Merritts, D.J. (2007) What is a natural river? *Geography Compass* 1: 871–900.

Wolfe, P. (2000) History of wastewater. Supplement to *PennWell Magazine*: 24–36.

Wolman, M.G. (1967) A cycle of sedimentation and erosion in urban river channels. *Geografiska Annaler* 49A: 385–395.

Wolman, M.G. (2002) The human impact: some observations. *Proceedings of the American Philosophical Society* 146: 81–98.

Wooldridge, S.W. and Morgan, R.S. (1937) *The Physical Basis of Geography: An Outline of Geomorphology*. Longmans, London.

World Bank (1993) *Using Water Efficiently – Technological Options*. World Bank, Washington, DC.

World Health Organisation (WHO) (1990) *Protecting and Promoting Health in the Urban Environment: Concepts and Strategic Approaches*. World Health Organization, Geneva.

Yatsu, E. (1966) *Rock Control in Geomorphology*. Sozoscha, Tokyo.

Yatsu, E. (1988) *The Nature of Weathering: An Introduction*. Sozosha, Tokyo.

Ye Grishankov (1973) The landscape levels of continents and geographic zonality. *Soviet Geography* 14: 61–77.

Zalasiewicz, J., Williams, M., Smith, A., Barry, T.L., Coe, A.L., Bown, P.R., Brenchley, P., David Cantrill, D., Gale, A., Gibbard, P.F. John, Gregory, F.J., Hounslow, M.W., Kerr, A.C., Pearson, P., Knox, R., Powell, J.,Waters, C., Marshall, J., Oates, M. Rawson, P. and Stone, P. (2008) Are we now living in the Anthropocene? *GSA Today* 18: 4–8.

Zube, E.H. (1999) Environmental perception. In Alexander, D.E. and Fairbridge, R.W. (eds), *Encyclopedia of Environmental Science*. Kluwer, Dordrecht, pp. 214–216.

INDEX

Major items are indexed, locations and individuals are not included. Page numbers in **bold** refer to tables those in *italics* to figures.